BIOINDUSTRY ETHICS

BIOINDUSTRY ETHICS

DAVID L. FINEGOLD
CÉCILE M. BENSIMON
ABDALLAH S. DAAR
MARGARET L. EATON
BÉATRICE GODARD
BARTHA MARIA KNOPPERS
JOCELYN E. MACKIE
PETER A. SINGER

ELSEVIER
ACADEMIC
PRESS

Amsterdam   Boston   Heidelberg   London
New York   Oxford   Paris   San Diego   San Francisco
Singapore   Sydney   Tokyo

o 57751490

Elsevier Academic Press
30 Corporate Drive, Suite 400, Burlington, MA 01803, USA
525 B Street, Suite 1900, San Diego, California 92101-4495, USA
84 Theobald's Road, London WC1X 8RR, UK

This book is printed on acid-free paper. ∞

Library of Congress Cataloging-in-Publication Data

British Library Cataloguing in Publication Data
A catalogue record for this book is available from the British Library

ISBN: 0-1236-9370-5

For all information on all Elsevier Academic Press Publications
visit our Web site at www.books.elsevier.com

Printed in the United States of America
05 06 07 08 09 10 9 8 7 6 5 4 3 2 1

CONTENTS

RESEARCH TEAM

David L. Finegold, D.Phil.
Associate Professor
Keck Graduate Institute for the Applied Life Sciences
Strategy & Organization Studies
Claremont, California

Cécile M. Bensimon
Graduate Student
University of Toronto Joint Centre for Bioethics
Toronto, Ontario, Canada

Abdallah S. Daar, D.Phil (Oxon), FRCP (Lon), FRCS, FRCSC
Professor of Public Health Sciences and
    Professor of Surgery
University of Toronto
Director, Program in Applied Ethics and Biotechnology
Director, Canadian Program on Genomics and Global Health
University of Toronto Joint Centre for Bioethics
Director of Ethics and Policy, McLaughlin Centre for
    Molecular Medicine
Toronto, Ontario, Canada

Margaret L. Eaton, Pharm.D, J.D.
Senior Research Scholar
Stanford University School of Medicine
Center for Biomedical Ethics
Palo Alto, California

Béatrice Godard, Ph.D.
Associate Professor and Director
Bioethics Programs
Faculty of Graduate Studies
Researcher, C.R.D.P.
Université de Montréal
Montreal, Quebec, Canada

Bartha Maria Knoppers, O.C., Ph.D.
Canada Research Chair in Law and Medicine Professor
Faculties of Law and Medicine
Senior Researcher, C.R.D.P
Université de Montréal
Montreal, Quebec, Canada

Jocelyn E. Mackie
Graduate Student
University of Toronto Joint Centre for Bioethics
Toronto, Ontario, Canada

Peter A. Singer, M.D., MPH, FRCPC
Sun Life Financial Chair in Bioethics
Director, University of Toronto Joint Centre for Bioethics
Professor of Medicine
University of Toronto and University Health Network
Canadian Institutes of Health Research Distinguished Investigator
Toronto, Ontario, Canada

PREFACE

This book is the end product of a study that examined how bioscience companies are incorporating ethics into their decision making. The study was prompted by our collective belief in the value of business ethics, and by our recognition of the special need for an ethics perspective, when companies develop and market biotechnology products and services which, in and of themselves, often raise strong moral, social, and religious concerns. Going into this project, we knew that the manner in which bioscience companies were approaching ethics varied, and that no set of fundamentals or uniform management practices guided it. Indeed, very little basic information has been available about the main ethical issues bioscience companies spend their time on, the policies and practices they are using to deal with ethical issues, and how effective these approaches appear to be. We decided, therefore, that lessons could be learned from comparing and contrasting the various approaches to ethics, and that these lessons, which appear in the final chapter, could inform the future development of this managerial activity.

The book contains detailed case-study descriptions of how 13 bioscience companies are approaching ethical concerns, primarily from the viewpoints of the executives and employees within the firms. As such, the cases are not intended to be comprehensive or highly critical accounts of company decisions and behaviors. Rather, the research teams served as outsiders, pulling together different perspectives within the firms; these, along with secondary material, provide compelling stories that can serve as the basis for stimulating discussion and debate about the proper roles and responsibilities of the private-sector companies who are taking the lead in developing practical applications of the bioscience revolution. It is also important to note that these stories are like snapshots, capturing a picture of the company at a particular point in time. While circumstances may change in unpredictable ways, the cases should still provide valuable background and context for interpreting subsequent events.

A primary audience for the book is the thousands of biotech, pharmaceutical and bioagricultural companies around the world who are confronting an ever-evolving set of ethical issues and looking for practical frameworks and examples of how to approach them. The cases should also be of interest to the many specialized service providers—venture capitalists, law firms, financial analysts, and contract research organizations—who work with or study the bioscience sector. The other main audience is the rapidly growing number of higher-education programs designed to prepare individuals to enter the bioscience sector. We feel that it is vital that these programs—whether they focus primarily on the business, scientific, legal, medical or regulatory aspects of the biosciences—contain a strong component on the ethical issues inherent in this industry. In addition, by not assuming any prior scientific knowledge and defining technical terms whenever they are introduced, we have tried to make the cases accessible to the general public, who are all affected by the developments in the bioscience industry.

The unusually large and diverse nature of our project team came about three years ago when we discovered that two independent groups of researchers—one at the Canadian Program on Genomics and Global Health (CPGGH) based at the University of Toronto's Joint Centre for Bioethics, and the other a network of researchers affiliated with the Bioscience Business Ethics Center at the Keck Graduate Institute in Claremont, CA—were about to embark on separate projects, each focused on ethical decision making in bioscience companies. Rather than compete, we decided to combine our resources and our different disciplinary and national perspectives to do a more comprehensive study than either team could have completed on their own.

David L. Finegold

Acknowledgments

We would like to thank the Seaver Institute and Genome Canada for providing funding specifically for this project.[1] We would also like to thank Anne McLoughlin, Deborah Flynn, Diane Isonaka, and Andrew Taylor who have provided invaluable support at different stages of the project. Susan Finegold helped with editing each chapter, while Luna Han provided editorial guidance and encouragement to make sure the book was completed. Most of all we would like to thank all of the companies and the many individuals within them who agreed to take part in this study. We hope the resulting collection of cases will help assist them and others to incorporate ethics into the business of bioscience.

[1]In addition, Bartha Knoppers and Béatrice Godard received support from Genome Quebec, and the CPGGH also received funding from the Ontario Research and Development Challenge Fund; matching partners are listed at www.geneticsethics.net. Peter Singer is supported by a CIHR Distinguished Investigator award and Abdallah Daar is supported by the McLaughlin Centre for Molecular Medicine.

1

INTRODUCTION

The genesis of the topic of bioscience-business ethics is the realization that DNA knowledge can be applied to produce breakthrough medical, agricultural, and other products. Just over 50 years ago, James Watson and Francis Crick unraveled the structure of DNA; 2 decades later, the first two biotechnology companies, Cetus and Genentech, were founded. Since then, there have been great advances in biotechnology: in the 1980s, the first biotech drugs reached the market, and in the 1990s, the first genetically engineered crops became commercially available. These new industries and products are beginning to have a dramatic impact on human health and agriculture. We are now well into a biotechnology revolution that has transformed both the ability to generate medical and agricultural products and the industries that produce these products.

Physics dominated the first half of the 20th century; the 21st century has already been dubbed the *age of biology.*[1] The sequencing of the genomes of the major crop plants, many animals, pathogens, and humans has laid the groundwork for a new, genetically driven approach to medicine, agriculture, nutrition, and many other fields. Biotechnology has so invaded the methodologies of the traditional medical, industrial, and agricultural companies that, as an industry, they collectively deserve a new name—what we are calling the bioscience industry. There are still relatively few major biotechnology drugs on the market, but the number of new biological drugs in clinical trials now exceeds the number of novel, small molecule, chemical-based drugs in development. And despite the European controversy surrounding their introduction, biotechnology crops (now called *genetically modified,* or *GM,* crops) are growing at a rapid rate, already accounting for a majority of the soybeans and cotton grown in the United States, with global sales projected to reach $25 billion by 2010. The potential impact of new biotechnologies

[1]Shahi G. *BioBusiness in Asia.* London: Prentice Hall; 2004.

extends well beyond health and agriculture to a variety of other industrial and environmental implications that may account for up to 50% of the gross domestic product (GDP) of world economies by 2050.[2]

There is another reason to group and rename the industry that is producing these new biotechnology products. Their very nature raises such profound social and ethical issues that managers in these companies now have to address concerns emanating from the relatively new field of bioethics, as well as the older field of medical ethics. Looking back, there were hints that this phenomenon was coming. In the 20th century, public concerns about safety and misuse caused major setbacks in the development of nuclear power and recombinant DNA technology that presaged similar attacks against genetically modified crops. Applying lessons from these three examples makes it plausible to predict that public concerns about issues such as cloning, stem cell research, genetic privacy, human genetic enhancement, and antibioterrorism technology can pose a greater risk to the future of the bioscience industry than the technologic challenges themselves.

Given this potential risk, some bioscience company managers have begun to address the social and ethical consequences of producing biotechnology products. These efforts are driven in part by an interest in protecting companies from harm. But other motivations also are at play, driven by the realization that the role of corporations in the use of biotechnology products is so vital that failure to address ethical concerns risks stalling scientific advances and wasting the potential for building a healthier society. Companies at the forefront of incorporating biomedical ethics into product development decision making are plowing new ground; there is, as yet, no well-established body of scholarship to guide them. This book, based on a research project involving in-depth case studies of 13 bioscience firms, aims to begin to close this gap.

ETHICAL ISSUES CONFRONTING BIOSCIENCE COMPANIES

The ethical issues facing bioscience companies are likely to vary based on their particular technology, product focus and stage of development. Some examples of ethical issues that firms are facing as they progress from research to bringing products to market include:

- Who owns genetic information? Is it socially equitable and feasible for commercial firms to own genetic information? Does this unfairly inhibit academic research? How do firms respond to those who claim that no one should patent and sell human genetic information? What consent, if any, do firms using this information need from individuals and communities?
- Is it responsible to develop and market genetic tests for conditions for which there is as yet no related prevention or treatment? Who bears responsibility for

[2]Ibid.

those individuals who lose medical insurance or employment when, using a company-developed gene test, they are identified as possessing disease-related genes?

• How should pharmaceutical and biotech firms price their drugs and other medical products in a global economy, to ensure that they can justify the major research costs and risks, and still provide the maximum access and benefit to those in need in both the developed and developing countries?

• What constitutes fair and balanced medical product marketing to physicians and the public? Do the benefits of direct-to-consumer advertising outweigh the corporate and social costs?

• How does a company manage the inherent conflicts of interest in its funding of the clinical trials of a medical product that it intends to market? At what stage should it be obliged to disclose publicly the results of its research, including negative findings?

• What are the economic, social, and health implications of the move toward more personalized medicine? (Targeting treatments much more specifically to individual underlying genetic conditions improves health outcomes but reduces the target population group, and hence the potential market, for drug development firms.) Will individuals and societies be able to afford it?

• How can the benefits of genomics and biotechnology be extended to billions of people in developing countries, in an effort to avoid creating a "genomics divide" between industrialized and developing countries?

The rest of this introductory chapter is in four sections.

1. We examine why bioscience firms should address these ethics questions. By using the term *bioscience business ethics,* we intend to include the ethical issues that attach to:

   • the technologies that these companies commercialize (medical, agricultural, etc.)
   • the manner in which these technologies are researched in animals and humans
   • the way that drugs and other products are marketed and sold

   Depending on the phase of product development, a bioscience business manager is likely to encounter ethical issues in all these areas.

2. We describe the development of the field of biomedical ethics, placing its interaction with bioscience firms in an historical context, and addressing some of the associated controversies.

3. We describe the methods used in the research that led to this book. We interviewed more than 100 senior decision makers to develop the 13 case studies from which we have attempted to derive lessons for the development of bioscience business ethics.

4. Finally, we provide an overview of the rest of the book, to explain the logic of how we grouped the company case studies. The cases can be read individually; however, we encourage readers to read this overview chapter first.

The concluding chapter pulls together lessons learned from the cases on how bioscience companies approach those aspects of their business that raise ethical issues.

WHY SHOULD BIOSCIENCE FIRMS CARE ABOUT ETHICS?

Although this work is a study of how companies are managing bioscience ethical issues, there are reasons why many have not adopted this managerial practice. To provide a broad understanding about the current state of affairs on our subject, we examine the most common arguments from new bioscience firms against the need for dealing with ethical issues. We follow with our counter-arguments about why it is prudent for bioscience firms to overcome the objections and address the ethical issues raised by the products they develop proactively.

Argument No. 1: *We're worried about raising enough capital to survive and to see if our technology works; we don't have the time or money to worry about ethics.*

Bioscience is one of the riskiest bets an investor can make. Of every 10 companies they back, venture capitalists generally expect only 1 to become a major success, hope that 2 or 3 others will yield some return on investment, and assume the rest will fail. To make money, given bioscience's inherent uncertainties, these investors and the companies they fund must become experts at managing many types of risk: technical, financial, regulatory, and managerial. They screen hundreds of business plans, weeding out any with risks considered unacceptably high. The last thing they want is to add additional risk—regarding either a bioethics or business ethics issue—that could undermine the future of the company. The near-impossibility of obtaining any funding in today's market for a business model that involves GM foods is an example of this unacceptable risk. This is not because of concerns about the feasibility of the technology nor the size of the potential market. It is because of the global controversy surrounding the initial applications of GM crops, which has made the prospects for regulatory and consumer acceptance of any next-generation product highly uncertain. Harsher critics would say that, as ethical issues are now known to be an inherent aspect of commercially developing biomedical technology, firms that do not plan for ethical analysis are guilty of a failure of vision. So it would pay an early-stage bioscience firm seeking to raise capital to anticipate ethical concerns and develop responses to them.

Argument No. 2: *We're too small. We'll worry about ethics when we get a product to market.*

Given that it typically takes a decade or more to go from laboratory research to getting a new drug on the market, it is not surprising that many bioscience start-ups do not feel a sense of urgency about addressing ethical concerns. Many argue that, because the failure rate is so high (only a small percentage of initial compounds make it through the clinical research phase), it is a waste of time and effort to focus on ethics at the early stage of drug discovery.

Our research and that of others suggests that bioscience companies miss important opportunities when they assume that they will need to address ethical issues only after a mature product has been developed. As the Genzyme, Millennium, Novo Nordisk, and Affymetrix cases in this volume reveal, the culture and values established in a company's early days by founders and leadership teams have a profound and lasting effect on the subsequent ethical decision making of the organization, even decades later when the company has evolved into a large global business. To cite an analogy that has become a general rule of organizational change, it is easier to change the course of a small speedboat than a supertanker. Likewise, it is far more difficult to change an organization's culture and approach to ethical issues at a later stage than to instill the ethical values and priorities from the outset. This is also true in the more narrow sense of individual products. Identifying at the outset which ethical issues are likely to attach to a particular product informs all subsequent decisions about that product and prevents getting blindsided by unanticipated reactions.

A solid foundation in ethical decision making will enable companies to see, at each stage in product development, where the minefields may lie. For instance, there are always business, technological, and medical aspects to the first decision most companies make—that is, what products to develop. Adding an ethical dimension to the decision can generate useful insight, such as when companies decide to forego short-term profits from an expensive "me-too" drug with only marginal benefits over existing drugs, and to execute instead a business plan to produce a first-in-class drug to fulfill an unmet medical need. Many would argue that society needs the latter and not the former. The company, however, may not be able to afford the socially responsible choice. An ethical analysis would force the company to consider the utility, the consequences, and the justice of each potentially feasible option. Whatever the decision, the company proceeds with an understanding of the potential ethical objections to that choice and how those objections can be mitigated. Assuming that most business choices in this industry have an ethical dimension allows managers to see and manage the ethical aspect of all activities, such as animal and human research, contracting with academic physicians, data publication, pricing, advertising, off-label use, or post-market monitoring, all of which have produced ethical problems for companies, with the consequent negative press and financial costs. Companies founded on a coherent set of values and an ability to incorporate ethics into decision making find it much easier to see where the ethical issues lie and how to address them.

Argument No. 3: *We're fine so long as we follow FDA guidelines.*

The Food and Drug Administration (FDA) and its counterparts around the world play such a powerful role in regulating the bioscience industry that it is not surprising many managers in these companies feel they are behaving ethically as long as they comply with what the FDA requires. But, as a general proposition, legal compliance alone never ensures that companies are shielded from allegations of unethical behavior. There are two specific difficulties associated with conflating FDA compliance and ethics. First, what the FDA requires changes often and, especially for cutting-edge technologies, it may be years before it supplies clear guidelines on how companies should proceed. For example, the FDA never required the producers of GM foods to demonstrate human safety, just equivalency to natural foods. Many herbal medicines are marketed with little, if any, FDA oversight. Genetic tests, if marketed as testing services, can be sold with no safety or efficacy data submitted to the FDA.[3] The FDA is almost inevitably at a disadvantage compared with the knowledge possessed by company experts working on a new technology. Hence, regulatory evolution often trails the evolution of technology. This is clear in several of our case studies of pharmacogenomics or nutrigenomics firms that seek to link drugs or foods to an individual's genotype. The FDA has not yet developed a full set of guidelines to govern how these products get to market. In the absence of regulation, having an ethical justification for key development and marketing decisions can only benefit, as it is more likely that the agency will approve what the company has done and might even adopt the ethical reasoning of the company to generate an industry standard. The Monsanto, Interleukin, and Sciona cases illustrate the ethical problems that can occur when developing technologies in a regulatory vacuum.

Relying solely on the FDA to determine the right actions to take creates a second, related problem for companies, which is what the FDA demands is not always considered ethical. This is most clearly brought out in the Genzyme case: the firm argued vociferously (but ultimately in vain) with the FDA over the design of a Phase IV clinical trial because it believed it would be unethical to ask doctors to continue to give some patients a placebo when the drug was now approved for sale and no other viable treatment alternatives were available.

Argument No. 4: *We are ethical. Our whole business is about helping people.*

Corporate scientists and managers may be fully convinced that the research they are doing, and the products they are developing, will be beneficial for mankind. However, it does not follow that all or even a majority of the public will agree. The leaders of Myriad Pharmaceuticals thought they were helping women when they marketed the tests for the breast cancer *BRCA* gene. However, the company encountered strong opposition from breast cancer activists who thought that the gene test was harmful because it resulted in women being told they carried a breast cancer gene, but there was nothing reliable they could do to prevent the disease. Companies also have been labeled "unethical" when they deny access to clinical trials to desperately sick patients who do not meet the research entry criteria.[4] A number of the cases in this volume demonstrate that interest groups and nongovernmental organizations (NGOs) can oppose the marketing of bioscience company innovations, no matter how beneficial the company believes the product to be. For instance, Monsanto scientists went from being convinced that

[3]Eaton ML. *Ethics and the Business of Bioscience.* Stanford, Calif: Stanford University Press; 2004.
[4]Eaton, 2004.

the GM crops they were creating could reduce environmental pollution and help feed needy populations in developing countries to being reluctant to identify themselves as Monsanto employees because of the opposition their firm faced. Likewise, Diversa Corporation expected public approval when it voluntarily decided to exceed the stipulations of the Convention on Biological Diversity for sharing the benefits stemming from its research. Nevertheless, Diversa has encountered opposition for seeking to profit from nature's genetic diversity.

Although the human health applications of bioscience research have generally been more favorably received, a number of issues remain, such as the use of animals in research, applications of embryonic stem cells, handling of genetic information, access to medicines once they are developed, which can cause significant problems for bioscience companies. To anticipate any issues that may arise, to shape policies and educational activities that may alleviate criticism, and to build awareness of the potential benefits of their activities, it is useful for companies to analyze the ethical context of their decisions long before their products reach the market.

In addition to the arguments above, our research and earlier works[5] have identified a number of other reasons why bioscience firms should bring an ethical perspective to their business decision making. Unquestionably, the most common reason we heard was, "It's the right thing to do." Managers saying this not only highlighted the strong normative dimension of taking an ethical approach to business, but also told us that there were some tangible business benefits, including:

- helping to attract, motivate, and retain some of the most talented scientists and other employees
- enhancing the image of individual companies and an industry that, at least in the case of bioagriculture and pharmaceuticals, has fallen in public trust
- making it easier to attract strong, high-quality business partners
- attracting the growing number of socially conscious investors who are factoring corporate and social responsibility (CSR) into their evaluation of companies
- reducing the likelihood of the very large legal settlements and personal criminal liabilities that some bioscience companies and their executives have faced for such things as overly aggressive patent protection, misleading data interpretation, poorly managed drug toxicities, deceptive marketing practices, or inducements to prescribe
- Finally, in an industry that is wholly dependent on the government for approval of its products and much of its revenue, adopting a strong approach to self-regulation as part of good governance practices may be the best means to avoid greater restrictions imposed by government on their license to operate.

[5]Dhanda RK. *Guiding Icarus: Merging Bioethics with Corporate Interests.* New York: Wiley-Liss; 2002; Eaton, 2004.

ETHICISTS AND THE BIOSCIENCE INDUSTRY

The previous section addressed why bioscience companies should care about ethics. This section addresses how biomedical ethics has evolved along with the bioscience industry. We document three approaches.

In the early days of the bioscience industry, there was a long period of distance and skepticism between bioethicists and the corporate world. During this time, most academic ethics experts, (along with interest groups and the media) criticized the industry from the outside, and there was relatively little direct interaction between the ethicists and the industry. More recently, some bioethics experts have moved to the other end of the spectrum, going to work for bioscience companies, either as consultants or employees, to help them deal with the complex ethical issues companies face. In our study, we suggest a third way: conducting detailed research *in*, but not *for,* bioscience companies, to try to improve our understanding of the ethical decision making process in the corporations that are pushing forward the bioscience revolution.

Bioethics and biomedical ethics have their roots in medical ethics, which has existed as a moral guide to the practice of medicine since antiquity. A particularly strong resurgence in medical ethics took place after the Nuremberg Trials revealed the atrocities committed by Nazi physicians in their experimentation on prisoners of war. The resulting Nuremberg Code of Ethics in Medical Research (1946–1949) sparked a movement in medical research ethics. Stimulated again by postwar revelations of research misconduct, such as the infamous Tuskegee Study,[6] another expansion in ethics concern resulted in the development of the national and international guidelines, regulations, and laws that set new standards for what is considered ethical conduct in human research.

Additional concern about the activity of medical scientists resulted in a new term, *biomedical ethics,* which started appearing after the consumer and protest movements of the 1960s. Amid the general questioning of power, the professions of medicine and science did not escape scrutiny for the ways medical treatments and technologies were used. Physicians were denounced for ethical failures, such as when "God committees" used social-worth criteria to decide which patients had access to scarce dialysis treatment. Scientists were condemned when it was discovered that they had exposed unwitting people to nuclear radiation. How to avoid the abusive application of new medical and related technologies became such a widespread concern that it led to the formation of several government commissions. (The President's Commission for the Study of Ethical Challenges in Medicine and Biomedical and Behavioral Research, and the Advisory Committee on Human Radiation Experiments, both sponsored by the United States government, are two examples.) Products of this era were bioethics centers (such as The Hastings Center), the introduction of ethics teaching in medical schools, and the formation of hospital clinical ethics committees that sought to ensure ethical treatment of patients.

---

[6]When a group of African American men were infected with syphilis without their knowledge as part of a medical experiment.

Soon thereafter, scientists recognized the need to engage in public debates on the ethical implications of new biotechnologies. A notable example is the 1975 Asilomar Conference, convened by the Nobel Laureate biochemist Dr. Paul Berg, where scientists involved with new recombinant DNA research agreed voluntarily to halt their research until the safety concerns about the technology had been addressed and resolved.[7] The lessons of Asilomar, however, did not penetrate deeply. In the 1980s and 1990s, new technologies (in vitro fertilization, embryonic stem cells, genetic testing, etc.) continued to spark debates about whether they were being used before their full impact on patients, medicine, and society was fully understood.[8]

The ethics climate changed drastically when, in 1988, the Human Genome Project (HGP) became the first large scientific undertaking to dedicate a significant portion of its research budget to associated ethical issues and created HUGO (the Human Genome Organization). HUGO's purpose was to foster international collaboration and provide study guidance for international ethics and intellectual property issues linked to the decoding of the genome. In the United States, the Department of Energy devoted 3 to 5% of its HGP budget to a program that would conduct prospective and concurrent research into the ethical, legal, and social implications (ELSI) of genome research.[9] Other funding agencies, such as Genome Canada, have built on the lessons, both positive and negative, of the HGP's ELSI program to devote an even larger percentage of their funding to large-scale bioethics programs. These lessons also are being incorporated into similar European Union initiatives and the United States National Nanotechnology Initiative. The ethics initiatives associated with the HGP set the important precedent that scientists must now address the ethical issues associated with their work as part of seeking funding for their research.

The need for these ethics initiatives was clearly demonstrated as the genetic advances that accompanied the HGP attracted growing public concern. This ethical concern then began to include industry: when companies such as Incyte and Celera began to seek patent rights for human genetic material, critics claimed that corporate ownership would limit the access of non-profit and academic researchers to valuable genetic tools. Some argued that no human genetic material should be owned by anyone, least of all a for-profit corporation, and thus challenged the very existence of bioscience companies.[10]

The new ethical challenges that accompanied innovations by bioscience companies were occurring against a backdrop of ethical issues that had existed for decades within the pharmaceutical industry. For reasons often related to the cost

[7]Reconsidering Asilomar. *The Scientist.* 2000;14:15. Available at http://www.the-scientist.com/yr2000/apr/russo_p15_000403.html.

[8]See, Rose H, Rose S. (2003, July 3). Playing God: Eggs from Foetuses, Artificial Wombs, Dead Men's Sperm—It's Not Only the Religious Right Who Object to Such 'Advances.' *The Guardian,* p. 25.

[9]Collins FS, Morgan M, Patrinos A. The Human Genome Project: Lessons from Large-Scale Biology. *Science.* 2003;300:286–290.

[10]Waldholz M, Stout H. (1992, April 17). A New Debate Rages over the Patenting of Gene Discoveries. *The Wall Street Journal,* Section B1.

and safety of prescription drugs, pharmaceutical companies and their device and vaccine cousins have been under more or less constant criticism from academia and the media since the consumer movement started in the 1960s. Groups that lobby against these companies are AIDS activists (ACT-UP was formed specifically to protest the price of the first anti-AIDS drug); breast-cancer advocacy groups that protest the denial of access to experimental cancer drugs; religious groups against companies that use embryos in research, or that market contraceptive- or abortion-related products; environmental activists against companies that market genetically modified foods and crops; and animal rights groups whose attacks against firms and their employees who use animals in scientific research have sometimes escalated from verbal abuse to physical violence. These anti-industry groups began to form alliances and to demonstrate at the annual Biotechnology Industry Organization (BIO) conferences to get their message across that biotechnology companies could not be trusted and would remain under scrutiny.[11]

The steady stream of allegations of corporate misconduct against a few high-profile companies bolstered the general perception that bioscience companies place profits before the welfare of the patients who use their products: media revelations of companies taking shortcuts with research protections, rigging trials, hiding negative research results, paying for scientific opinions, providing kickbacks to physicians to prescribe certain brand-name drugs, and misleading advertising gave protesters more ammunition, as did challenges to the justice of high, and rising, drug prices when many companies spend more on sales and marketing than on research and development of new drugs.[12] Such publicity has severely eroded public trust in the industry.[13] It has been exacerbated by the environment of declining trust in corporations in general[14] and in conjunction with the prevalence of the enduring joke that business ethics is an oxymoron.

Starting in the 1990s, a few bioscience companies and BIO responded to this mounting criticism by seeking ways to include ethics as a part of their business strategy. Under Carl Feldbaum's leadership, BIO helped define an ethics agenda and mission statement for the industry, convened an ethics committee, and published an ethics brochure.[15] Ethics publications, papers, and books highlighted the benefits that could be obtained when business managers in this industry realized the ethical pitfalls associated with their products and included a consideration of

[11]Priest SH. US Public Opinion Divided over Biotechnology? *Nature Biotechnology.* 2000;18: 939–942.

[12]Angell M. *The Truth About the Drug Companies: How They Deceive Us and What to Do About It.* New York: Random House; 2004.

[13]Marsa L. *Prescriptions for Profits: How the Pharmaceutical Industry Bankrolled the Unholy Marriage Between Science and Business.* New York: Scribner; 1997; Grady D. (1998, April 15). Study Says Thousands Die from Reaction To Medicine. *The New York Times,* Section A1; Williams N. Food Safety: U.K. Government Tries to Reassure Wary Public. *Science.* 1998;282:856; DeAngelis CD. Conflict of Interest and the Public Trust. *JAMA.* 2000;284:2237–2238.

[14]Trust in Corporate American Has Hit Rock Bottom. *Harris Interactive.* February 24, 2004. Available at http://www.harrisinteractive.com.news.

[15]Bioethics: Facing the Future: There's More to Biotechnology than Science and Business. Available at http://bio.org.

ethics as a part of product development and marketing.[16] As individual companies sought to build an ethics capability that they often lacked internally, some hired outside ethics experts as consultants or employees. Others established ethics advisory boards to tap into a diverse spectrum of ethical perspectives—medical, religious, legal, and philosophical. Our collection of company cases features each of these approaches. The cases also feature ethics experts employed by bioscience companies, such as Steve Holtzman, who studied moral philosophy at Oxford before heading business development and ethics for Millennium Pharmaceuticals. Interleukin Genetics is another company with an affinity for hiring ethics experts with managerial responsibilities: Rahul Dhanda held a business-development position there while completing his book, *Guiding Icarus: Merging Bioethics with Corporate Interests*. Perhaps the most prominent example of such an individual is Phil Reilly, who began as an ethics advisor and board member for Interleukin and eventually became the firm's CEO.

The efforts by bioscience companies to address ethical issues, however, were greeted with a high degree of skepticism by the industry's critics. Allegations included the assertions that industry was "allowing corporate conundrums to masquerade as ethical problems,"[17] that companies were "buying a conscience,"[18] hiring ethicists to keep them honest,[19] merely enhancing the company's public image,[20] and using ethicists to preempt critics.[21] Biomedical ethicists who consulted for these companies drew similar criticisms. They were accused of allowing themselves to become coopted by the companies, of whitewashing for corporations,[22] of being "like watchdogs but...used like show dogs"[23] and not freely addressing complex issues.[24] Others wondered if the bioethicists were really performing as legal compliance officers.[25] Their motives were impugned and they were accused of profiteering as consultants,[26] and of being primarily interested in acquiring power, prestige, or money.[27,28,29,30]

[16]Brower V. Ethical Culture: Millennium Pharmaceuticals, Inc. *The HMS Beagle,* December 24, 1999; Dhanda, 2002; Eaton, 2004.
[17]Devries R. (2004, February 8). Businesses are Buying the Ethics They Want. *The Washington Post.* Section B2.
[18]Elliot C. Pharma Buys a Conscience. *The American Prospect.* 2001;12:16–20.
[19]Kerr K. (1999, June 5). Research Corporations Hire Ethicists to Keep Them Honest. *San Jose Mercury News,* Section 1E.
[20]Park P. Note 11, citing Virginia Ashby Sharp.
[21]Caruso D. Note 16, 1.
[22]Turner L. Note 12, 947.
[23]Elliot C. Note 14, 20.
[24]Michaels D, et al. Advice Without Dissent. *Science.* 2002;298(5594):703.
[25]Magnus D. Note 2, 385.
[26]Blumenstyk G. Bioethics Criticized for Lack of Policy on Their Own Industry Ties. *The Chronicle of Higher Education,* July 19, 2002.
[27]Younger SJ, Arnold RM. Who Will Watch the Watchers? *Hastings Center Report* 2002;32:21.
[28]Brower V. Biotech Embraces Bioethics; But Are They Being Exemplary or Expedient? BioSpace.com, 1999. Available at http://www.biospace.com/b2/Articles/061499_bioethics.cfm.
[29]Boyce N, Kaplan DE. And Now, Ethics for Sale? *US News & World Report,* July 30, 2001; Stolberg SG. (2001, August 2). Bioethicists Find Themselves the Ones Being Scrutinized. *The New York Times,* p. 1.
[30]Turner L. Bioethic$ Inc. *Nat Biotechnol.* 2004;22(8);947–948.

There are clearly instances in which corporate ethics programs have served as mere window dressing, but in most cases, companies appear to be making good-faith efforts to seek better ways to understand and handle the ethical issues they face. Within the industry and the organizations that represent them, there is a growing consensus among executives, physicians, and scientists about the wisdom of addressing the ethical aspects of their work. To address these issues competently, it seems wise to work in cooperation with biomedical ethics experts. As Dhanda has noted,[31] bioethicists and biotechnologists have much to learn from each other. As with any field of expertise associated with business activity, companies benefit most when the experts do not avoid them or capitulate to corporate interests, but rather assist in their operation, either from the inside or from the outside as consultants, by asking hard questions, and exploring the firm's underlying assumptions. Consequently, just as an appreciation for biomedical ethics has improved and enlightened the practice of medicine, human research, and hospital care, it now seems that a broader ethical perspective should also become a standard part of doing business in the bioscience industry.

A NEW APPROACH: BIOSCIENCE BUSINESS ETHICS

The goal of our project was to reveal and comment on the ways in which ethics is being incorporated into bioscience business ethics decision making. The research approach developed for this book, and the wider efforts of this project team,[32] are an effort to walk a middle path between the extremes of the critic safe in an ivory tower, too distant to offer practical guidance to the industry, and the entrenched in-house ethicist or consultant who may have a narrower experience that does not lend itself to forming wider conclusions. We tried to maintain research sensitivity through the use of a well-accepted qualitative research methodology for developing rich case-studies. At the same time, our independence from the firms and the ability to look at experiences across companies was intended to enable us to adopt an independent view, neither unduly critical nor too conciliatory. This balancing act is difficult, yet it seems to offer a way to provide in-depth analysis of the actors—bioscience companies—that are developing the new technologies that are raising the most challenging ethical issues.

To conduct our research, we assembled a highly interdisciplinary team of business school professors, legal scholars, sociologists, scientists, physicians, and graduate students. Our team is made up of researchers affiliated with the

---

[31]Dhanda, ibid.

[32]Among other activities, Stanford cases (Eaton, 2004) and KGI cases have been used to teach a business version of biomedical ethics to employees, business, and bioscience students. The CRDP at the Université de Montreal helped to develop an ethical approach to DNA databanks (http://humgen.umontreal.ca), and the Canadian Program at the University of Toronto Joint Centre for Bioethics has emphasized the potential benefits of biotechnology for developing countries.

Biosciences Business Ethics Center (BBEC), which includes professors from the Keck Graduate Institute (KGI) of Applied Life Science in Claremont, Calif., Stanford University, and the Université de Montreal, and those with the Canadian Program in Genomics and Global Health (CPGGH) based at the University of Toronto Joint Centre for Bioethics. The study was funded primarily with grants from the Seaver Institute to the BBEC and from Genome Canada to CPGGH,[33] with additional support from Genome Quebec to the Centre de Recherche en Droit Public (CRDP) of the University of Montreal. What follows is a brief description of our methodology.

Each case study focused on the ethical decision making processes in the company, and the ethical issues and mechanisms surrounding these processes. Our goal was not to identify companies that were representative of the bioscience sector, as many firms have no systematic process for dealing with ethical issues. Instead, we included companies that were adopting innovative and distinct approaches to ethical decision making, from which the rest of the industry might be able to learn valuable lessons. In addition, we sought a diversity of sector and geography, and of companies at different stages of growth. In all we invited 19 companies to take part in the project, and 13 agreed (see next section for brief descriptions of each of the 13 cases). Four pharmaceutical, one biotech, and one bioagricultural company declined. The participating companies covered the range of large and small biosciences companies in health and agriculture in the United States, Canada, and Europe. In the interest of full disclosure, we note that researchers on the project team had had prior relationships with some of the companies studied: the Centre in Toronto has received support from Merck and Co.; Peter Singer has received consulting funds from Merck Frosst and, subsequent to this study, from Genzyme Canada; and Margaret Eaton had accepted a small grant from Affymetrix for general research in business ethics 5 years prior to her work on this project.[34] To minimize potential conflicts of interest, researchers who had ongoing or past significant relationships with companies were not directly involved with that case's data collection, data analysis or write-up.

Data were collected between autumn 2002 and spring 2004. The primary source of case information was face-to-face interviews with more than 110 individuals in the companies. We worked with our main contact at each company to identify a list of participants who would cover most functional areas of ethical decision making in each firm: regulatory affairs, research operations, marketing, corporate affairs, corporate communications, human resources, clinical and laboratory operations,

---

[33]CPGGH also receives funding from the Ontario Research and Development Challenge Fund and other funding partners listed at www.geneticsethics.net.

[34]Bartha Knoppers and Abdallah Daar sit on Ethics Advisory Boards for other companies that did not participate in the study.

bioethics management, and corporate social responsibility management. Most interviewees were senior business leaders—the people most familiar with how ethics featured in the firm's strategic decision-making process—but frontline workers and lay people were also included. Wherever possible we used these individuals' own words to describe how they approach ethical issues. We aimed to have enough interviews to reach a point where we were hearing repeated information about the issues and mechanisms central to ethical decision making in that firm. Additional phone interviews were conducted when information was incomplete or when geographic constraints made face-to-face interviews impractical. The cases thus represent a "slice-of-time" look at company activity and are as accurate and current as we could make them up until the date written.

The interview data were supplemented with archival material gathered during the visits and from general research. Corporate sources included annual reports, ethics and values statements, press releases, official company policies and any evaluations conducted of ethics-related initiatives. We also obtained publicly available medical, professional, and lay material about the companies. These data were then analyzed, using a modified thematic analysis: that is, we identified key themes in why and how these companies had addressed ethics and what ethical issues they faced. Material and data were then incorporated by one team-member into a case write-up, which was then reviewed and edited by the other members of the team, as described in the first footnote of each case study.

Such detailed case studies require extensive cooperation from participating companies. To secure their agreement to take part in a project dealing with a highly sensitive subject, we allowed the company contacts to review the case write-ups to correct any perceived inaccuracies and remove any information they deemed to be confidential before publication, a common practice in business school case writing.

STRUCTURE OF THE BOOK

The case studies form the heart of the book. They show, in the kind of detail that current and future leaders in the industry should find useful, how 13 bioscience companies approach ethical decision making. Each case includes:

- an introduction to the primary ethical issues facing the firm
- a brief company history and business overview
- background explanation of the science or technology involved
- detailed discussion of the ethical decisions the firm must make as part of its strategy and operations
- a range of approaches the firm uses to make ethical decisions
- a conclusion identifying issues and questions for the future

Although the cases were written with narrative flow in mind, we attempted to address four key questions in each case: Why does this company address ethics?

What ethical issues does the company address? What mechanisms does the company use to address ethics? How effective does the company perceive these mechanisms to be, including whether any efforts are made to assess their effectiveness? The cases are arranged in five thematic sections. Part 1 has four cases that introduce a wide array of ethical issues and show the vital role that a company's early leadership plays in shaping the organization's approach to ethical issues. The first case is *Merck,* one of the world's largest pharmaceutical companies, and one that has been a leader in adopting an ethical approach to this competitive global industry. The case examines how Merck, at a time when the corporation and its leaders were under growing pressure from the investment community, was able not just to sustain, but to significantly increase its ethics and corporate social responsibility programs. The Merck case study explores the variety of mechanisms and initiatives—an internal ethics office, the creation of Ethics Centers around the world, the charitable programs of the Merck Foundation—that Merck uses for ethical decision making.

We then examine *Genzyme,* the world's fourth-largest biotech company, and a leader in the development of ultra-orphan drugs that treat rare genetic disorders suffered by 10,000 or fewer patients worldwide. This case considers how Genzyme's strategic focus on "putting patients first" has shaped its approach to a wide array of ethical challenges, including drug pricing and access for treatments that can cost $200,000 a year for the rest of the patient's life; how to conduct an ethical, placebo-controlled trial for a drug that is already on the market; how to deal with personal conflicts of interest when managers or employees have relatives who need treatment for these rare disorders.

The third case is *Millennium,* one of the most successful of the second generation of biotech companies, and arguably the leading voice for ethical behavior in the biotechnology industry. The importance of ethical issues was first articulated by the CEO and evident from the outset of this Cambridge, Massachusetts-based firm when it hired Steve Holtzman to be both its Chief Business Officer and in-house ethicist. The case study traces the approach to ethics decision making over the firm's 10-year history, and the many innovative strategies that it has used to make ethics a key focus. Among them are a movie-discussion series on ethics; instilling the company's core values into every corporate program; an ethics helpline; and hiring ethics experts to run ethics education workshops within Millennium.

This section concludes with *Maxim Pharmaceuticals,* a San Diego-based bioscience firm that develops drugs from naturally occurring histamine. The case highlights the vital role that the firm's CEO, Larry Stambaugh, has played in shaping the firm's core values and innovative approach to good corporate governance. The case concludes by exploring how the CEO and board deal with a series of business and ethical challenges posed by efforts to raise financing that are complicated by the passage of the Sarbanes-Oxley Act.

In Part 2, we look at companies that are facing ethical issues specific to the environment and to the use of animals in research. *Diversa,* a San Diego-based company, has been a pioneer in the development of benefit-sharing partnerships that align with the goals of the Convention on Biological Diversity. As a leader in this area, however, it also has become a lightning rod for critics of public and private sector partnerships, inhibiting its ability to collect genetic samples. The case examines the rationale for Diversa's biodiversity collaborations and how these agreements operate. This is followed by *Pipeline Biotech,* a contract research organization (CRO) located in a rural part of Denmark that specializes in conducting animal studies for drug companies. The case focuses on how this start-up enterprise has grown, thanks in part to the controversy surrounding animal testing in England and the eagerness of UK-based companies to outsource this work to another part of the European Union with a friendlier regulatory and societal climate for such research. It also illustrates the distinctive approach to ethical issues that Pipeline has adopted—delegating decision making to the frontline laboratory technicians who are empowered to end any experiment if they feel an animal is not being treated ethically or humanely. The last case in this section deals with *TGN,* an early-stage Canadian company trying to build a sustainable approach to ethical decision making in preparation for the challenges to its controversial transgenics business: producing human proteins in, and then extracting them from, pig semen.

Part 3 focuses on an emerging field in the bioscience industry that bridges health and nutrition. Nutrigenomics has developed since the sequencing of the human genome, as companies seek to analyze individuals' genetic information to provide guidance on disease prevention through diet and on effectiveness of medication when treatment is necessary. *Interleukin Genetics* is a Waltham, Massachusetts-based company that made the unusual decision to appoint an ethics expert, Dr. Phil Reilly from Harvard, as CEO. Reilly and his leadership team were faced with an ethical challenge when their firm was in danger of running out of money: should they accept a takeover offer from Alticor, the parent company of Amway, to transform their firm into a nutrigenomics firm? *Sciona* is a UK-based company that advertises and sells direct-to-consumer nutrigenomics services on the Internet. This case study, instructive in several ways, highlights ethical issues related to genetic testing in the very new field of direct-to-consumer marketing of genetics services; a bioethics-consultant model of addressing these issues; and the interaction between a firm, NGOs, and a government commission.

Part 4 examines how companies have sought to enhance their decision making by seeking the advice of diverse ethical experts. *Affymetrix,* a Silicon Valley-based company that has been a pioneer in the development of gene chips, has created an ethics advisory committee to advise them on the many ethical issues involved in the appropriate use of human genetic information. *PharmaSNPs* (a pseudonym we used because the company was taken over during our research), was a genomics firm focused on identifying genetic linkages with major diseases; it gathered human genetic data from around the world through CROs and tried to develop clear ethical guidelines on how to conduct this research. The case study focuses on the ethics advisory board it created to aid in this effort.

Part 5 examines one of the greatest challenges that any company faces in dealing with ethical issues: assessing whether its particular approaches are effective.

Monsanto, the leader in the commercialization of GM crops, became the focal point for critics of GM products. This criticism was so strong that it had a major impact on the company's ability to operate in Europe and other parts of the world. The case study describes how Monsanto made major changes in its approach to ethical issues in response to these critics. These changes included the adoption of a new, detailed reporting system to track its performance against a new ethics guide, the *Monsanto Pledge.* The final case is *Novo Nordisk,* a Danish firm that has been a world leader in the treatment of diabetes, which has relied on stakeholder involvement and the use of its "Triple Bottomline" to assess both its social and its business performance.

The concluding chapter analyzes the findings across the 13 cases. It observes the patterns in the ethical issues facing bioscience firms and how these vary across sectors and stages of firm development. It presents a framework that integrates the different mechanisms firms use for dealing with ethical issues. The analysis highlights lessons learned from the experiences of these leading companies and some of the issues that the bioscience industry may face in the future.

2

MERCK: STAYING

THE COURSE[1]

INTRODUCTION

Driving slowly to avoid the deer that roam the 1000-acre nature reserve in Whitehouse Station, New Jersey on which Merck has built its impressive corporate headquarters, there is no sign of financial stress. During the 1980s and 1990s Merck had discovered a series of novel, blockbuster drugs to become the world's largest, most profitable pharmaceutical company and continue its leadership in corporate citizenship and philanthropy. Starting in 1985, *Fortune* magazine ranked Merck as America's "Most Admired Corporation" an unprecedented seven straight times.

By early 2004, however, the news was not as favorable.[2] The firm had just experienced one of the most difficult years in its proud, 113-year history. For the second year in a row, it had reported a decline in earnings, down to $6.8 billion, a drop of 3% from the year before. The stock price was down close to 50% from its peak in 2000, twice the decline of the pharmaceutical industry average (see Exhibit 2.1A). Its best-selling drug, Zocor®, which accounted for 18% of the firm's $22.5 billion in revenues, was due to come off patent in the United States in 2006.[3] Furthermore, the development programs for two of the drugs that the company had been counting on to replace Zocor's revenue were canceled, after hundreds of

[1]The case study was written by David Finegold based on interviews he conducted with Margaret Eaton and Jocelyn E. Mackie at Merck on March 30-31, 2004 at company headquarters in Whitehouse Station, NJ and Merck's Research Laboratories in Rahway, NJ. Unless otherwise specified, all quotes attributed to these people were obtained during interviews. This case study is one in a series produced under grants funded by The Seaver Institute and Genome Canada to study models for ensuring ethical decision making in bioscience firms. Everyone's participation, suggestions, and assistance are greatly appreciated. Special thanks are extended to Samir Khalil for arranging the visit.
[2]Simons J. Merck's Man in the Hot Seat. *Fortune.* 23 February 2004:111–114.
[3]Once the patents on prescription drugs expire, other companies can compete with generic forms of the drugs, causing prices to fall precipitously and dramatically.

EXHIBIT 2.1A   **Merck Stock Market Performance and
Charitable Contributions**

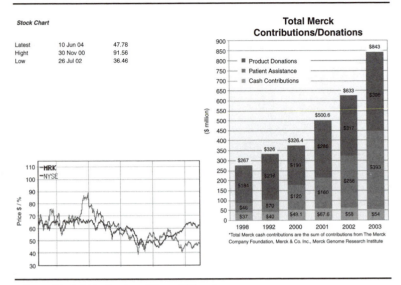

Source: Merck & Co., Inc. Annual Report 2003, p. 6.

EXHIBIT 2.1B   **Merck Company Foundation Spending
and Endowment**

Year	Spending	Assets
1999	$24 M	$360 M
2000	$29 M	$317 M
2001	$40 M	$267 M
2002	$45 M	$191 M
2003	$43 M	$173 M

millions of dollars in research and development, because of disappointing results
in late-stage clinical or preclinical toxicology studies.

At the end of 2003, in response to the lower-than-expected earnings, Merck had
announced the largest layoff in the firm's history, cutting 4400 jobs, or about 7%
of its global workforce. Although Merck stuck steadfastly to its strategy of seeking to grow through discovering new, first-in-class drugs for large, unmet medical
needs, rivals Pfizer and Glaxo SmithKline (GSK) had overtaken Merck's sales
through a set of major mergers and acquisitions. Merck's reputation for attracting
and keeping the best researchers in the pharmaceutical industry had also suffered
a blow with a rise in its historically very low turnover rate and the departure of
several of its stars. Roger Perlmutter left the number two position in Merck's
Research Laboratories to head Amgen's R&D efforts and took three senior

colleagues with him, whereas Emilio Emini, the head of Merck's AIDS program, left to join the International AIDS Vaccine Initiative (IAVI).

As a consequence of these events, Merck's Board was coming under growing pressure from the investment community to make a change in leadership and strategic direction. CEO Ray Gilmartin, 63, who had been running Merck for a decade since being recruited from Becton Dickinson in 1994, was due to retire in 2006. "We believe that resetting the management clock is preferable to letting it run out," wrote Deutsche Bank analyst, Barbara Ryan, at the end of 2003.[4]

In such a difficult environment, it might be expected that one of the first casualties would be Merck's large charitable activities and contributions to fighting disease around the world. Perhaps the most prominent example of this commitment was Merck's unprecedented decision, starting in 1987, to donate Mectizan® in perpetuity to try to eliminate river blindness as a public health problem. Over the last 17 years, Merck has given away over 1 billion tablets of this drug, which, with a once-a-year dose, continues to prevent blindness in millions of African and Latin American people. Rather than cut back on these efforts, however, Merck has increased them substantially in the last few years. Although the company's stock price and the value of the endowment for The Merck Company Foundation, established to fund its charitable programs, were declining, Merck had more than doubled its charitable cash contributions and significantly increased the size of its drug donation programs between 1999 and 2003 (Exhibit 2.1A).

How has Merck been able not just to sustain, but to grow, its global drug donation programs and other charitable efforts in the face of growing pressure from competitors and investors? To answer this question, and to examine more broadly Merck's approach to ethical issues, this case looks first at the history, culture, context, and internal structures that underpin Merck's global health initiatives. It then describes Merck's internal approaches to ethics—how it evaluates the effectiveness of its ethics-related activities and the benefits to the company of this approach. It concludes by examining some of the key ethical issues facing the pharmaceutical industry and how Merck's values and ethical processes have shaped its response to them. These issues include: what medicines to develop; how to price them to balance a fair profit and access to drugs for those who need, but cannot afford them; when and what data to publish regarding the development of new drugs; and how to market and promote drugs ethically.

FOCUSING ON PATIENTS AND THE LONG TERM

THE MERCK CREDO

Any understanding of how Merck operates and makes key strategic decisions begins with the firm's credo, articulated by George W. Merck, a long-time Merck

---

[4]Ibid., 112.

chairman and son of the firm's founder, in a speech at the Medical College of Virginia in 1950:

We try never to forget that medicine is for the people. It is not for the profits. The profits follow, and if we have remembered that, they have never failed to appear. The better we have remembered that, the larger they have been.

George Merck's words contain two powerful elements that continue to shape the company's culture more than half a century later (see Exhibit 2.2 for Mission Statement). First, they lay the groundwork for the company's strategy of focusing on the development of first-in-class or best-in-class drugs to address major unmet medical needs and the long-term commitment to research this entails. This long-term orientation is reflected in executives' views on Merck's recent performance

EXHIBIT 2.2   **Merck Mission Statement**

Merck & Co., Inc. is a leading research-driven pharmaceutical products and services company. Merck discovers, develops, manufactures and markets a broad range of innovative products to improve human and animal health, directly and through its joint ventures.

Our Mission

The mission of **Merck** is to provide society with superior products and services by developing innovations and solutions that improve the quality of life and satisfy customer needs, and to provide employees with meaningful work and advancement opportunities, and investors with a superior rate of return.

Our Values

1. **Our business is preserving and improving human life.** All of our actions must be measured by our success in achieving this goal. We value, above all, our ability to serve everyone who can benefit from the appropriate use of our products and services, thereby providing lasting consumer satisfaction.
2. **We are committed to the highest standards of ethics and integrity.** We are responsible to our customers, to Merck employees and their families, to the environments we inhabit, and to the societies we serve worldwide. In discharging our responsibilities, we do not take professional or ethical shortcuts. Our interactions with all segments of society must reflect the high standards we profess.
3. **We are dedicated to the highest level of scientific excellence and commit our research to improving human and animal health and the quality of life.** We strive to identify the most critical needs of consumers and customers, and we devote our resources to meeting those needs.
4. **We expect profits, but only from work that satisfies customer needs and benefits humanity.** Our ability to meet our responsibilities depends on maintaining a financial position that invites investment in leading-edge research and that makes possible effective delivery of research results.
5. **We recognize that the ability to excel — to most competitively meet society's and customers' needs — depends on the integrity, knowledge, imagination, skill, diversity and teamwork of our employees, and we value these qualities most highly.** To this end, we strive to create an environment of mutual respect, encouragement and teamwork — an environment that rewards commitment and performance and is responsive to the needs of our employees and their families.

and the appropriate response. "We're not going to change the culture of the company because of two hard years," said Dr. Adel Mahmoud, President of Merck's Vaccine Division. "We know it is due to pipeline, market, and regulatory issues. We have to address them directly, not by running around [and cutting long-term commitments]. We've said we'll invest more in research and sustain it." In 2003, Merck increased its R&D expenditure from $2.7 to $3.2 billion, a growth of 19% over the prior year, an investment equal to 14% of its sales and nearly half (47%) of its profits.[5] As several business leaders noted, including Merck's Senior Vice President and General Counsel, Ken Frazier, the small drop in Merck's net income "doesn't mean we are not profitable. Recent trends in our EPS (earnings per share) growth are not indicative of our long-term prospects." Zaki Hosny, Vice President of Marketing and Operations in the Europe Middle East and Africa (EMEA) region, concurred. "I'm not sure what correlations one should see between charitable donations and business performance. I feel that the company's charitable level of giving should be at a certain level—based on our size and not on the income statement of a few years."

The second element of the credo is the recognition that Merck, as a leading pharmaceutical company, has a responsibility not just to its shareholders, but also to the wider society of patients around the world. Perhaps the best illustration of this societal responsibility came during World War II, when Merck tested potential antibiotics isolated by Rutgers University Professor Selman Waksman, who discovered streptomycin, a powerful antibiotic that could be used to treat tuberculosis and a variety of infectious diseases. Merck made the decision to give up its potentially very lucrative exclusive license to the wonder drug so it could be made freely available.[6]

Samir Khalil, who, as Executive Director for HIV Policy and External Affairs for the EMEA region, plays a key role in many of Merck's charitable initiatives, stresses these two elements of the credo in discussions with fellow employees:

> When I go into the cafeteria, I will often get questions from my co-workers, asking why we're spending so much in Africa when the stock price is way down and we're laying people off. I tell them we have to focus on the long-term. Merck has always supported efforts to improve global health and we now face a huge (HIV) problem and have an opportunity to make a difference. We have a responsibility to do something.

CHARITABLE PROGRAMS AND THE MERCK COMPANY FOUNDATION

Merck institutionalized its commitment to societal stakeholders in 1957 by creating the not-for-profit Merck Company Foundation as a mechanism to fund most of its charitable activities (although the drug donation program is funded and administered separately by the Corporation). The endowment for the Foundation,

---

[5]Merck had historically prided itself on the productivity of its research units, spending less than its main pharmaceutical rivals on R&D, yet producing many of the initial drugs in new blockbuster categories.

[6]Hawthorne F. *The Merck Druggernaut*. Hoboken, NJ: John Wiley & Sons; 2003:13.

which stood at $176 million in June 2003, has been periodically replenished through the sale of business units. Merck's annual charitable spending comes predominantly (over 80%) from the Foundation's endowment, supplemented with cash contributions from Merck & Co., Inc. operating funds. Although all its funding comes from Merck, the separation of the Foundation from the operating company shields it from the ups and downs in Merck's performance, inevitable in a business in which the average cost of bringing a drug to market is estimated at $800 million, and in which only 1 of 10 drugs that enters clinical trials eventually makes it to market. "That is why we have the endowment, so that (the Foundation) is not influenced by short-term fluctuations," said Frazier.

The major growth in Merck's charitable spending and donations can be traced to a shift in the Foundation's strategy in the mid-1990s and the accompanying long-term commitments that Merck has made to global health initiatives. Brenda Colatrella, Senior Director of the Office of Contributions, explained the shift in the Foundation's focus:

> In the last ten years, it has changed very significantly. At one point, about 70% of our cash-giving was focused on education in the US, including support for universities and other non-profits. The last few years have seen a major shift, as Merck took a careful look at our global priorities in corporate and social responsibility. This was part of becoming a more global corporation. We spent a lot of time four to five years ago looking at the different issues and strategies for how to address them. We identified four strategic priorities (see Exhibit 2.3). The other thing we've done is, instead of funding many little projects, we're now focusing on a few larger ones where we can really make an impact and evaluate that impact. That's harder when your resources are spread over many small efforts.

This strategic change appears to have occurred because of the confluence of a large and urgent, unmet medical need—the AIDS pandemic—and Merck's new capability to address this need. "Once we had our HIV/AIDS medicines, Crixivan and Stocrin,® [Merck's two AIDS medicines that came to market in the mid-1990s] on the market, then we started to work on partnerships with other organizations to get drugs to the people who need them," said Brian Healy, Vice President for Economic and Industrial Policy. "You could see that the money you donated was actually doing something, so we could get support for it. Same as countries who

EXHIBIT 2.3 **Merck's Corporate Philanthropy Priorities**

Merck's corporate philanthropy programs are guided by four strategic priorities, aligned with our business capabilities and mission:

- Improving access to quality health care and the appropriate use of medicines and vaccines.
- Promoting environments that support innovation, economic growth, and development in an ethical and fair context.
- Building and strengthening long-term capacity in the biomedical sciences through advancement of education and research.
- Contributing to local communities where Merck has major facilities

Source: http://www.merck.com/about/cr/policies_performance/social/philanthropy.html

can see that their foreign aid is having an impact." And Hosny commented, "What we've been able to do is be opportunistic in the sense of contributing and partnering where we've had the opportunity to make a difference."

As part of this new strategy, Merck and its Foundation have made a number of new, major, multiyear commitments,[7] including:

- extending the Mectizan program, which now reaches 40 million people a year in 34 countries and enables the delivery of additional drugs[8]
- treating children with HIV in Romania. Merck donated $1 million in 1999 to create a network of seven AIDS treatment centers to care for the 6000 children who were infected and reduced prices on HIV/AIDS drugs to non-profit levels in 2001
- joining the UN/Industry Accelerating Access Initiative (AAI), in which Merck, along with seven other major pharmaceutical firms and five U.N. organizations, has been helping to improve access to HIV care and treatment in developing countries since May 2000. By the end of 2003, AAI estimated that as many as 150,000 patients in Africa now have access to antiretroviral therapy, a more than 15-fold increase from when the initiative began, and drug prices have been reduced by as much as 95%
- creating the African Comprehensive HIV/AIDS Partnerships (ACHAP), a 5-year partnership with the Bill & Melinda Gates Foundation, launched in July 2000, in which each committed $50 million to a large-scale demonstration project in Botswana to help the government of Botswana develop and implement a national and comprehensive HIV/AIDS program, covering prevention, care, treatment, and support. Merck also agreed to donate its two antiretroviral drugs to Botswana for the 5-year initiative
- supporting the Gates Foundation's Global Alliance for Vaccination and Immunization (GAVI) by donating $100 million worth of Recombivax, Merck's hepatitis B vaccine.

The Foundation program priorities and funding levels are set by an oversight committee that is chaired by Merck's Vice President of Public Affairs, Joan Wainwright, and consists of five heads of Merck's strategic business units. Wendy Yarno, Executive Vice President, Worldwide Human Health Marketing, explained how the process works:

> As far as setting our priorities for the Foundation, there are different ways, depending on the issue. It may be discussed by the management committee or sometimes at a policy level. There is a set of guidelines and principles around what the Foundation will support and won't support. We've looked for areas and examples where we can teach people to fish and not just give them the fish. That's why we've focused on major demonstration projects like Botswana, or Mectizan, where we had to partner. Providing access to healthcare is a major issue we try to support with the Foundation.

[7]For more details on these programs see: http://www.merck.com/about/cr/policies_performance/social/philanthropy.html.

[8]For details see www.mectizan.org.

In setting its priorities, Colatrella notes that the Foundation tries to be very careful in assessing, "Are we making decisions for the right reasons?"

> To be funded by the Foundation, the initiative needs to be philanthropic in nature while being consistent with our company mission and our business. A couple of times on the product-donations side, we would have someone say, 'We'd like to make a donation of product x, so we can get a head start on a market when we launch the product in 6 months.' Legally there may be nothing wrong with it, but ethically it is not OK. We want to do things that are done for the right reasons, not for a self-serving motive.

Although the cash commitments from the Foundation have leveled off after a period of steep growth (see Exhibit 2.1B), Colatrella says:

> Overall, our commitment has not changed, despite what's going on commercially. I would argue it has gotten stronger in the last few years. I have not been asked to cut the budget. We put in our budget requests based on what we'll need for our multiyear commitments. Product donations also continue to grow, driven by previous commitments. This sustained commitment is due to Mr Gilmartin's philosophy and leadership in this area. He's made it very clear he doesn't expect us to cut back.

EXTERNAL PRESSURE

The role of the Foundation's Oversight Committee in keeping Merck aware of constituents' viewpoints suggests another important reason why the company has avoided cutbacks in its charitable efforts: the strong external pressure on Merck and other pharmaceutical companies to help combat global health problems. Indeed, whereas some financial analysts and shareholders may have criticized Merck and its CEO for not focusing enough on its bottom line, some activist groups and non-governmental organizations (NGOs) were accusing Merck and the industry as a whole of not having done enough. "We were the most admired industry and now we're one of the most criticized; it's as if we are marketing cigarettes rather than life-saving medicines," observed Michael Quinlan, Vice President and Assistant General Counsel, Human Health International. "In retrospect, we may have brought some of this upon ourselves. Perhaps, as an industry, we didn't sufficiently understand what the public was thinking and we may have insulated ourselves until it came upon us... In the US, the industry is viewed by some as greedy and out for a profit. I don't agree with that view, but the press is publishing articles pointing to this."

One example of this came in the winter of 2004, at the time when this case study was being prepared. The *Wall Street Journal* ran a story that portrayed Merck as being slow to carry through on its promise to make its AIDS drugs available to the world's poorest nations at no profit.[9] Khalil, of Merck's EMEA External Affairs group, was clearly perplexed at the article and the attack from Médecins Sans Frontières (MSF, or Doctors Without Borders) that had prompted it.

[9]Zimmerman R. (2004, March 3). Merck Still Draws Fire for HIV Drug for Poor Nations. *The Wall Street Journal*, Section D3.

MSF criticized Merck for not following through on a promise of providing its newest AIDS drug, Stocrin, at $0.95 per day, because it was charging more for three 200-mg tablets ($1.37), than for one 600-mg pill ($0.95), and for failing to submit all the information needed for approval of the drug in different countries, citing a specific delay in South Africa. Merck wrote a letter of response explaining first that the price of the pills differed "due to different manufacturing processes for the capsules and the tablets…but each is set at a level at which we do not profit in the poorest countries." Merck refused to "sell the lower-dose pills at a loss…because it would not be sustainable." Second, Merck noted that, in South Africa, it had applied for "fast-track registration of Stocrin…the day we made the announcement about our pricing policy for the tablet" and that it had complied quickly with any requests for additional data.[10] "I'm not sure what motivates some NGOs, like MSF, to attack the companies that are trying to do the most in the region," said Khalil, recounting his impression of the sequence of events that led to the article:

> Recently they contacted us and asked us for a list of all the countries where we had registration for our new AIDS drug. We told them we couldn't release that outside the company, but that if they faxed us a list of countries they were interested in, we could then tell them in which of these we had approval. They faxed us a list of 25 countries and we faxed back the 10 countries where we had approval. But many of those where we didn't yet have approval were not our fault. In South Africa, for example, we'd applied for approval for the drug two years ago. We already had approval in Europe and from the Food and Drug Administration (FDA), but they kept asking us for more data. We've applied for approval everywhere in the world we can get it, but we can't make these governments go faster. MSF didn't reply to my fax, but instead sent a letter directly to our CEO protesting that we hadn't done more to get approval around the world and issued a press release with their allegations. I think they got almost everything within it wrong. The *Journal* then ran a story that got most of the facts right.

THE BOARD OF DIRECTORS

One of the key factors that has allowed Merck's management to sustain its long-term focus on charitable programs has been the composition and structure of its Board. In keeping with leading practices in corporate governance, Ray Gilmartin is the only insider on Merck's Board. An unusually high number of outside directors, however, come from academia, and they have placed pressure on the company to focus on wider issues of corporate and social responsibility. According to Frazier:

> We have a Board that recognizes the long-term nature of our business. Of course we want to minimize the ups and downs, but they are, to some extent, inevitable. Just because our rate of growth has slowed down, it doesn't change the fact that people are still dying of AIDS in Africa.… Our issue from the Board regarding donations has been whether we are doing enough—never that we're doing too much.

[10]Correspondence between Merck and MSF, letters dated March 8, March 15, and April 1, 2004.

An example of the Directors' concerns regarding Merck's wider role was the Board's decision to create a new Standing Committee on Public Policy and Social Responsibility. This Committee, composed predominantly of academics, was formed at a time of growing public debate about drug pricing to examine the full range of social, political, economic, and environmental issues that affect or could affect the company. As Frazier observed:

> The Committee doesn't only deal with the drug-pricing issue. Some in our society feel that drug prices are too high, so they [the Committe] will do anything that is within the bounds of the company's means to get them down. This is not an easy industry to run a company in. We are not exempt from any of the performance standards of other companies, yet we feel an obligation to do more. This committee functions as a sounding board; they feel quite empowered in their role as overseers of Merck's reputation to challenge us. Their point is to make sure we are always cognizant of what other constituents think of our drug-pricing policies.

Although the Board is thus firmly behind Merck's commitment to combating global health problems, Frazier admitted that "from shareholders, we sometimes get questioned why are we giving their money away at all."

ACADEMIC ORIENTATION

Another element of the Merck culture that aligns with and reinforces the long-term, societal orientation is its similarities with academia. Although Merck is a for-profit company, over time it has been able to attract top scientists by offering an environment where they have the resources and freedom to work on major challenges in the discovery of new medicines. In drawing a contrast with some of Merck's competitors, Mahmoud pointed to the close parallels between Merck and a university:

> Merck is closest in culture and value to academic medicine. Dr. Vagelos, Ray Gilmartin's predecessor as CEO, Scolnick (Edward, former head of R&D), Kim (Peter, Scolnick's successor) all came from academia... In many ways the coin of the realm at Merck is knowledge. And that is really the coin of the realm in academia. We exist here primarily because of our knowledge, not just because we are the best deal-makers or product marketers.

Critics of the pharmaceutical industry have argued that a more cynical reason that companies like Merck make drug donations is for the tax benefits, but according to Colatrella, this is not the case:

> There is a popular misconception in the press and public that we're claiming a huge tax write-off on these drugs. For public reporting we value the drugs we give away at the US wholesale value or list price (which totaled $789 million in 2003, up from $575 million in 2002). But for tax purposes, we determine the value using a complex formula set by the IRS. It is based on inventory cost or the cost of production—it is not based on the US wholesale or market value. A good ballpark estimate of the tax credit is 2 to 2.5 times the inventory cost, which takes into account overhead, research, and other costs. Generally this is much lower than the list price per tablet. Then, on top of this, you have to take into consideration the corporate tax rate, which can be estimated at 42%.

ETHICS RESOURCE CENTERS

In addition to its growing portfolio of global health initiatives, The Merck Company Foundation has also undertaken a major new business ethics initiative in partnership with the Washington, DC-based Ethics Resource Center (ERC). Ray Gilmartin served as a chairman of the ERC both prior to joining and in his early years at Merck and, according to Frazier, Gilmartin's work with the ERC "completely transformed his life in a business sense," providing a "view of how to instill an ethical culture informed by ethics practices." Thus, when Khalil began exploring an initiative to address ethical issues affecting how business is conducted in countries where Merck operates, he recalls that Gilmartin "was very supportive":

> I first got the idea in the mid-90s that industry was spending a lot of time regulating itself, but not discussing ethical decisions with policymakers. If government is corrupt, then there is only so much that firms will be able to do. The ultimate outcome I wanted was a two-way dialogue between industry and government, but I wasn't sure how to pursue the idea.

A colleague at the Foundation suggested he contact the ERC, which agreed to partner with Merck. With financial backing from the Foundation, they have now helped to establish four ethics centers around the world, beginning with the United Arab Emirates (UAE) in 1998, followed by South Africa in 2000, Colombia in 2001, and Turkey in 2003. These centers are independent, and each has its own operating model and activities, but all "provide education and advocacy on health-care and business-ethics issues."[11] "We started with the UAE," said Khalil, "because like Botswana (for the AIDS initiative) we wanted a demonstration project with a high likelihood of success. It was a small country where it was easy to get to know the main players. But we were met with lots of skepticism initially. The leaders kept asking, what's the real reason Merck is doing this?" Merck eventually overcame this skepticism and convinced the government to become a partner in creating the Center. Among its biggest achievements has been creating a code of ethics for the UAE Ministry of Health.

Khalil explained how the ethics centers fit into Merck's overall strategy and mission:

> Ethics is one of Merck's strategic objectives. Our role as a company is to contribute to the world. Our primary way to do this is through innovation to create new drugs and vaccines, products that will save lives and improve health care. A second part of our role is to facilitate access to those in need, through differential pricing, donations, etc. The third part is something we've learned to do over time. We've learned from Mectizan and Botswana that we can have an additional important role to improve partnerships and bring care and hope to those living with HIV. Building infrastructure has been essential. Because of the limited capacity in these countries, they are not able to start it on their own. The best

---

[11]How Merck is Helping to Build an Ethical Business Environment Worldwide. *MerckWorld*. First Quarter 2004:8.

thing to start with is a public-private partnership to build capacity. Finally, I see a further role over those three major ones: to ensure that the environment we operate in is ethical, both internally and externally.

INTERNAL ETHICAL PRACTICES

ETHICS OFFICE

When Gilmartin took the job as Merck's CEO, one of his first initiatives was to create an internal ethics office to match Merck's high-profile external efforts to promote business behavior that was more responsible. Merck had had an Ethical Business Practices Policy on the books since 1976, but Gilmartin wanted to put in place more formal programs to increase awareness of ethical issues. He chose Jacqueline (Jacquie) Brevard, an international lawyer who had worked with all the major areas of Merck's global operations, to be the Chief Ethics Officer heading up this effort. Recalled Brevard:

> Ray decided he wanted a lawyer to head the Ethics Office. Given the issues that arise, it is helpful to have a legal perspective… And with the nature of our business, Ray wanted the Office to be global, so he wanted an international lawyer with experience in developing nations. By the time he added up all the desired criteria, the list narrowed to me.

Brevard, who had no formal training in ethics, explained further the rationale for choosing a lawyer to head the Ethics Office and how the new effort differed from Merck's past practices:

> Ethics has always been part of a lawyer's responsibility at Merck. We used to have ethics training conducted by Legal and Finance—a compliance kind of effort. Lawyers here have always had commercial responsibility; you can't just opine on legal issues. You're responsible for negotiating the deal and informing people of the risks and potential exposure, including ethical business practices. If a deal is happening in a part of the world where ethical business practices may be an issue, then a lawyer has to be part of the negotiations.

The Ethics Office was established in May 1995, with Brevard reporting to the CEO. The Office has now grown to include Brevard, plus four ethics officers serving different parts of Merck's operations, two support staff, and administrative support, with a budget in 2004 of $2.1 million. Its original charter was to draft a new Code of Conduct and create a comprehensive ethics training and awareness program. Brevard hired the International Business Ethics Institute (IBEI) to help with this effort, beginning with an organizational ethics assessment. "It was a way of taking the temperature of the firm in the ethics area," she said. "I wanted to use an outside consultant to do it, to get information that was as candid as possible. We took a cold, hard look at ourselves and asked, 'Is the audio matching the video?' or, 'Are our deeds consistent with our words?'"

IBEI interviewed senior leaders and conducted focus groups of 8 to 15 midlevel managers, production personnel, and administrative staff. This was followed by a business ethics survey of a random sample of 22% of the workforce in every country where Merck had operations. Given the scope of Merck's operations, the survey had to be translated into about 20 languages, sometimes for as few as six people in a country. The goal, Brevard explained, was to design an approach to ethics that would reflect Merck's international and cultural diversity. "We didn't want to repeat the mistake of a lot of US firms by creating a program designed for the US that we would then take on the road. We wanted to be global from the outset." The internal assessment produced three main findings:

1. Merck's employees generally have a high awareness of the company's expectations regarding a high standard of ethical behavior.
2. Actual practices didn't always match the official policies, which could be, as Brevard noted, "because certain policies had become obsolete, which resulted in employees fashioning creative solutions."
3. Employees often didn't know where to go for help if they faced an ethical dilemma.

Along with the internal assessment, Brevard used contacts from the Ethics Officer Association to conduct external benchmarking on how other firms, such as General Electric, United Technologies, Levi Strauss, and Texas Instruments, handled ethical issues. These visits were very informative, but Brevard noted the differences in the industry environments:

> Some, like GE and United Technologies, were in the defense industry and were heavily regulated. Levi's had a fabulous program, but they made blue jeans, which are not exactly a life-and-death product, so if they didn't approve of the politics in Burma, for example, they could just leave. Our business is different; and a decision to withdraw from a given market would be a lot more thought-provoking.

Merck did not benchmark any firms in the pharmaceutical industry, because Brevard perceived them to be focusing on compliance, rather than ethics. She explained the distinction:

> We take it as a given we will comply with all the relevant laws; the legal standard is the minimum. But there are a lot of things we don't do that are legal, but not consistent with our values and standards. For example, if a person who has been here 25 years is going to retire in November and their skills no longer match what's needed, it is unlikely that we would fire them in May for non-performance. We continually ask ourselves, "What kind of company are we? How do we want to be perceived by our employees and the medical community?" We would like it to be a place that believes in full disclosure. When our reps walk into a doctor's office and they are asked a question, they can answer in a narrow way, or they can give full scientific information. We want them to do the latter.

This combination of extensive internal data gathering and external benchmarking ultimately resulted in the drafting of Merck's first Code of Conduct, entitled *Our Values and Standards: The Basis of Our Success,* together with the design of an accompanying training program. The new 25-page Code, translated into 27 languages, was launched in 1999. It was written in a question-and-answer format and posted on Merck's intranet as well as its company external website, making it easy for employees to consult when dealing with everyday ethical issues (see Exhibit 2.4 for a sample page).[12]

To emphasize its importance, Merck conducted a carefully staged roll-out of the new *Our Values and Standards,* in which the firm's most senior leaders discussed the new Code with their employees and made all employees aware of the toll-free 800 number they could use to contact the Ethics Office. Based on

EXHIBIT 2.4   **Standards and Values – Sample Page**

Source: http://www.merck.com/about/conduct.html

[12]For the Code of Conduct see: http://www.merck.com/about/cr/policies_performance/pdf/ code_of_conduct.pdf.

feedback and questions from this roll-out, the Ethics Office then created a scenario-based training program, tailored to different countries and individual job functions. All employees received 2 hours of training online, by video or in person, with two additional hours of training for those in key decision making positions. This training, which was rolled out over a 2-year period, produced additional feedback on what was and was not working with the Code. "That's when we learned that no one was happy with the policy regarding personal use of the company computer," said Brevard. "When the Code first came out, it said you could only occasionally use the Internet for anything personal. Based on the negative feedback that we received, we changed it. Now the policy is more sensible: if you've got a computer on your desk, use it like a phone. You may make a personal call, but you wouldn't stay on for hours talking with your mother. You should do the same thing with the Net. The criterion is productivity."

After the initial training, the Ethics Office has reinforced the importance of the company's *Values and Standards*. Employees are regularly reminded of the company's key corporate policies, and all key personnel are required to certify annually as to potential conflicts of interest. All new employees are required to undertake ethics training within 6 months of joining Merck. Although some may view the review as a bureaucratic formality, Ken Gustavsen, the Mectizan Donations Program Director, said he found it helpful. "We get an e-mail as a reminder to review policies yearly to show what we stand for. I've had lots of examples where people said, 'Come stay at this hotel or take these baseball tickets.' I'd love to go to a Yankees game, but we just can't do it." Beyond the activities associated with the *Values and Standards*, the other main function of the Ethics Office has been to create channels for employees to share concerns about ethical issues. They can do this in several ways. One is to call a free advice line that is run by an outside vendor; this advice line gets 50 to 75 calls a year. More common is for employees to contact the Ethics Office directly. They can then ask to speak to an ethics officer or an ombudsman. Brevard explains the distinction for the staff that play both of these roles:

> The ombudsman process is marketed as a safe haven; the office is meant to be neutral, somewhat outside the company, not reporting to any line manager, but straight to the CEO. Employees can have a conversation that is completely confidential... There is an exception—we will keep the discussion anonymous unless it could cause harm to the company or an individual, such as physical harm, sexual harassment, or illegal conduct. We can talk for an hour about why it was unfair that someone didn't get tuition benefit and that's no problem; we do not have to do anything. If it is a case of sexual harassment, then we tell them we have to act.

About 85% of the 1000 or so calls that the Ethics Office gets each year are classified as part of the ombudsman process, that is, kept confidential, with the main distinction being that all records of this conversation are destroyed once the case is complete. Brevard commented, "I measure how we're doing based on the willingness of employees to raise issues and the comfort they have in making allegations. When I look at the advice line, I notice that at one time 100% of the calls were anonymous; now it is about 50:50. I see this as a sign of having an effective program. People are more open and have no fear of retaliation."

LEADERSHIP AND PERFORMANCE MANAGEMENT

Although the Ethics Office activities have clearly been important in raising awareness of ethical issues and how to deal with them, a number of the interviewees commented that the key to making the Merck credo real was having leaders send consistent messages about what types of actions are desirable. Observed Yarno:

> I think that at the end of the day, you can have credos and training programs, but if the behavior is not modeled and not rewarded—for example, if you're told to act ethically, but if you don't get the business because you did act ethically and then you get punished—it sends the wrong signal. Management has to back people up when they make decisions, not just by paying them accordingly, but also by supporting their behaviors as managers... We have a set of leadership principles that are part of every person's performance evaluation. We've had them now for 7 or 8 years—they are fairly well integrated. There are three dimensions: 1) strategic ability, which is knowing what to do, 2) can they get it done? and 3) do they get it done the right way? and that's the leadership aspect—do they collaborate and treat people with respect. We've terminated people for leadership issues—people who have been very good performers, but lousy leaders. It is meaningful.

TREATING EMPLOYEES ETHICALLY

Employees are one of the key stakeholders that Merck has always tried to treat in a supportive and ethical manner. Long before Silicon Valley startups had popularized the practices of free food and flexible work practices, Merck's huge complex in Rahway, New Jersey, had sought to provide everything that its star scientists needed to be happy and productive. This Merck minicity contains multiple eateries, a day-care center, boxed take-home suppers, a bakery, a health club, a dry cleaner, and a credit union, to go along with world-class laboratory and production facilities. These amenities, plus above industry average pay and benefits, have helped Merck rank consistently in *Fortune's* list of the 100 best places to work.

Merck has pursued two specific initiatives to try to live up to the ethical principles of creating equal opportunities for all of its more than 60,000 global employees. The first is a major focus on promoting diversity, which began in the early 1980s as an effort to comply with new regulations regarding affirmative action and equal opportunity. Now, these programs go beyond compliance to a commitment "to expand the efforts globally to make all of Merck an inclusive work environment," according to Deborah Dagit, who joined Merck in 2001 from Silicon Graphics, Inc. to head up its diversity programs. These diversity programs include:

- providing training to employees, such as on micro-inequities, which Dagit describes as focusing on "verbal and non-verbal cues that convey favoritism or have the effect of marginalizing contributions." An example, according to Dagit, is, "You are the only woman at a meeting and you make a suggestion and none of the men respond to it. Then a man at the meeting makes essentially the same suggestion and it is well received by the others. It's your idea and he gets the credit for it. This type of phenomenon is not usually intentional, but it is also not uncommon at most companies. At Merck we are looking at a creative way to address this by raising awareness and teaching new skills."

- flexible work arrangements that make it easier for employees to balance work and personal commitments
- development of leaders, in particular the "talent pipeline" for upper management

One signal of the importance that Merck attaches to its diversity efforts is the CEO's decision to create a Diversity Worldwide Strategy Team, the only one in the company that does not deal with a product or disease area. This emphasis appears to have worked, because Merck has, according to Dagit, "the most diverse management team and board in the pharmaceutical industry and one of the most diverse among Fortune 500 companies." Merck's goal is "to be a recognized leader" of companies on diversity. Dagit sees Merck's diversity efforts as closely connected with its broader approach to ethics:

> When we get questions from communities (employees, shareholders, customers, the public) about our commitment to values and whether what we say is the same as what we do, diversity is critical in this regard. From an ethical standpoint, I can say we do it. But we also need an authentically inclusive environment so that what we say and what we do are the same as reported by our employees.

Dagit is careful to point out that diversity is not just an ethical issue, but also a core part of the talent strategy for the business. "If you put all of our employees together who are from disenfranchised groups (women, African Americans, Hispanics, American Indians, Asian Americans, people with disabilities, the GLBT community, older workers, etc.), the number exceeds the population of white males in the company."

The other, more recent human resource initiative that relates to Merck's global health efforts was the adoption of an "HIV/AIDS, Tuberculosis and Malaria Workplace Policy." As Merck's global operations expanded into areas where there was a much greater risk of infectious disease, it wanted to ensure that all its employees and their dependents had access to prevention and treatment programs. Explained Dagit, "Merck has played a leading role in making these drugs affordable in the developing world, so we wanted to make sure that we had the same access policy internally for our employees in all countries—and we wanted consistency and to appropriately address the stigma that can be associated with these diseases."

One sign that employees appear to share Merck's concern for how people are treated came during the most recent annual employee donation program at the end of 2003. Recalled newly appointed Director for Corporate Responsibility Communications, Maggie Kohn, "It was around the time of the lay-offs and we still had a record number of donations... The stock was at 50% of where it was three years ago and people gave even more donations than ever before, and this amount was matched by Merck."

ASSESSING THE IMPACT

Although there appears to have been little demand within Merck to cut back on its ethics initiatives and charitable giving in the face of disappointing financial results, there has been recent emphasis on demonstrating the impact of the substantial resources Merck is devoting to these programs. For the Foundation, Colatrella observed:

> What has changed is there is now more pressure on us to show value. One thing we haven't done well in the past is evaluation, but that's true for any company with corporate giving. Measurement is tough. There's a lot of anecdotal evidence that what we do matters, but hard data are difficult to come by; most programs have very long-term results. There is a lot more pressure now to become more focused and to show that our activities both make sense for the company and make a difference for stakeholders and the community.

Colatrella described how Merck has gone about assessing the effectiveness of its different programs, noting that the focus of this assessment is not on any benefits for Merck, but rather whether the initiatives have achieved their intended objectives:

> We've tried in the last couple of years to put more emphasis on credible third-party evaluation. The Mectizan Donation program is in its 17th year; we think it has achieved good results, but we don't really know. The World Health Organization (WHO) and the World Bank have done studies that have appeared in journals like *The Lancet* showing the positive effects. [To conduct a formal evaluation] we decided to engage the Johns Hopkins-Bloomberg School of Public Health to look at impact. We've asked them to look not just at the reduction in river blindness, but also what effect the program may have had on the wider infrastructure and building of public-private partnerships.[13] We've done a similar thing with several of our cash donation programs. For example, we've been working with the United Negro College Fund for the last decade to try to expand the pool of African American candidates in the field of science. We've just started an assessment to see whether this funding has made a difference.

Evaluations are also underway for Merck's Enhancing Care Initiative with the Harvard AIDS Institute, which will end in 2004, and the ongoing program on Pharmaceutical Policy Issues that supports academic centers engaged in research on health-policy issues. Colatrella conceded:

> We still struggle with the evaluation. With many of our programs, we didn't build evaluation in from the start. But now with any discussions of new programs, we say that you can't move forward without designing-in evaluation, which can be difficult.

---

[13]The generally very positive results of this evaluation were published in a March special issue of *Tropical Medicine and International Health.* 2004;9(3):A1–A56. See http://www.mectizan.org/impact.asp.

Folks here are so eager to get things off the ground; they see a problem and they want to move quickly. But we try to tell them that it will do a world of good five years from now if we build evaluation in from the start. That was not true [of Merck's approach] in the past.

Khalil argues that while the Botswana HIV treatment program may not have been designed with a formal evaluation from the outset, it has always focused on setting targets and assessing progress against them:

> We have clear objectives: by the end of 2004 we seek to have 30,000 people in treatment. We measure to see if we're meeting criteria... There's now more and more of a focus on measurement. We're evaluating some of the programs from ACHAP and becoming more sophisticated at it. For example, they have a huge program for condom distribution that costs a lot of money. We have a specialized team working with the organization we fund to do that. We're looking at how many were distributed, rate of utilization, and what impact they've had on infection rates.

In the case of the longer standing Mectizan program as well as AIDS programs, Gustavsen pointed to a wider range of criteria:

> An easy one is to measure the number of people treated. We can see how many are treated and check the transmission rates as well. Another way is to see how committed our existing partners are, and the new partners that want to get involved. It's exciting to see the new organizations that want to get involved. For example, when I was just in Africa, several NGOs came to Kenya to see how they could get involved in the Mectizan program (that is beginning) in southern Sudan.

When it comes to Merck's internal ethics efforts, Brevard noted the importance of being very cautious in what types of evaluation data are collected and how they are interpreted:

> You have to be very careful to avoid any relationship with the bottom line. Finance might look at it that way, but I don't. If we get an increased number of contacts [on the help line] it could be because employees are disgruntled or because they view us as very open... Our database is all we have to assess the ethics issues that come up. For example, conflict-of-interest calls shot through the roof because we issued new guidance to employees, and people finally figured out what the policy was [and were calling with questions].

The Ethics Office recently conducted a follow-up, company-wide survey to its initial organizational assessment to see what effect its training and other programs have had on awareness of ethical issues and behavior.

For Merck's diversity programs, Dagit uses a variety of methods to assess their effectiveness including:

- frequent, short surveys of the workforce to assess job satisfaction, how engaged people are, and how well Merck is doing at employee retention
- employee groups consisting of historically disenfranchised people (black, Hispanic, gay and lesbian, women), who are asked to give a scorecard of how well Merck is doing in diversity and how they view the company
- external benchmarks, such as the rankings by Diversity Inc.com, *Working Mother,* and *Fortune*

• longitudinal analysis of diversity among top management and the rest of the workforce, as well as access to training and flexible work arrangements

BENEFITS OF AN ETHICAL APPROACH

Although Merck employees clearly believe that supporting global health initiatives, and behaving in an ethical manner, are the right way to do business, they are also careful, like Brevard, not to draw a direct connection to the company's financial performance. When asked about the benefits to Merck of an ethical approach, Yarno seemed to sum up the sentiments of her colleagues: "It's not something you can say really adds to your bottom line. I think it's pretty convoluted how you get there. It's the right thing to do and it's something we want to do. Some of these things are fulfilling George Merck's credo."

Despite these reservations, Merck employees were able to identify a variety of ways in which the company benefits both directly and indirectly from an ethical approach to doing business. These benefits include:

• attracting, motivating, and retaining top talent
• performing good science
• creating a positive brand and company image
• enhancing regulatory compliance
• providing access to policymakers
• improving relations with stakeholders
• promoting a good climate for global business

AN EMPLOYER OF CHOICE

Although the way Merck treats its workforce clearly contributes to its ranking as a top place to work, so do the company's wider efforts to give back to the community (see www.Merck.com for a list of awards). "Surveys show that the things we do for people around the world are part of what makes employees feel good about working here," said Yarno. Frazier noted that this had played a key role in his decision to move from a law firm to in-house counsel at Merck. "Merck was my favorite client… More than any other company, Merck seemed to be animated by a larger conception of what it could do for people's health worldwide. It had the sense of a social cause." This was reflected in Merck's decision to allow Frazier to continue with a very time-consuming *pro bono* death penalty case after he had joined the company's payroll.[14]

For more junior employees such as Kohn, Merck's wider stakeholder orientation is a clear motivator. "I remember when I was in a bookstore with my parents and brought them over to show them the *Merck Manual*," said Kohn. The Merck Manuals are a well-recognized set of medical reference books that Merck publishes and distributes on a non-profit basis to doctors and patients around the world. Kohn added,

---

[14]According to Frazier, about half of Merck's attorneys do *pro bono* work.

"I am so proud to work here… People are so committed and are the top people in their field. I've been here for 8 years, which I think is unique for people my age (to stay with one firm)." According to Hosny, an ethical approach to business is becoming an increasingly important part of attracting and retaining a new generation of workers:

> Defending your ethics with your employee base is a new dimension. More and more intelligent young people are willing to challenge and push back, and they need to feel that they are working for a company whose values they can subscribe to. I think you can fail *The Wall Street Journal* test, as stock prices can go up and down. But if you don't have the confidence and trust of your employee base, then you are really in trouble.

GOOD SCIENCE

Dr. Elizabeth Stoner, the Senior Vice President in charge of Merck's global clinical development efforts, believes "that when ethics is ingrained in research and business practices, in the end you see better science, which is better for the company. So by being straightforward with our data and doing it right, it produces a strategic advantage." Merck's decision to make Mectizan available, despite the lack of a paying customer, was also closely connected to the company's belief in pursuing good science. "If we would have dropped it," said Gustavsen, "it would have been so demoralizing for the researchers who had worked so hard on it."

BRAND IMAGE AND MARKET ACCESS

In today's ever more crowded pharmaceuticals marketplace, where sales representatives compete for a few minutes of doctors' time, Merck's strong reputation for doing good work, both scientifically and socially, can sometimes help it gain access to customers. Gustavsen said, "I've heard anecdotally from reps that docs will say to them, 'I'm letting you into my office because of the *Manuals'* or other things we've done." This can be particularly important in the international marketplace, noted Healy, where many doctors have spent time in a developing country as part of their medical training or national service:

> There's a lot of social awareness on the part of doctors in Europe. Doctors look at reps based on their company's reputation. Our company reputation helps our sales reps see more doctors, more so than our competitors… It won't sell more Vioxx directly, but indirectly it helps the company sell drugs because of our enhanced reputation when the doctors see the company doing programs in Africa. The reps can say, 'We are working to develop a vaccine—just thought you'd be interested to know.' So it does work to improve the relationship between the rep and the doctor.

Gustavsen noted, however, that the company has to strike a fine balance when deciding how much to publicize its global health initiatives. "I think sometimes we've done too good of a job of not tooting our own horn. There's a fine line between doing good and promoting it."

MANAGING RISK AND IMPROVING COMPLIANCE

Although Merck clearly views its ethics and corporate and social responsibility (CSR) efforts as going well beyond what is needed to ensure regulatory compliance, pursuing these standards and increasing employee awareness of its values and standards may have helped reduce the number of warning letters that it receives from the FDA and other regulatory agencies. "I think the Ethics Office has had an impact on the compliance numbers," said Brevard. "But this opinion is not scientific. As an example, recently one of our lawyers wrote a memo noting that for some unknown reason our compliance numbers are going down. I could tell him that the reason is because of our initiatives, but couldn't prove it."

ACCESS TO POLICYMAKERS

In addition to customers, Merck's approach to CSR may also facilitate access to influential global policymakers. "It certainly helps our corporate reputation with at least some audiences," said Yarno. "It gives credibility with policy-setting and health organizations, and helps put Merck on a different footing." Khalil gave a concrete example from when Tony Blair and George W. Bush were meeting in London in November 2003 to discuss global AIDS initiatives. "They invited our CEO to fly out to join them. They didn't choose other companies who are closer or have more AIDS drugs than us. There were several reasons for this, but one of the main ones was wanting to hear about what we're doing in Botswana." Gustavsen described a similar experience: "The reason to donate Mectizan wasn't for the PR value… But there have been benefits—employees have talked with government officials that have heard about it. In Mexico, they wanted to talk with me because of the Mectizan program."

IMPROVING RELATIONS AND DIALOGUE
WITH STAKEHOLDERS

Although making a strong commitment to ethics and CSR clearly has not insulated Merck from criticism by some NGOs, activist groups, and the media, it can help improve the company's relationship with those critical of the pharmaceutical industry. Gustavsen gave an example: "Last summer I was at a meeting with personnel from an NGO in Tajikistan who told of receiving a box of Merck medicine there. The docs were so happy when they saw that it was an actual Merck medicine (and not a counterfeit) because it will actually work." The public-private partnerships that Merck has been building as part of its global health initiatives can also lead to dialogue that improves the effectiveness of these initiatives. "Through discussions with our six NGO partners," said Gustavsen, "we were told that the products we had available for donation weren't necessarily the ones they wanted or needed. So now we allow those organizations to choose medicines from our product line to meet the needs of their ongoing humanitarian programs—we give them a menu so they can choose, as long as it fits in with what we can produce.

We can't always give them exactly what they want, but at least it gives them more opportunity to direct what is given to them."

PROMOTING AN ATTRACTIVE BUSINESS ENVIRONMENT

Longer term, Merck hopes that some of its initiatives will create more transparent, ethical, and growing economies in developing countries where it can do business. "When there is a lack of ethics in a country," said Frazier, "it makes it more difficult for us to do business there while still adhering to our standards." As an example, he cited an ethics center that was working to fight corruption in one "integrity-challenged" country, where "it is difficult for our employees to know what to do when they are asked to pay off someone to get the product in." Likewise, Healy notes that one of the reasons for Merck's participation in the global health initiatives was the growing recognition that the AIDS epidemic and other infectious diseases were preventing these economies from growing:

> We now emphasize that healthcare is important to help developing countries actually develop, and that countries should not wait until they can afford better healthcare. Before, we weren't really into that. And I don't think the international organizations were really into that either. They finally saw that healthcare is really a prerequisite to development, instead of the other way around.

ETHICAL ISSUES

As one of the leading global pharmaceutical firms, Merck faces the same set of ethical issues—global drug pricing and access, appropriate promotion practices, what products to develop, conduct and publication of clinical trial results— that have brought a growing amount of unwanted scrutiny to this industry (see Exhibit 2.5 for a categorization of the types of issues).[15]

PRODUCT PROMOTION AND GLOBAL BUSINESS PRACTICES

When asked about the main ethical challenges that Merck faced, the leaders and employees interviewed pointed most frequently to the marketing and promotion of pharmaceuticals, and the challenge Merck faced trying to compete successfully to sell its products around the world while adhering to high ethical standards. In general, promotions such as gifts, lavish dinners and entertainments, trips to resorts for education sessions, etc., have been criticized as having an undue influence on physician prescribing. Merck has policies and practices in place that prohibit this

[15]Merck took on an additional set of challenges in 1993 when it purchased Medco, which was then the United States' largest pharmaceutical benefits manager (PBM). Because of space limitations and the fact that Merck no longer owns Medco, we elected not to make it a focus of this case. For more information on the ethical issues associated with Medco, see Note 6, Hawthorne, pp. 183–189.

EXHIBIT 2.5   Merck's General Counsel, Ken Frazier Description of
Types of Issues He Handles

- "Corporate governance – is about having the controls and systems in place to run the company in the best interests of the shareholders
- Compliance – is much narrower. It is rule, instead of value-oriented. Compliance is run out of legal and is about obeying the letter and spirit of codes that govern our business.
- Ethics – I think of ethics as a more value-based approach to fair-dealing with various constituents – shareholders, employees, customers, suppliers, regulators, communities. … It's what we teach our children (about) behaving the same way when their patents are gone as when they are there.
- Corporate and social responsibility – touches on compliance and ethics but goes beyond these concepts. Because of our field, we have the ability to reach out to disadvantaged people around the world and make a difference – simply because we are an ethical company. In some cases, lowering our prices is the responsible thing to do. CSR is distinct from ethics (though they) certainly overlap. We have a responsibility to behave ethically and comply with laws, but that is not enough for us in order to adhere to the (credo). If we have the capabilities to help and reach out to needy, sick people, hewing to Merck's values, we ought to do more. CSR is going beyond ethics. It's why Merck developed the Mectizan program. We should do this because it's a force for good for people's health in the world. Ethics doesn't require us to do this but our internal and values and commitments do."

kind of inducement, but this ethical approach has caused its own difficulties. Healy explained that, "generally speaking, product promotion is not a level playing field, since not enough firms adhere to the voluntary codes of conduct and guidelines of the AMA and of industry associations in the US and Europe, as Merck does."

The problem is compounded, according to Healy, because when Merck refuses to promote its products with gifts and other perquisites, it poses a motivational challenge for the Merck sales representatives. Hosny gave one example of a common practice that Merck eschews: "Many of our competitors fly physicians around the world to meetings in exotic destinations. We refuse to do so—we just think that is wrong." He argues, however, "that in the long run the practice of ethical standards gives us a competitive advantage. It's not driven by what the traffic will bear in any country. We apply the same standard at Merck in Bangladesh as in Boston."

Merck also has had to contend with critics who deplore the high cost of drugs and who claim that these prices are a result of excessively high advertising and promotion expenditures. Merck does invest heavily in promoting its products, spending over twice as much on marketing, sales, and administration as it does on R&D ($6.2 billion vs. $2.3 billion in 2000).[16] In 2001, it added over 1000 sales representatives to keep pace with its main industry rivals, such as Pfizer, which were growing more rapidly through acquisition and heavy promotion of blockbuster products. In 2000, Merck had the second largest advertising budget in the industry, and it spent more to advertise its blockbuster pain medication,

[16]Hawthorne, p. 70.

Vioxx—$161 million—than any other drug that year.[17] And the company is an unabashed supporter of the direct-to-consumer (DTC) advertising of prescription drugs, which is currently only legal in the United States, Costa Rica, and New Zealand (and in Canada, but with stricter guidelines). "We're trying to change the law in Europe," said Hosny. "We think people have the right to know. Why can't we give appropriate health information to the consumer? There must be a way to get health information to people. Now, people just go to the Internet and get information that may not be accurate."

Despite the vigorous promotion of its products, Merck so far has not been subject to the high-profile legal challenges and large fines for improper promotional practices that have sullied the reputation of many large pharmaceutical firms.[18] Merck's DTC ads are typically more conservative than its competitors', and it has done better than average in avoiding warning letters from the FDA, although it was asked to pull a controversial ad for one of its AIDS products in 2001, along with seven other pharmaceutical firms, because the FDA considered the young men in the ad "too healthy-looking" to accurately represent an AIDS patient, even one in treatment.[19]

Yarno gave a first-hand account of how Merck's ethical marketing approach compared with one of the firm's major competitors:

> I left Merck at one point and went to another major pharma company… I truly found a difference in the way that they made decisions. Let me give you an example— it was around the brand name of a new product that they introduced. The name they chose was very close to two drugs already on the market. Within a couple of weeks there had been some mix-ups at pharmacies. The FDA approached them and said it was a problem. They still haven't done anything about it. Their approach was to try to convince the FDA it wasn't a problem. We had a product called Losec that was confused in a couple of cases with a common drug, Lasix. We said right away this is a problem and we've got to fix it, so we changed the name to Prilosec. It costs us millions of dollars but it's the right thing for the patient. I was not used to even thinking about these things at Merck. We just make decisions based on that value system and I was uncomfortable with the way things were decided at this other company.

Ethical processes such as the credo and the strong culture associated with it, a clear code of conduct with concrete examples of what's permissible, training for all sales people to reinforce the code, and leaders who demonstrate clearly what behaviors are acceptable are all intended to help Merck maintain consistent global standards of ethical behavior. Even with these in place, however, it is not possible to ensure that every Merck employee around the world will behave in an ethical

---

[17]This case was completed before the high-profile controversy over Vioxx that followed Merck's announcement it was recalling the drug on September 30, 2004. Hawthorne, p. 156.

[18]In May 2004, Pfizer pleaded guilty to criminal charges and paid a $430 million fine for the way it marketed its drug, Neurontin. In 2003, AstraZeneca paid a $355 million fine. Schering-Plough and other drug companies are now under investigation for, among other practices, paying doctors large "consulting" fees to prescribe their drugs. See Harris G. (2004, June 27). As Doctors Write Prescriptions, Drug Companies Write Checks. *The New York Times,* Section A1.

[19]Hawthorne, p. 157.

manner. "Our commitment to ethics doesn't always succeed—of course, human frailty does exist at Merck like everywhere else," said Frazier. "But the repetition of these mantras really is a part of the culture and the way it is lived."

To try to minimize any such occurrences, Merck has put in place several additional mechanisms to encourage the desired behaviors. These include:

Hiring people with the right values and then reinforcing them—"I think we look for people who we believe have a similar value system, including how they make decisions, and what they've done earlier in their career," said Yarno. "The sales organization uses a fairly consistent type of questions and scenarios… In the interviewing process, it is the softer-fit issues that sway whether we hire someone or not." Kohn gave an example from Merck's South American operations of how this plays out once someone is hired:

> "The reps will come back and say, 'My competitors in that country take doctors out and spend money, and I know it's not legal in the US but it is here and I'm not going to make my quota if I don't do this.' And we just have to say no, that's not how we do things here. In our communications we talk about the fact that a doctor isn't going to prescribe a medicine because you take them on an outing; they'll do it if you show them it is successful… We have good products and they are able to hold their head up high because our product is backed by science. And management will say, 'if this environment isn't good for you, you are free to go.' Healy added that they sometimes lose people who are attracted by companies with higher salaries and looser standards."

Strong centralization of marketing materials — "Every promotional claim we make throughout the world has to be cleared medically and legally," said Hosny. "We have very strict global control over what we say about our products, pricing, promotion, etc.," agreed Healy. We're centralized across the board…unlike competitors that allow local managers to make decisions."

Prominent role for lawyers as business partners — "I help my [company] clients achieve business goals by reducing risk and increasing compliance with the laws and guidelines in a highly regulated environment," said Quinlan. "They need lawyers to help make the business work. For example, we need to train the sales force about providing fair and balanced information."

Performance-management system that doesn't create incentives to bend the rules — "We try not to put people in situations where they might have a conflict of interest. For example, having to make a sales number but also having the ability to award a $10,000 research grant to a doctor. We separate these functions," said Yarno.

Working at industry level to raise marketing standards — In October 2001, at a time when the pharmaceutical industry was under fire for excessive promotional efforts aimed at getting doctors to prescribe its products, Merck took the bold step of announcing it would no longer allow "anything that might give the impression that you might try to buy somebody's business," said Casey Weber, a Merck sales rep.[20] Nine months later, with Merck's encouragement, PhRMA, the industry association, followed suit, putting in place standards very similar to Merck's.

[20]Hawthorne, p. 141.

"I think the Achilles heel of the industry—certainly in Europe and the emerging markets—is promotional practices," said Healy. "The challenge for us is to lead the industry... If we can practice what we preach, if we give out medical information and if it is scientifically based, then doctors will want more informational materials."

DRUG PRICING, ACCESS, AND INTELLECTUAL PROPERTY

Over the last several years, the pharmaceutical industry has come under growing pressure because of these closely related issues of drug pricing: access for those who cannot afford the drugs; the protection of intellectual property (IP) that enables firms to charge the high prices that allow them to continue to invest in R&D for their products. In the United States, this debate has focused recently on the question of drug reimportation—whether organizations or individuals should be able to purchase drugs from countries, such as Canada, where drugs sell for a significantly lower price owing to government-imposed price controls. This practice is currently prohibited by the FDA for safety concerns, but not rigidly enforced, especially when individual patients cross borders to buy drugs. In the poorest developing countries, even heavily discounted drugs are often beyond the means of much of the population. Although Merck and other major pharmaceutical companies have agreed to sell their AIDS drugs and some other medications at no profit, they have generally been loath to give up their IP rights to allow generic manufacturing of their drugs. Companies fear the precedent this may set and predict that the cheaper generics will create a gray-market supply of drugs that will be resold in their core markets. Such competition would not only reduce firms' profits, but also, they claim, hurt society by impinging on their ability to fund future research. For Merck, approaching these issues goes to the heart of the credo that "medicine is for the people, not for the profits," and as Hosny observed:

> The real ethical challenge that the pharmaceutical industry faces is juggling these competing forces—staying true to one's corporate mission, not just adopting any practices to compete, [while being] competitive and profitable at the same time. In a pharmaceutical environment where you have the fast followers, no matter what the innovation is, you're never alone for long before someone comes in with a similar [drug]. That creates a set of tensions or pressures...The crux is knowing what pieces of the puzzle are so important to you that you have to have non-negotiable standards. Protection of IP is one. Maintaining the highest standards of information to doctors is another.

According to Frazier, however, "price and patents really are not the main barriers" to providing access to drugs to those who need them. In Africa, he contends:

> Infrastructure is the main problem. When Merck started the Mectizan Donation Program, all we needed to do was to distribute one treatment per person per year; however, the effort it took to accomplish this, with the networking, the delivery

systems, the human capital, the infrastructure—made it very difficult. The drug is just a component of the process. Now, when we are dealing with AIDS drugs (which require a much more complex and frequent dosing regimen), it is even more difficult. The argument that some generic company can make it significantly cheaper, and that cost is what will make a difference regarding access, falls short. When we lowered the price of drugs to the point where we didn't make a profit, there was no spike in demand... [Even] in Botswana, where we've committed to providing antiretroviral drugs at no cost, it is still very difficult. When you ask what is the one thing many of these populations need, the answer is clean water. There is no patent on chlorine tablets. Chlorine tablets are cheap and the system can't even manage that... So, it's not a patent or price barrier.

In the United States case, Frazier makes a different argument:

> The demand that we are seeing here in the US for lower pricing is not the result of the price differentials [between the US and many other countries where the government sets the price for drugs]; it's the fact that so many people in the United States don't have insurance coverage for prescription drugs. Those who are covered don't ask what the price is in Canada. We think the issue is more to do with coverage and not pricing.

It is for this reason that Merck was one of the first pharmaceutical companies to support the recently enacted prescription-drug coverage for Medicare patients, which PhRMA and many firms initially resisted, because of the risk that it could involve government price-setting, and why Merck has pledged to make its drugs available for free to current eligible Medicare recipients once they have exhausted their interim $600/year benefit.

Merck has recognized, however, that there may be some extraordinary circumstances in which the ethical arguments and public pressure are so strong that it cannot maintain its steadfast defense of IP protection. As Khalil noted:

> In the latest WTO negotiations, Merck was the first of the large pharmas that also brought others along in support of the initiative to allow countries to do compulsory licensing in the face of an international health crisis. When 25% or more of a population has HIV in a country, like you find in Botswana, you have to do something. In the US, seven people died of anthrax and the government was ready to consider a compulsory license for Cipro [ciprofloxacin, the antibiotic used to treat anthrax]. We still believe strongly in the protection of IP, but we have to know what battles to fight. This position wasn't winnable. Even though most of our competitors opposed Merck's stance initially as too liberal, now they are much more supportive.

An example of this new approach occurred in April 2004 when Merck announced its intention to grant a royalty-free license to South Africa's Thembalami Pharmaceuticals to produce Merck's AIDS drug Stocrin, for the countries in the Southern African Development Community (SADC).[21] Merck will continue to supply its AIDS drugs, Crixivan and Stocrin, at no profit in South Africa, but by creating an additional generic source in one of the nations most affected by AIDS,

---

[21]MSD (Merck's international name) press release, "MSD Announces Intent to Grant a License for HIV/AIDS Drug *Efavirenz* to South African Company, Thembalami Pharmaceuticals.

Merck reiterated its commitment to improve access to the medicine in this dire public health crisis.

CLINICAL TRIALS AND PUBLICATION
OF RESULTS

Recent changes in Merck's research publication practices are also illustrative of how Merck's decisions are influenced by its dedication to the credo and ethics. When Dr. Laurence Hirsch, who had headed product development for Fosamax®, Merck's blockbuster osteoporosis drug, was offered the chance to run a newly formed department that would coordinate all the clinical trial-related publications emerging from Merck's Research Labs (MRL) and other areas, he thought "it would be somewhat low-key, with less stress than drug development. I thought it was also a great chance to be involved with a broad array of products and programs and I eagerly took the job." He knew that in 1933, at the founding of the Merck Institute for Therapeutic Research, George W. Merck had made the revolutionary promise that Merck scientists would not only be permitted, but also encouraged to publish the results of their research in the scientific literature, and that this culture remained strong within Merck 7 decades later.

Dr. Hirsch began work as head of Medical Communications in MRL on September 1, 2001, overseeing a staff of 25 (since grown to nearly 40). It includes doctoral scientists and support staff who work closely with Merck researchers in MRL and in Marketing, as well as with their outside investigators, to publish the results of their clinical studies in leading peer-reviewed medical and scientific journals. Before the creation of this new department, the company's medical and science writing was dispersed across departments at Merck, with no centralized planning or consistent resource allocation. The rationale for the new department, according to Hirsch, was logical from an organizational perspective. Beyond that, he notes, "Publications really are Merck's currency for communicating with the medical community, academic thought leaders, and the general public." Timely and high-quality publications about the scientific and medical aspects of company products furthered the company's purposes of educating physicians for the ultimate benefit of patients and promoting its products in a responsible manner. Although all research data had to be submitted to the FDA and other agencies, standard industry practice was for those data to remain confidential, and a company usually was free to decide which data and studies were published in the medical literature. As a consequence, there had been instances in which companies acted to prevent publication of certain trials. A small number of these were egregious and had been widely publicized. Little did Hirsch expect how big an issue this was to become in his new position, which turned out to be anything but low key.

Nine days after starting his job, a blow-up occurred in the medical publishing field. The International Committee of Medical Journal Editors (ICMJE), which comprises the editors of 12 leading publications, among them the *New England*

Journal of Medicine (NEJM), *The Journal of the American Medical Association* (JAMA), and *The Lancet,* published an editorial in all 12 journals, amending the Uniform Requirements for Manuscripts Submitted to Biomedical Journals. This was done because of growing controversy about economic and competitive pressures in pharmaceutical company-sponsored clinical research. Critics of this research claimed that companies hired outside researchers to conduct clinically valid research, but then undermined the integrity of the data by exerting too much control over the study. The improper controls by trial sponsors included manipulating trial design, preventing investigator access to raw data, overmanaging the interpretation of study findings and writing of manuscripts, and failing to publish important studies with less-than-positive findings for a sponsor's product.[22] All of this activity, critics claimed, was intended to promote only favorable conclusions about a drug's safety and effectiveness and to suppress publication of other data. ICMJE addressed some of these complaints by strengthening its guidelines to promote principal investigator accountability and independence, sayings its editors would not accept articles for publication unless the investigators could attest to having had access to the study data, as well as control over the writing of the manuscript and the decision to publish the data.

The ICMJE's new guidelines, and the allegations that led to their adoption, were widely published and radically changed Hirsch's work life. Over the next 15 months, a multidivisional task force at Merck developed formal Guidelines for Publication of Clinical Trials, implemented in late 2003, and posted January 2004.[23] These guidelines codified many long-standing Merck practices and addressed a number of sensitive issues, including what the company would publish and when, how it interacts with investigators, investigator access to data, and authorship issues (to prevent corporate ghost-writing, guest authorship, and failure to include those who contribute). He began spending much of his time at conferences and interacting with editors, academics, and the media as a pharmaceutical industry representative on this issue. During these encounters, he described Merck's publication practices, justified the reasoning behind them, and explained how they differ from the widely perceived abuses.

Merck's publication policies and practices were designed to serve both the company mission and the medical professional's need for accurate information. Over the years, Merck has sought to reconcile its corporate needs and the legitimate need of physicians for information about its drugs responsibly, by trying to provide complete, accurate, and balanced information from its clinical trials, and to make public all information related to both the safety and efficacy of its marketed products. It has accomplished this balancing act by dividing its studies into two classes with respect to publication: exploratory and testing. In the clinic, exploratory (or pilot) studies are usually early-stage trials designed to provide preliminary information about an investigational agent, the disease or condition,

---

[22]This last practice became known as the "file drawer phenomenon," meaning that unfavorable study data were sent to the file drawer and never published.

[23]See Merck's Publication Policies at http://www.merck.com/policies/clinicaltrialspublication.

possible endpoints to measure, etc. These trials are highly proprietary to Merck and only published if they produce medically important data (they may be published at a later time, depending on the progress of the development program). Later stage human trials that are statistically designed to test a hypothesis (e.g., this drug will be effective in reversing a medical condition) are published whether the results are positive or negative. Confidentiality of the early studies is preserved for competitive reasons. According to Hirsch, "Drug development is extremely competitive. Even knowing that Merck (or Pfizer or Lilly) is studying a type of compound for a particular medical condition can be of great value to another company. Premature disclosure of such information can be very damaging." At the same time, Hirsch feels comfortable about the policy, because no physician or patient is harmed for lack of knowing about the early studies on a molecule that is far from (and is at high risk of never) becoming a marketed drug, and because the Guidelines do call for publication of exploratory trial results when medically important.

Nonetheless, Merck, as part of the pharmaceutical industry, has been facing growing criticism on this issue, with demands that drug companies publish all drug trial data. Hirsch has appeared at meetings to discuss Merck's practices and he describes his encounters as follows:

> I sometimes feel like a sheep before wolves. The industry is blasted at these meetings for suppressing publication of negative data, hiding data, etc. These episodes, while disturbing, are very rare, and I do not feel they represent the normal conduct of most companies (nor are they limited to research sponsored by companies). However, I speak on Merck's publication guidelines and try to convey that we have a long history of publishing all of our hypothesis-testing clinical trials—positive or negative—and that we understand (and accept) the ethical obligation to publish clinical data regardless of outcome. I did not think my new job would be like this... I'm actually getting used to the debate but my role is certainly very different from what I expected.

These issues show no signs of calming down. In June 2004, the controversy surrounding clinical-trial data flared again when New York Attorney General Eliot Spitzer decided to sue GlaxoSmithKline (GSK) for failing to disclose to physicians information about its drug Paxil® in children. This was followed by the American Medical Association (AMA) passing a resolution calling on the United States government to create a public registry of all clinical trials and their results. And *The New York Times* ran a front-page story reporting that the top medical journals represented on the ICMJE may adopt a new rule that would only publish studies based on clinical trials that had been entered into a public registry at the time the trial began, and which would be updated to show the results.[24] Immediately following these events, Merck became one of the first pharmaceutical companies to declare its support for a public registry of late-phase (hypothesis-testing) clinical trials.[25]

[24]Meier B. (2004, June 17). Group Said to Seek Full Drug-Trial Disclosure. *The New York Times,* Section A1.

[25]Meier B. (2004, June 21). A Medical Journal Quandary: How to Report on Drug Trials. *The New York Times,* Section C1.

WHAT DRUGS TO DEVELOP

Merck has also led the industry in setting a standard regarding the value of the drugs it chooses to develop. Many other companies have been the target of criticism for focusing their drug-development efforts on "me-too" drugs (those that are slight modifications of existing blockbusters and hence less risky to develop) or lifestyle drugs (such as Viagra and its new competitors), rather than drugs to treat the most serious global health problems. With few exceptions, Merck has prided itself on discovering and developing novel compounds with new mechanisms of action. Arguably, its drug Vytorin®, which was approved by the FDA in the summer of 2004, could be considered a "me too," as it is a combination of two existing cholesterol-lowering drugs, Merck's statin Zocor and Schering-Plough's Zetia (ezetimibe). Yet Vytorin is the first product to reduce cholesterol by targeting both its absorption in the intestines and its production in the liver. Merck hopes Vytorin will enable it to take market share from the world's best-selling drug, Lipitor, and to eventually replace the sales of Zocor, which will come off patent in the United States in 2006.

Merck also has what some call a lifestyle drug in Propecia®, marketed as treatment for baldness. But according to Yarno, "Propecia was an accident, not the focus of a major development program. It is the same chemical ingredient as our drug Proscar, which is a treatment for benign prostatic hypertrophy (or BPH), a serious medical condition in men. It was noticed in clinical trials that it helped men retain their hair." Hosny adds:

> "Serendipitous products are appropriate for companies to pursue as long as (1) They do not burden the reimbursement budget for payers—for Propecia we decided not even to seek reimbursement. That is part of being consistent with what we feel is important; we feel that health budgets should not be used for this; and (2) within the research labs, they do not take priority over other more important health needs."

Other than these examples, Merck's drug development program can be distinguished from many of its large rivals in two respects. First, as the title of its 2003 Annual Report suggests, Merck has historically concentrated a high proportion of its R&D budget on "novel medicines that make a difference," or as Dr. Hirsch put it, "Our mission is to develop the first-in-class, and/or best-in-class medicines... that address unmet medical needs. We don't do the 5th beta blocker or 8th ACE inhibitor."

The other distinctive feature of Merck's R&D pipeline is the heavy weighting toward vaccines. "The vaccine emphasis is not new," said Hosny. "It's been a part of our heritage for many years." Under the leadership of Maurice Hilleman, Merck established itself as a leader in the vaccine field between the 1960s and mid-1980s. It now has four potential vaccines in late stages of development, along with a major AIDS vaccine program that is at an earlier stage of clinical trials. In contrast "many pharma companies abandoned vaccines because of the risk of litigation, and the lower margins," said Hosny, leaving only 4–5 major global players.

Although vaccines have been profitable for Merck in the past, and are projected to be a much more central part of its business model going forward, a number of other characteristics of the vaccine market have deterred other companies from developing them. Among the characteristics that Dr. Mahmoud, head of Merck's Vaccine Division, points to are:

- the fact that the government is typically the sole customer for vaccines and sets the prices. Adds Hosny, "What's tricky is that it's a tender business… You win it or you lose it. The government decides [who wins the competition for a national contract]—you can lose it for the next three years and then that's that. There's no coming back the next month with your sales reps to try again."
- high capital demands, such as for manufacturing, with investment that must be made early in the development process, when there is still a high risk the vaccine will not work
- a stricter regulatory environment than with small molecules—"the FDA has to release every batch you make."

These characteristics result in an uncertain business model for vaccines, which are often a one-time preventative use, rather than the more standard blockbuster model of frequent pills for a chronic condition. "There are some calculations that say return on investment (ROI) is close to the cost of capital, which is about 13%. If that's all we get, then there's not going to be a lot of money coming our way."

Despite these drawbacks, Mahmoud says that it is the global health burden and the challenge of solving it that attracted Merck [to develop vaccines for HIV and other retroviruses]. Vaccines, Mahmoud adds, lend themselves to discovery and require "top-notch science in a very direct way—you have to use the best of biochemistry, biology and immunology. They pose a challenge for scientifically minded people." In addition, Merck is hoping that the financial benefits of the large customer base associated with vaccines, where ideally 70 to 80% or more of the whole population will receive the product, will offset the lower price/patient associated with vaccines. Although Merck might be expected to be praised for its willingness to commit major research dollars to vaccine programs to try to prevent serious diseases in the United States and around the world, even these programs are not without controversy. According to one senior executive, even before its human papilloma virus vaccine completes clinical trials "some religious groups are criticizing Merck for promoting sexual promiscuity in young women by developing a vaccine that will hopefully prevent this sexually transmitted cancer (of the cervix)."

CHALLENGES FOR THE FUTURE

As Merck looks to the future, it faces a number of questions regarding its business strategy and how best to stay true to the mission laid out by George W. Merck over half a century ago. Perhaps the most pressing issue is whether Merck can remain a leading pharmaceutical firm by going it alone, at a time when all its

major competitors are growing through mergers and acquisitions. And if it is
compelled to merge, can it sustain its distinctive culture? Merck's leaders appear
united in the belief that continuing to focus on the discovery of breakthrough
medicines is a better way to create value, and they argue that this, rather than
preserving the credo, is the reason to remain independent. "The primary deterrent
to pursuing a merger strategy is not that it will erode our standards," said Hosny.
"If we were to merge, it would be with a company with compatible strategies and
cultures." A key to Merck's maintaining its independence will be demonstrating
that its new vaccines not only work technically, but also that they form the basis
for a viable business model. As Mahmoud summed it up, "There is theory and
there is practice. The theory is that if you have a product and you prove that it is
medically and cost-effective, then the funding mechanism in [different] countries
should cover it. But there is only one payer [in each nation] and you can't test that
[until you have a proven vaccine]." If these vaccines are approved, then Merck will
face the additional challenge of how to make them available where the needs are
greatest around the world. Mahmoud continued:

> The solution is not only in the hands of Merck. The solution is in the hands of the
> total global community. What has been working in the last three to four years is
> the concept of public-private partnerships, [such as] the Global Alliance for
> Vaccines and Immunization (GAVI). The attractiveness is that by having that kind
> of mechanism, you are making it available to the poorest of the poor. The prob-
> lem is that it is much more complicated for vaccines to reach the kids in Africa
> than buying them and giving them to the countries. [We have to work next on] the
> mechanisms of delivery and the infrastructure. I'm not in charge of delivering
> vaccines to all those born around the world. But I am in charge of the enabling
> conditions: making sure that the Merck vaccine-manufacturing sites are approved
> by the WHO; that Merck practices differential pricing according to the value of
> the product and the ability of a country to pay; that Merck participates in the
> tender process for treating the poorest of the poor.

The vaccine case is an example of perhaps the greatest ethical challenge facing
Merck for the future: can it continue to sustain its commitments to global health
initiatives while still satisfying the short-term demands of shareholders? In the
case of its flagship philanthropic initiative, the Mectizan Donation Program, Merck
may be forced to raise its contribution if it wants to meet its pledge of eliminating
river blindness around the world. Not only has the program been expanding, but
as Gustavsen notes:

> Our commitment with the Mectizan program has been to ship the medicine and
> let others take it over from there. Because of high involvement of NGOs in Iraq
> and the resources that AIDS is taking, a lot of countries and NGOs are finding
> themselves short of financial and other resources to run these programs. So our
> decision will be whether to supplement these donation programs with donations
> of cash. People don't realize that giving away the products requires substantial
> financial commitment as well.[26]

---

[26]In May 2004, Merck announced a donation of $1 million to The Carter Center to fund river blind-
ness elimination programs in the Americas, in addition to its ongoing donation of Mectizan.

And Merck is facing these growing demands at a time when the resources in its Foundation are diminishing and there is increasing pressure on the company's bottom line. How Merck will fund these commitments when the Foundation's endowment runs out is a pressing question. According to Colatrella, the company has already begun to think about these issues:

> We've discussed the decline in level (of the Foundation's endowment), and whether we're spending it too quickly. We're looking at alternatives, to extend the life of the endowment. Once depleted, we need to look at whether we will re-endow or fund the Foundation annually. We have done some benchmarking of other corporate foundations and see a split among them [in how they are funded]. We still have a good eight years left of money.

The decision on how best to fund the Foundation, and how much Merck can afford to spend, is part of a wider issue that will shape the future of Merck as well as millions of desperately ill people around the world: how can pharmaceutical companies balance satisfying the short-term demands of their shareholders with their longer term commitments to a broader, global stakeholder community?

3

GENZYME: PUTTING PATIENTS FIRST[1]

INTRODUCTION

In the land of the giants, where large pharmaceutical companies are increasingly focused on the largest disease areas and drugs have the potential to be multibillion dollar blockbusters, Genzyme has become one of the world's most successful biotech companies by focusing on the very small. The United States and the European Union offer special incentives to encourage firms to develop orphan drugs for rare diseases, and Cambridge, Massachusetts-based Genzyme has shown that producing novel therapies, even for "ultra-orphan" drugs (for diseases with 10,000 or fewer patients), can be a successful business model.

Genzyme has become a recognized leader in delivering medicine to carefully targeted populations, a strategy that many feel is the future of healthcare, by adopting a business philosophy of putting patients' interests first. The firm's commitment to this philosophy has been rigorously tested as it has grown into a large public company, facing diverse ethical challenges associated with its business model, and with the need to satisfy shareholders. Some of the difficult ethical questions include:

1. How to conduct an ethical, double-blind, placebo-controlled clinical trial for a drug that the Food and Drug Administration (FDA) has already approved for sale, when there are no other effective treatments available for this condition.

[1]This case study was written by David Finegold and Jocelyn Mackie based on interviews conducted at Genzyme on September 30 and October 1, 2003 at Genzyme's headquarters in Cambridge, Massachusetts. Unless otherwise specified, all quotes in the case were obtained during interviews. Interviews were conducted by Finegold, Mackie and Cæcile Bensimon. This case study is one in a series produced under grants funded by The Seaver Institute and Genome Canada to study models for ensuring ethical decision making in bioscience firms. We would like to thank all those who partcipated in this study, especially Elliott Hillback for his help in arranging the case study and detailed feedback on the earlier drafts of the study.

2. How to balance the desire to provide treatment to everyone who needs it around the globe with the need to charge a high price to earn an adequate return on its investment, thus to continue expensive research and development and produce drugs to treat small patient populations.

BEGINNINGS OF A STRATEGY: SOLVING PROBLEMS

Genzyme was founded in June 1981 by Henry Blair, an enzymologist, and Sherry Snyder, an entrepreneurial investor. Henri Termeer became Genzyme's second president in October 1983 and today is Genzyme's President, CEO, and Chairman. Over a number of Saturday mornings in November and December 1983, Blair, Termeer, and members of Genzyme's advisory board of scientists sat around a table at Genzyme's office and laboratory in a run-down, old loft building in Boston's "Combat Zone." As many start-up biotechs' personnel probably did, they animatedly discussed the strategy of their firm, usually over Chinese takeout. What they didn't realize at the time was that the strategy they were putting in place would set the company on its path to become one of the largest biotechnology companies in the world, or that the strategy would still be in place more than 20 years later.

The core strategy set out in those sessions was to "seek unmet medical needs and find opportunities there, rather than starting with technologies and seeking applications to define markets."[2] Genzyme's continuing scientific focus has been on finding therapies and cures for debilitating orphan diseases. Employees of the company describe the strategy as being focused on the patient. In a recent internal publication, the company explains, "We have diversified by focusing within health care on promising technologies, clear applications, and well-defined markets where we can have a major impact on patients' lives."[3] Senior Vice President of Corporate Affairs, Elliott Hillback, put it simply: "Our approach is to identify a real problem, then solve it."

AN UNUSUAL BUSINESS MODEL: ORPHAN DISEASES

Genzyme's approach to solving health problems has been to use a mix of internal development, partnerships, and acquisitions. These arrangements have allowed Genzyme simultaneously to expand its research platforms, manufacturing capacity, and product pipeline. In 1986, Genzyme became a public company with a successful initial public offering. In the same year, it reached a financial break-even point with 75% of its revenues generated by product sales.

What makes Genzyme unusual is that the largest share of its revenue comes from treating orphan diseases. Orphan diseases are classified in the United States as diseases that afflict fewer than 7.5 out of 10,000 people, or roughly 200,000 Americans, and ultra-orphan diseases are generally defined as those that afflict

[2]Internal Company Document. *A Different Vision. The Making of Genzyme.* p. 20.
[3]Ibid. p. 9.

fewer than 10,000 people globally. The United States Congress passed the Orphan Drug Act in 1983 in the recognition that companies needed a financial incentive to invest in the high-risk process of developing drugs for rare diseases, for which the potential customer base was very small.[4] On top of patent protection, the Orphan Drug Act allows the FDA to grant 7 years of market exclusivity to a company that develops a therapy for an orphan disease. During this time, other firms are prevented from offering a similar treatment for that disease, unless it is clinically superior. The Act also allows the FDA to create a streamlined, quicker approval process, and offers firms a tax credit of up to half their clinical trial costs for developing the drug. European Union rules that went into effect in 2001 provide similar incentives, granting an even greater window of exclusivity (10 years), but also defining orphan drugs in a more limited way (<5 cases per 10,000 people).[5] These laws have encouraged a number of companies to invest in developing drugs for orphan diseases.[6]

Genzyme has probably focused more than any other firm on the diseases covered by the Orphan Drug Act, with a core focus on ultra-orphan lysosomal storage disorders (LSDs). LSDs are rare genetic diseases that afflict only a few thousand people globally, but have devastating consequences for those affected. Genzyme's first orphan drug to be approved was Ceredase, a treatment for type 1 Gaucher disease, which is an inherited, chronic LSD disease that causes children to have enlargement of the liver and spleen, anemia, bone deterioration, bleeding and bruising, and eventual premature death. Ceredase was produced through a very costly and time-consuming process of extracting tiny amounts of the missing enzyme from human placentas; roughly 22,000 placentas were required to generate enough of the drug to treat a single patient for 1 year. Recognizing that this approach was not scalable, Genzyme developed Cerezyme, a genetically engineered version of the drug that is currently its most successful product.

Two recent successes for the company in this area were the FDA approvals for ultra-orphan therapies Fabrazyme and Aldurazyme in the summer of 2003. Fabrazyme is an enzyme-replacement therapy for Fabry disease and Aldurazyme was developed in a joint venture with BioMarin Pharmaceutical for mucopolysaccharidosis (MPS) 1.[7] Genzyme has other ultra-orphan drugs in their pipeline, in various stages of development.

GENZYME TODAY

Genzyme currently has a number of products on the market and several more that are expected to reach the market in 2005. In addition to ultra-orphan diseases,

[4]The Orphan Drug Act (as amended). U.S. Food and Drug Administration. www.fda.gov/orphan/oda.htm.
[5]Wess, L. Orphans in Neverland. *BioCentury.* 2004;12(6):A1–A5.
[6]Roja, P. Genzyme Finds the Target. Nasdaq, Sept/Oct, 2003.
[7]Mucopolysaccharidosis 1: "A rare, genetically determined disorder caused by the body's inability to produce certain enzymes… MPS belongs to a larger group of genetically inherited disorders known as Lysosomal Storage Diseases." www.mpssociety.org.

it has business units that focus on renal disorders and biomaterials. Renagel is one of their products currently sold in the United States for the treatment of kidney disease, and it was approved for sale in Japan in early 2003. Synvisc is Genzyme's leading biomaterial product and is used to treat osteoarthritis.

Genzyme is currently focused on five major therapeutics areas: (1) LSDs; (2) renal; (3) biosurgery; (4) transplant and immune diseases; (5) cancer, plus a Diagnostic Products and Services business. The company's revenue for 2003 was around $1.574 billion and 2004's revenue is expected to be around $2.1 billion. Genzyme has approximately 6,500 employees worldwide with offices across North America and Europe, and in Israel, China, Korea, Taiwan, Japan, and South America. Genzyme's corporate headquarters, which has just moved into a new building in Cambridge, Massachusetts, is one of the most environmentally friendly in the United States. As Genzyme continues on its path of growth, the ethical challenges and dilemmas it faces will only become more complex.

PUTTING THE PATIENT FIRST: GENZYME'S APPROACH TO ETHICAL ISSUES

When asked why they joined Genzyme, almost all the employees interviewed responded in a way similar to the Vice President, Global Medical Programs in the Therapeutics Business, Geoff McDonough: "Because of the over-the-top devotion to patients." Roger Louis, Chief Compliance Officer, described Genzyme as "a very patient-centric place. Our values and company mission are just so much a guiding light for ethics… It's just about the patients." Jan van Heek, Executive Vice President, was even more emphatic: "People who are not willing to go out and talk to patients and face the reality of the patient we treat will not work out well here."

From the outset, this patient focus appears to have played a key role in determining where Genzyme invests its resources, and how it resolves potential conflicts in its strategic decision making. Genzyme's product strategy has been to focus predominantly on life-threatening diseases for which no effective treatment is available, in contrast to the trend within the pharmaceutical industry to develop "me-too" and lifestyle-enhancing drugs. Jan van Heek explained, "One thing is that we only want to work on things that make a difference. We don't want to work on the fifth product on the market. If we don't see a medical benefit, we won't do it. It's instilled. It's not a slogan." McDonough elaborated on Termeer's vision for Genzyme: "Henri's answer is, 'How far do you move the bar?' In cases where therapeutics exist, unless we can revolutionize the standard of care, [he's] not interested. Of course, secondarily there are other important tests like, can we get there technologically, financially, etc?"

Louis explained how the patient-centric approach has guided Genzyme's decision making: "Some might assume that there is a tension between different constituencies in a for-profit company, shareholders being one. But if you do what is right for the patients, sustainable value will accrue to shareholders." McDonough added, "The commitment to providing therapy to patients is not driven by a business rationale," but rather that this commitment comes first, and serves as a basis for the business decisions that are made.

ETHICAL ISSUES

Genzyme's effort to make ethical decisions has been subject to a series of strenuous tests, as the cutting-edge nature of Genzyme's technology has led the firm into some uncharted ethical waters. These challenges have intensified as Genzyme has grown into multiple business units with a larger global employee population and a growing number of competitors. The main ethical issues identified by Genzyme's employees fall into the following categories: drug pricing, marketing and access; clinical trials; conflicts of interest; and product mix and priority issues.

DRUG PRICING, MARKETING, ACCESS

At a time when the price of prescription drugs is perhaps the most controversial issue facing pharmaceutical and biotech companies, Genzyme is a natural target for industry critics: it produces some of the most expensive drugs in the world, costing $150,000 to $200,000 per year for chronic conditions that may require life-long treatment. It is worth noting that the overall cost to the healthcare system of expensive drugs used for very small populations is less than the system costs generated by less expensive drugs used by large populations. In addition, unlike most pharmaceutical and biotech companies, Genzyme generally charges about the same price for its drugs wherever they are sold. Hillback explains the rationale: "People are people, wherever they may live. We make the argument that our drug creates a certain value by its excellent performance and that this is how the price should be determined. For countries that can't afford that price, we try to step in in other ways."

Genzyme has taken a multipronged approach to the pricing issue: to justify how the price is set, it makes both a "cost of development" and an economic argument to drug payers, and strives to provide the drug to as many people as possible globally, regardless of their ability to pay. Genzyme's economic justification for its drug prices has three parts. First, the costs and risks of developing any new drug are high, and because the patient-population for Genzyme's ultra-orphan drugs is so small, the price per patient must be high if the company is to recoup its investment, continue research and development on additional drugs, and satisfy its shareholders by continuing to grow. McDonough said simply, "[We] charge what we do based largely on what it costs to develop and produce the drugs." To make this case, Genzyme has adopted an unusual policy of full disclosure of its product-development costs. Caren Arnstein, Vice President of Corporate Communications, explained:

> Early on we took some verbal bashing but we just kept going. When Ceredase first came out, a *Wall Street Journal* reporter launched an investigation because he thought the price was exorbitant. Henri just said, "We're going to open the books and show them. This is what it cost us." In a separate investigation, a congressional OTA (Office of Technology Assessment) reviewer also concluded that it wasn't excessive. We got past it. But periodically it comes back. We are the poster child [for high-priced drugs]. We are one example they always dig out.

The second part of Genzyme's economic argument to payers is that, in contrast to many "me-too" drugs on the market that offer only marginal improvements to some patients, its drugs make a huge difference to the lives of severely affected individuals, including children who have no other treatment available. Hillback said that these factors must be taken into account: "The cost of a person who can't work, drive, take care of kids—whatever that's worth, to get that person productive—should be worked into the value of the drug." The rarity of the conditions that Genzyme's drugs treat helps Genzyme in making this case, because each country or insurance firm is likely to have only a few patients to cover. So far, 55 countries have agreed to reimburse the costs of Cerezyme, and Genzyme expects similar numbers will reimburse its more recently approved drugs. Evidence for the strength of Genzyme's pharmacoeconomic case, according to Hillback, comes from the fact that whereas many large payers have substantially reduced the amount they will reimburse for many expensive drugs, Genzyme's orphan drugs have seen only small price cuts.

Genzyme recognizes, however, that many individuals with these rare diseases, in the United States and around the world, will not be able to afford these drugs. In response to this problem, Genzyme has created a compassionate-use program for the uninsured in the United States, as well as working with the non-profit organization, Project Hope, to make its drugs available in countries that cannot afford to pay. At the end of 2003, 17 countries had programs to distribute Cerezyme for free, some to just a single patient. Tomye Tierney, Vice President of International Therapeutics, explained that Genzyme works with Project Hope on "getting to very poor developing countries. We wanted to make sure patients would not have access to drugs limited because of an inability to pay." She acknowledged, however, that there may be some tensions between the two goals of the program: to provide access to patients based on medical need, and to make sure all possibilities of reimbursement were exhausted first. As Jan van Heek put it:

> [We] can't do everything ourselves. So we work in partnerships (with governments and insurance companies) to try to make it available…We will get the drug approved but they have to work out who is going to pay. We can't ultimately pay. In some places we can pay for a while, but not forever. The health care systems always start with concerns…even though numbers are small.

Tierney described how the partnership between Project Hope and Genzyme operates:

> Project Hope's main mission is education and training. [Genzyme] provides them funds for this. Project Hope has the connections with the ministries to bring the drug in [to the country]. We now have a separate field rep from Genzyme working with Project Hope in several countries. The role of people on that team is to build support programs for patients, including reimbursement; they have meetings with the minister of health, and will put an application in for approval. They are trying to work on a process so the drug will eventually get accepted. We help patient advocacy groups, who can lobby the government to get funding. Project Hope is there to make sure that no matter how long it takes for that to happen, the drug will still get to the patients.

This approach can be very time consuming and difficult, as many of these countries have had no prior experience with treating these rare conditions. Tierney explained, "We'll spend a year or two to get one patient on. In Korea they wanted a full drug trial…but they don't understand that this is an orphan-drug population. We hired someone there who was well respected. We helped create orphan-drug legislation in Korea. Now we're trying to do that in Egypt and China." This approach has already succeeded in gaining reimbursement for Genzyme's drugs in three or four Eastern European countries, as well as in Tunisia. Tierney noted that in these situations there is often initial criticism that paying for these patients is "draining hospital resources. But the results are so dramatic that it really gets people excited."

Louis was asked how the company has dealt with the whole issue of compassionate use, especially in cases in which the supply of a drug is limited. He answered, "Carefully and thoughtfully. It's all about resource allocation. We are sometimes faced with hard decisions."

CLINICAL TRIALS

Genzyme has a number of drugs in various stages of clinical trials and must address the standard medical ethics issues of informed consent and trial safety. The unusual nature of orphan and ultra-orphan diseases and their very small patient populations, however, has posed some unique ethical challenges for Genzyme. Nowhere was this more evident than in the recent case of Fabrazyme, for which the FDA gave an expedited approval because the results from the Phase III clinical trial were positive, and no other treatments were available. As a condition of the approval, however, the FDA required that Genzyme complete a double-blind, placebo-controlled Phase IV trial. This meant that even though the drug was now available on the market and no other good alternatives were available, some of the very sick patients who were involved in the trial would not be receiving treatment until the Phase IV trial ended. Dr. Mark Goldberg, Senior Vice President in charge of Genzyme's Clinical Research said:

> This was the most difficult ethical issue I have faced. The standard I've always used to determine whether what we're doing is ethical is if I would be willing to

do it myself or recommend it to a patient. If I wouldn't be willing to participate in the trial, how can I ask someone else to? This is the first time that we've been in a situation that didn't fully pass my personal test.

He went on to explain that the issue was "not black and white"; that Genzyme's first priority was to do what was in the interests of patients; that it also recognized that it would be hard to get healthcare payers to reimburse for a drug that would be very expensive; and that, without definitive data from the trial, this task would be even more difficult. Said Goldberg, "Each investigator may say to their patients, 'It is in your personal best interest to drop out of the trial and get the drug,' but if everyone drops out, then the trial will fail and the drug won't be as broadly available." He pointed out that it puts pressure on patients to think about their peers and their children and created some difficult choices.

Genzyme tried to dissuade the FDA from putting the trial participants, their doctors, and the company in this situation, but the Agency stuck with its position. Alison Lawton, Senior Vice President of Regulatory Affairs and Corporate Quality Systems explained:

> We kept proposing different options. But placebo-controlled trials are the gold standard at the FDA—we can't get away from it. At the end of the day…we didn't have a choice. Sometimes the ethical issues may be in conflict with what is needed for scientific validity from the FDA's perspective.

Faced with the FDA's decision, Genzyme had long internal discussions on the best way to proceed. In the end, Genzyme decided it had no choice but to complete the trial as ethically as possible. This meant, said Goldberg, making sure that "everyone gets consent, that the patient is aware of what the alternatives are, and that they can drop out." It also meant offering assurances that the company would see to it that any patient who completed the trial would have access to the drug, even if they could not get third-party reimbursement.

Another ethical issue that Genzyme encounters more frequently in its clinical trials is how to determine who gets access to its drug, as the targeted patient population is typically in a dire medical state, and no effective alternatives are available. This dilemma was acute in the Myozyme drug trials for Pompe disease, in which the number of very ill young children seeking to enroll in the trial far exceeded the available supply of the drug. Louis says, "The worst part of my job is handling the occasional calls from parents who want their kids in the trial; it's heart-wrenching."

After working with a panel of outside experts to determine the eligibility criteria for the trial, Genzyme then had to grapple with whether they had a responsibility to provide treatment to those who ultimately did not meet the criteria. If they did not provide treatment, some of the children might die. But supplies were limited. Most, if not all, of the drug that could be produced was needed for the trial, and treating patients who didn't meet the criteria risked delaying completion of the trial, which could affect approval of the drug. Jan van Heek added

> When there is not enough product at first, like with Pompe, it is very tough. We try not to make judgments ourselves…we use multi-disciplinary teams to look at

patients and make these decisions. [When it comes to getting kids on trials,] parents and politicians are very vocal and don't want to take "no" for an answer. We try to do the best we can, but we know we are not perfect.

PERSONAL CONFLICTS OF INTEREST

The difficult choices involved in patient access to clinical trials can be compounded when the patients are related to Genzyme employees. Considering the very small populations affected by these disorders, a surprisingly large number of employees have personal connections to these diseases. Some examples include one employee's sister having Gaucher disease, a scientist's wife who has multiple sclerosis, and two people who have Huntington's disease in their family. "I think we struggle with the ethics of it," said Lawton. "People know the patients at the individual level and it becomes so emotional." To try to take the emotions out of the process, Genzyme has relied heavily on the guidelines developed by the groups of medical advisors in its disease areas. Goldberg explained the process:

> Either the patient is eligible according to protocol, or not... In every instance, we have tried to deal with these in the most appropriate fashion. If they are eligible, we direct them to an investigator to decide. If a patient does not qualify, we consider compassionate use. We want to make sure we don't treat someone because of who they are, at the exclusion of someone else, so we keep the name confidential so as not to obscure decisions...We use a mechanism that is unbiased, so anyone who would meet criteria would get treatment.

Perhaps the greatest test this process has faced was in the case of John Crowley. Crowley was originally a financial consultant, but when two of his three children were diagnosed with the debilitating, and eventually fatal, Pompe disease, he set out to find a cure. He quit his job, borrowed money, and teamed up with scientists working in the area who had started a biotech company. As the young company made progress developing an enzyme that they felt could be promising, Crowley and the rest of the Board of Directors felt they "needed the muscles of a big company to get a drug into production and testing."[8] Even though it meant giving up total control of the program, John, the rest of the Board, and the other shareholders voted to sell the company, Novazyme, to Genzyme for $137.5 million[9] in September 2001. John then became the head of Genzyme's Pompe research-program.

In the end, a molecule developed internally by Genzyme proved better than the enzyme that Novazyme was developing in its preclinical studies. One of the two pivotal clinical trials designed for this drug targeted infants younger than age 6 months and who had the severest form of Pompe disease; the other targeted children aged between 6 months and 3 years, and who had a less severe form. By the

---

[8]Anand G. (2003, August 26). For His Sick Kids, A Father Struggled to Develop a Cure. *The Wall Street Journal,* Section A8.

[9]Ibid.

time the trials were underway, Crowley's children did not qualify. Crowley continued to seek ways to get treatment for his children, a situation that was very complex and sensitive.

Late in 2002, Crowley left Genzyme to focus solely on his children, whose health continued to decline. As drug production capabilities increased, Genzyme was eventually able to provide the treatment to a number of additional children around the world, including Crowley's. Lawton said then:

> We all struggled with dealing with these issues. At the end of the day we were pleased that our rapid production scale-up allowed us to expand our program and treat more affected children, including the Crowleys. This was a complex moment but in the end we were comfortable about both our decision making process and the decisions we made and we were better for having gone through it.

PRODUCT MIX

The complex array of businesses that Genzyme manages can itself create interesting ethical challenges. This is particularly true in the emerging area of personalized medicine or "theranostics," in which genetic diagnostic tests are combined with targeted therapeutics. Genzyme has separate genetic testing and drug-development business units. As Louis explains, "When dealing with genetic information, we have to be exquisitely careful about the appearance that testing is done to benefit therapeutics." He says, "Obviously we avoid providing testing-based patient information to the therapeutics group. Testing needs to give that back to the doctor." The doctor would tell the patient the results. The patient would probably approach Genzyme for treatment later anyway; but, notes Jan van Heek, "[We] have to be very careful not to get accused of behaving inappropriately to generate a market for our products."

Genzyme's core strategy of developing "novel therapies that offer dramatic health improvements for underserved populations" might also potentially come into conflict with the imperatives for growth and profitability in each business unit; for example, when Genzyme has made discoveries or acquired another firm with some products that don't fit its core strategy criteria. In one recent case, Genzyme bought BioMatrix which, in addition to its lead products, produced a compound that helps reduce wrinkles. Rather than market the product itself, Genzyme is licensing it to another company, Inamed. Explained Hillback, "It's really not our type of product. BioMatrix had already agreed to supply the raw material to Inamed, and they did the FDA filing and got it approved. For us it is basically just a supply agreement... It's not a Genzyme-like product. We don't feel we should play in that ball game." Even this supplier partnership, however, made some employees uncomfortable. Lawton said, "It's hard to have facial-care products on one side and these orphan drugs on the other. The decision to work with this partner had been made years earlier and was seen by the business unit as a way to leverage their technology to generate some income from a partner to fund important research."

For a period of time, the Biosurgery business unit had a separate tracking stock with its own profit and loss statement. Genzyme was the first biotech company to create tracking stocks for different parts of the company, to allow outside investors to buy shares in separate businesses. In 1994, the Genzyme Tissue Repair division was formed and began trading separately from the Genzyme General (GENZ) stock. This was followed in 1998 by the creation of separate tracking stocks for Genzyme Molecular Oncology (GZMO) and Genzyme Biosurgery (GZBX) in 2000. Genzyme announced in July 2003 the dissolution of this structure. Termeer explained, "These tracking stocks...really became highly inefficient and very confusing to the market-place. So we eliminated them."

ETHICAL MECHANISMS

AN INFORMAL APPROACH

The focus on the patient has played a particularly vital role in how Genzyme has approached these diverse ethical issues. From the founding of the firm, the leaders have tried to encourage a flexible and creative corporate culture, and strongly resisted more formal processes and mechanisms. This strategy was summed up by the Vice President of Workforce Development, Joan Wood. She explains her early experience at Genzyme:

> When I first joined, I talked to the CEO. One thing he didn't like was how my predecessor interviewed him and then tried to distill down his message into three words [that could serve as values for the company]. That practice was offensive to the CEO. He explained that "Life is so complex and keeping an open mind is so important"; distilling something down to three words closes so many doors for a corporation like this, it robs the organization of all the possibilities...The whole subject of ethics as a formal pursuit—we haven't done a ton on that—we just live it.

Hillback described why Genzyme doesn't have a corporate slogan or values statement: "We were nervous to put a list of key traits on paper. First of all they are only words, and exactly what they mean ought to be constantly evolving and changing. It isn't stagnant; it's got to be a living situation." Arnstein agreed: "If I have to pull a laminated card out of my wallet to find our mission, we have failed."

It is for these reasons that defining ethical mechanisms at Genzyme is something of a paradox. Despite the lack of formal mechanisms for creating or disseminating values, "...it's fascinating to realize," says McDonough, that Genzyme has "a common corporate culture and that people have gone to different business units and found the same things that are important." In place of formal mechanisms, Genzyme's employees mentioned other ways in which the firm builds and sustains its distinctive culture: leadership, the hiring strategy, story-telling and communication tools, the business structure, and the use of external advisors. Some of these are starting to become more structured and defined, whereas others still occur somewhat naturally and without much planning.

LEADERSHIP

In his 21 years running Genzyme, Termeer has struck a balance between providing strong central leadership and encouraging the people to be entrepreneurial and make their own decisions. This approach can sometimes be challenging for new employees. Arnstein described it as "…a top-down communication infusion. There really is a Genzyme way and you learn what it is through a socialization process. No one is telling us the rules; we learn and live the rules, but also help redefine them and keep them alive."

Termeer conveys these rules and the patient-first philosophy through his own leadership style. According to Goldberg, "Henri's compassion, passion and consistency really set a tone." It appears that Termeer's management approach directly reflects his stance on ethics. Arnstein says, "Partly why I'm still at Genzyme is because I'm 100% confident that the CEO always takes an ethical approach." She explains that Termeer is committed to the patients and says that "when presented with tough issues he gets people from diverse areas together to brainstorm about what should be done. At strategy meetings, they try to figure out how to work better with patients." And Louis said, "Henri will sit in Genzyme's indigent-care program meetings, reviewing individual patient cases and make an effort to understand the patient perspective. I don't think there has ever been an instance where someone was denied clinically appropriate treatment."

Termeer, however, is not alone in providing ethical leadership. Hillback has played a variety of senior management roles since following Termeer from Baxter to Genzyme in 1990. He has informally taken on corporate culture, values, and ethical issues as one of his job responsibilities. He has represented Genzyme externally on this subject, serving as cochair of the BIO (the Biotechnology Industry Organization) Ethics Committee, and giving talks at business schools on values and ethics. In addition, Sterling Puck, the Medical Director for Genzyme Genetics, has played a key role in how Genzyme approached the controversial area of genetic testing. According to Dr. Phil Reilly, one of the international ethics experts in this field, who has provided advice for Genzyme, "Puck has been a huge influence on their high standards of genetic testing. She's not trained in ethics, but she puts patient privacy and other ethics issues first." Likewise, he observed, "Elliott [Hillback] really educated himself on ethics and has carried that into the company. These are just a couple of examples, because in the end everyone is part of the process."[10]

STAFFING

Given the lack of formal process controls, Genzyme focuses a lot of attention on hiring individuals who fit the firm's strong, yet unstated, values and ways of making decisions. "If they're here just for the money, they won't fit in," said

[10]Interview with Phil Reilly, October 1, 2003. Waltham, Mass.

Eric Tambuyzer, Senior Vice President Corporate Affairs Europe. "The decision-making process involves a lot of consensus, not hierarchical demands... You will not stay long if you are not comfortable with a consensus approach." Arnstein added, "People from more structured environments may find it hard [at Genzyme]," because entrepreneurship and flexibility are so highly valued.

As noted earlier, one of the most direct ways in which Genzyme has generated a strong connection between its employees and patients is that a number of employees have a family member who suffers from an ultra-orphan disease. The stories of these employees and their families are shared through the newsletter (discussed below) and in informal conversations, magnifying their impact. This link between employees and patients appears not to be a conscious staffing strategy, but rather the case that these individuals are attracted to Genzyme by the firm's mission.

STORY-TELLING AND COMMUNICATION TOOLS

Genzyme has used various ways to communicate what the company is all about to employees. The most common method is to tell the Genzyme story: "We go back to the very early history and make people part of where that history is today," Jan van Heek explained. "We tell stories—we make people part of who we are." McDonough recalls being told one such story when he joined Genzyme. "[It was about the] central founding event of the company; which involved lugging placentas around in ice vats to make [the first drug, Ceredase, that helped] this company go forward." Christi van Heek, former Senior Vice President of Therapeutics and Global Marketing, explained that these stories are especially good to share with newer employees and with those who are more removed from patients. She describes how the message gets across:

> ...by introducing all corners of the organization to patients' stories. While they are often heart-wrenching, they do define our mission. [For example, we] even show the shipping group a video about patients—to help them understand and get a feel for the importance of their work. It's very motivational, very people-oriented.

Arnstein explained:

> We really had no formal internal communication program until 3 years ago. What we implemented recently is an email newsletter...a monthly newsletter plus a hot-story newsletter (where you get an email with one story). The objective is to familiarize people about different areas of the company, with the range of what we do, and help them connect with patients—since that is the most important thing.

Some business units have also implemented a hotline/helpline as a mechanism to allow employees to communicate and get guidance in difficult situations.

BUSINESS STRUCTURE

Another way in which Genzyme has sought to embed the importance of ethical issues while encouraging entrepreneurialism in its global businesses has been through the matrix structure of the organization. The heads of strategic business units have been given a high degree of autonomy to manage their own businesses. At the same time, the clinical, regulatory and legal functions all report directly to the CEO. Goldberg explained the importance of this structure from his Clinical Research Division perspective: "The reporting structure is designed to minimize conflict of interest; in other words, we do not report directly to the business units. We work closely with them but remain independent. It allows us to do what we think is right regardless of the business repercussions." This matrix structure entails cross-functional meetings in which the different areas come together to try to make the best decision for the business. This is when really tough decisions are made about which initiatives to continue to invest in and which ones to cut. It is up to top management, and especially Termeer, to perform the balancing act of trying to meet all the stakeholders' needs. Jan van Heek noted that the different areas are each "…businesses in themselves and driven by their own circumstances." Lawton explained the benefit of this centralized, matrix structure from her perspective as Vice President of Regulatory Affairs:

> The advantage of the central structure is that I know everything that is happening at the FDA. As a hypothetical example, suppose we get a letter from the FDA with some little concerns about a marketing issue. And I know that one business unit is thinking of taking a bit of a risk with a marketing strategy. I can bring these points of view together [so that decisions will be made in the best interests of the firm overall]. And of course I can go to Henri directly if I need to—but I've never really had to. The FDA looks at us as a whole—not as different units—but as one Genzyme. So the matrix design recognized that and we manage appropriately.

EXTERNAL ADVISORS

While Genzyme has elected not to have a company-wide Ethics Advisory Board, it has asked Reilly occasionally to provide some expert consulting, such as a day-long training on bioethics for Genzyme's Board. In keeping with its focus on patients, Genzyme has made greater use of external review boards, made up of medical professionals, to assist several of their divisions. Jan van Heek explained how this works in the therapeutics division:

> We have an advisory board of physicians who are disease specific. They advise us not for individual patients, but for general issues for that disease or treatment. We are trying to develop guidelines and standards of whom to treat and how to treat them. We work with that group to develop clinical trials.

Goldberg described how these guidelines work in the clinical research process:

> For all the bad press about clinical research, in general there are tremendous checks and balances in place. If you follow the rules that are in place, it's very hard to do something unethical and find yourself in an ethical dilemma. I get annoyed with journals that criticize industry trials and question the integrity and the ethics of industry clinical trials. With the checks and balances in place we are often more in line than academic ones.

Another part of Genzyme that uses outside experts is the Project Hope program that was established to make sure that patients around the world have access to, and can get coverage for, Genzyme's drugs. Tierney, who is the director of this program, explained how they have used an ethics expert:

> We wanted to be at arm's length, so we established a medical expert committee with leaders in the field. On this, we included an ethicist—a Jesuit priest. He is Assistant Dean of Genetics at the University of Cincinnati. There were interesting dynamics in pulling the group together: as we were discussing the medical criteria, the ethicist kept coming back asking, would you do this for this case, etc.? He is our conscience. We took the expert committee to Egypt a few years ago. There was a good exchange between doctors in both countries. The committee has had to make tough decisions in dealing with patients who are not going to make it. We would be creating false hope by keeping the patient on the drug. The ethicist had to describe to the parents of the sick child why we were ending treatment.

SUSTAINING THE GENZYME CULTURE

Genzyme's patient-driven culture and core values appear to have served it well in confronting many challenging issues over its first 2 decades. As the firm has grown—through acquisitions, partnerships, and internal development—into a global corporation with more than 6500 employees, its leaders have begun to question how it can ensure that the diverse workforce and portfolio of businesses understand the firm's core values, and that employees continue to behave in an ethical manner. When the firm was small, it was "very easy for a large percentage of the population at Genzyme to identify themselves directly with patients. You could see it and hear it," said Jan van Heek. "For a large part of Genzyme, that is still the case. But now there are some groups that are more disconnected, for example, with the anti-adhesion product, where you don't know directly who the patients are."

"While we do think about it, we don't yet know exactly how we are going to manage it," continued Hillback. "We regularly take some time out of the day-to-day [operations] to work on how we are going to keep the crucial 'strength of character' in the company 10 to 15 years from now. Our approach is not very formal but it has worked well so far." While continuing to resist overly formal processes that could stifle entrepreneurship, Genzyme has recently adopted several new approaches to try to institutionalize its values and approach to ethical issues. As McDonough put it, "As we become larger, we can't just assume we are all on the same page because Henri wants us to be."

One example that illustrates the tension between formalization and maintaining informality was the decision to write a short book, *A Different Vision: The Making of Genzyme,* to commemorate Genzyme's 20th anniversary. The book highlights key events, tells inspiring employee stories, and chronicles heartfelt patient triumphs. Although organized around a set of concepts—A Pioneering Spirit, An Entrepreneurial Ethic, A Compassionate Approach, A Passion for Impact, A Collaborative Environment—that could easily serve as the company's values, these are never identified as such. And while copies were given to every employee, Genzyme made the decision not to share the book externally. In part, this was to protect the confidentiality of the many employees and patients pictured, to allow inclusion of some of the fun or funny events in their history, but also, explained McDonough, because "we didn't want to go so far as to say that these, and only these, are our values."

Another recent example of this tension was the decision to develop a 14-page Code of Conduct (Exhibit 3.1), which was seen as a step in a different direction for Genzyme. Louis, who was charged with facilitating the development of the Code, which covers areas such as internal and external business relationships and compliance, explained his own ambivalence:

The Code of Conduct will become a principle mechanism (defining some of the rules of ethical decision making)—it's a quantum leap. It will be posted on the web, plus we will disseminate it to every employee. I'm working with HR to have

---

EXHIBIT 3.1   **Genzyme Code of Conduct**

Our Commitment

This code formally states our commitment, as a company and individuals, to conduct our business in conformity with appropriate legal and ethical standards. This commitment supplements our dedication to overall excellence in the way we develop and deliver our products and services. The Code expresses our common understanding of what we at Genzyme mean when we talk about a set of fundamental principles for the conduct of business. **It means that:**

- We respect the laws of all places where we operate, as well as our own company policies and procedures;
- We are honest and treat people with respect and courtesy;
- We strive to make Genzyme a great place to work and a company that is respected for the quality of its people and products;
- We are all role models. Every one of us can influence and lead our fellow employees when it comes to behaving appropriately in the conduct of our business. This means acting responsibly, fairly, and honestly in our dealings, and exercising sound judgment in performing our jobs; and,
- None of us should face challenging situations alone. When you are faced with concerns about legal and business ethics issues, discuss the matter with your supervisor and immediately report any business conduct that appears to be illegal or unethical. If reporting such conduct to your supervisor is impractical, each of us has alternatives. Consider discussing your concern with your supervisor's manager or a member or your site's, region's, or division's senior management. You may always call your Business Unit or Functional Area Compliance Officer. Human Resources representatives, the Legal Department, and the Chief Compliance Officer are also available.

This code is the cornerstone of our Corporate Compliance Program. Each of us on the worldwide Genzyme team is expected to comply with the spirit as well as the letter of this Code. Since no Code can anticipate every situation, many of the concepts described in this Code are further explained in our corporate policies and procedures. The broad guidelines of this Code can help us make appropriate decisions and act with confidence in performing our jobs.

this be a living document...the code came out of a consensus from different areas—individual business units...who felt they needed a common theme...while recognizing new external pressures caused by actions and events at a few other companies.

Wood explained the importance of the proper dissemination of the Code to have any impact for Genzyme: "I'm working with Roger [Louis] to make sure that it isn't just a piece of paper, but is actually part of an ongoing process. We need to discuss it on a regular basis. I think it is structured to encourage dialogue, which is important."

CONCLUSION

These initiatives suggest that Genzyme has reached another critical juncture in the company's development. The connection with the patient, which has served as the guiding light for the firm's strategic and ethical decision making, has become harder to maintain in a larger, more diverse, global company. And while Genzyme veterans remain comfortable with the traditional way of working, many newcomers seem to feel a need for more structured ways to get connected and socialized to the culture, and for less reliance on culture and individual self responsibility. Jan van Heek summed up some of the issues:

> As the distance to the individual patient becomes greater, how do people feel about our standards? Do people still believe in them? What other means that are more formal need to be applied? We do bring doctors and patients in [to discuss the issues], but it is becoming more complex to do that effectively. When you look at a Gaucher patient after treatment, the impact is obvious. When you look at a surgery patient, our impact is much harder to see.

Another strategic issue for Genzyme is what new orphan drugs to develop to sustain its growth: it has already addressed the larger LSD patient populations; the remaining diseases affect even fewer people. McDonough said that there is "no clear level" as to what is too small a patient population. Jan van Heek raised the challenging question, "How low to go? I don't think anyone can say. If a product works, there is an obligation to make it available. Then the question is, at what cost? We have to generate a return. If you can't generate a return, you can't run as a commercial company." Finding an economic way to develop drugs to treat such small groups will require new approaches to clinical trials and to regulatory approval that Genzyme is just beginning to explore with the FDA. As Lawton described it:

> And now, we are getting even smaller populations. [There are] longer term discussions that I and others have to start to think about concerning approval issues. Pressure to find new ways to get drugs to patients, and to change what have been routine FDA requirements. [We need a] model for trying to recover investments, especially with smaller and smaller diseases. There is also more pressure for collecting more information in the post-approval phase. The question is, is it shifting the cost of development [of a drug] or is it increasing it? If the cost is increasing [to develop a drug] and the trials continue longer...then what?

To satisfy its investors and generate the profits to fund future development of these ultra-ultra-orphan drugs, Genzyme has sought to expand into new areas that share many characteristics with its LSD franchise. In 2003, it bought SangStat Medical Corporation, which enabled it to enter the immune–disease marketplace with a profitable antibody product used in organ transplants. In 2004, it announced two further acquisitions in the cancer field: ILEX Oncology (pending receipt of regulatory approval), whose lead product is an antibody approved for treating the most common form of chronic leukemia, and IMPATH's cancer diagnostics business "which will bring to Genzyme a leading array of solid tumor and blood-based cancer tests."[11] These, together with DNA tests from Genzyme Genetics, might be used to identify and develop more personalized treatments linked to particular forms of cancer and an individual's genetic profiles. These acquisitions significantly contribute to Genzyme's revenues and potential future growth. They also are expected to add more than 1000 employees and several new business leaders, all of whom will need to be socialized in Genzyme's distinctive culture while also altering it with their own ideas and experience. Despite these recent additions, Hillback stressed that the firm's commitment to putting patients first, by treating serious conditions in which the treatment can have a major impact, is as strong as ever:

> We've reiterated this message more strongly lately. We could call it a decision rule, even though nothing is really a rule at Genzyme. This focus also is a central idea in the concept of a value-based, rather than a cost-based, healthcare system. Genzyme's patient-oriented approach would help the company do well in that kind of world.

As he looked toward this future, and to the more immediate task of integrating these new acquisitions, Hillback pondered some of the key challenges Genzyme would face: "Can we perpetuate the culture without formalizing it? And can we continue to depend on culture, consensus and personal self-regulation, rather than formal rules, as we diversify and grow?"

---

[11]Genzyme. *2003 Annual Report*, p. 4.

4

MILLENNIUM

PHARMACEUTICALS, INC.:

CREATING AND

SUSTAINING CORPORATE

VALUES[1]

INTRODUCTION

Every time he tells this story, Millennium's former Chief Business Officer (CBO), Steven Holtzman, smiles and shakes his head as if he still can't believe it. The story deals with the conditions under which he was hired, in 1994, by the Cambridge, Massachusetts biotechnology company. The company was a small start-up at the time, and its founder and interim chief executive, Mark Levin, was looking for a permanent chief executive officer (CEO) from established pharmaceutical companies with research and development backgrounds. Levin's wife, Becky, a corporate recruiter, met Holtzman during this search. Holtzman had founded, built, and taken public a biotech company,[2] and he had other qualities that Becky thought would interest Levin. She convinced the two of them to have a glass of wine together, just to talk. As Becky predicted, Levin and Holtzman hit

[1]The case study was written by Margaret L. Eaton based on interviews with 10 Millennium employees on September 25, 2002 at company headquarters. Interviews were conducted with David Finegold, who also gave valuable editing assistance. Rahul Dhanda provided important insights and Emily Taylor collected and provided helpful background materials. This case study is one in a series produced under a grant funded by The Seaver Institute and Genome Canada to study models of informed ethical decision making in bioscience firms. Grateful appreciation is extended to all those who participated and assisted, especially to Millennium's Gary Cohen, without whose efforts this case study could not have been written.

[2]Holtzman had held several management positions at the biotech company, DNX Corp., including the presidency of its wholly owned subsidiary DNX Biotherapeutics.

it off. Levin was attracted to Holtzman's experience in bioscience-business, and to this former Rhodes Scholar's philosophy degree, his interest in ethics, and his dedication to unleashing the potential of innovative technology to do social good. For his part, Holtzman admired Levin's commitment to excellence and his beliefs about corporate integrity. Levin asked Holtzman to become Millennium's CBO. Levin's thinking was that Holtzman, the biotech entrepreneur, would have sufficient clout to negotiate alliances with large pharmaceutical companies, essentially David-and-Goliath deals. In addition, Levin was captivated by Holtzman's "non-science" side. Levin subsequently sent Holtzman an offer letter that was unique in the hiring of any corporate business officer. As Holtzman remembers it, the offer letter enumerated his various responsibilities (perform the job of CBO, execute certain deals, build specific corporate programs, etc.) and then stated that he was to spend 10 to 15% of his time dealing with company values and ethics, so as to "make Millennium a recognized leader in socially responsible deployment of genomics and genetic technologies." This is the part that makes Holtzman smile. But he was not the first, and by no means the last, employee who was hired or who sought employment at Millennium because of its two qualities of *scientific excellence* and *corporate values*.

MILLENNIUM PHARMACEUTICALS INC.[3]

Incorporated in January 1993, Millennium Pharmaceuticals was among a group of second-generation biotechnology companies whose business plan was devoted to genomics: translating the knowledge gained from the human genome-sequencing projects[4] to identify genes associated with human disease and, from there, discovering drug targets important in combating these diseases. This work was highly valuable to the large pharmaceutical companies that would be Millennium's first customers. Millennium's business proposition was created at a time when the large pharmaceutical companies were having new drugs approved at a dwindling rate and when a significant number of big money-making drugs were losing patent protection. Therefore, large pharmaceutical companies were seeking new avenues

---

[3]Unless otherwise specified, company information was obtained from financial reports and interviews with the company employees listed above. Further information about Millennium's business plans and strategies can be obtained from: Champion D. Mastering the value chain: an interview with Mark Levin of Millennium Pharmaceuticals. *Harvard Business Review.* 2001 June:109–115. See also Harvard Business School case study. Millennium Pharmaceuticals, Inc. (A) (9-600-038).

[4]The human genome is the entire genetic information contained in a human being. The sequencing projects were primarily the US government-sponsored Human Genome Project in collaboration with the International Human Genome Sequencing Consortium begun in 1990, and a similar rival sequencing program run by the private biotechnology company Celera Genomics, Inc. The goal of these two projects was to determine the sequences of the 3 billion chemical base pairs that make up human DNA. Substantial completion of both projects was announced in February 2001. With this sequencing information, scientists could identify the approximately 30,000 human genes, work on discovering their function, and translate that knowledge into important discoveries about human physiology and disease.

and sources of drug development to fill their product pipelines, and to replace income from the generic competition they would face when their drug patents expired. The promise of gene-based drugs lay in their potential to exploit the genetic component of most diseases, which could result in thousands of new drug candidates.[5] The second major commercial promise of genomics came from the likelihood that drugs could be developed that could alter the effects of disease genes. Because such drugs would operate by altering the physiologic causes of diseases, they were expected to be much more effective and less toxic than drugs that ameliorated only the symptoms of disease. A third expected benefit of genomics-based discovery of drug targets was that the technologies would speed up the process, and reduce the cost of their identification and, hence, development of the drug. Anything that could speed up drug development would be a tremendous boon to any pharmaceutical company.[6] Lacking in-house genomics expertise, many large pharmaceutical companies naturally looked to smaller genomics firms like Millennium to provide new drug targets and their associated benefits.

Millennium started by associating with key scientists[7] such as Massachusetts Institute of Technology's (MIT) Eric Lander, the head of the largest genome center in the United States. The company continued to be very successful in hiring or consulting with top-level scientists and acquiring or licensing state-of-the-art technologies, all of which facilitated the process of gene discovery, gene-function analysis, and identification of drug targets. In addition to venture capital, Millennium raised money by forging technology transfer, research and development (R&D), and commercialization partnerships with a series of biotechnology

---

[5]Scientists calculated that human genome sequencing information would yield an estimated 3000 to 10,000 gene-based therapy targets. This number comes from an estimate of about 100 important diseases that are caused by 5 to 10 genes each (most common diseases are associated with a number of genes), yielding perhaps 500 to 1000 disease-related genes, each of which probably interacts with 3 to 10 proteins associated with a disease process. Cohen J. Drug Companies, Biotechs, and Wall Street Investors are Putting Their Money Down on Efforts to Unlock the Secrets of Human DNA. *Science.* 1997;275:767–772.

[6]According to the Tufts University Center for the Study of Drug Development, in 2001 it was taking 12 years and costing up to $800 million for a pharmaceutical company to develop and market a new drug. These figures included the cost of failure. For every 5000 chemical entities tested, only 5 made it to human trials and, of these, only 1 was eventually approved by the FDA for marketing. Any company that could deliver viable drug targets would allow pharmaceutical companies to improve on these numbers. See http://csdd.tufts.edu/.

[7]In addition to Mark Levin, the founding scientific advisors included Daniel Cohen, MD, PhD, a Director of the Jean Dausset Foundation/Centre d'Etude du Polymorphisme Humain (CEPH) in Paris; Eric Lander, PhD, a member of the Whitehead Institute for Biomedical Research and Director of the Massachusetts Institute of Technology Center for Genomic Research; and Raju Kucherlapati, PhD, of the Albert Einstein College of Medicine. Other original scientific advisors included David Baltimore, PhD, a Nobel Prize winner and professor at Rockefeller University; Patrick Brown, MD, PhD, of Stanford University and the Howard Hughes Medical Institute; Neal Copeland, PhD, of the ABL-Basic Research Program, National Cancer Institute, Frederick; Nancy Jenkins, PhD, of the ABL-Basic Research Program, National Cancer Institute, Frederick; and Richard Wilson, PhD, of the Washington University School of Medicine.

and large pharmaceutical companies. The R&D and commercialization partners were firms with no genomics expertise, and they paid to acquire access to Millennium's technology and/or to have Millennium produce a series of high-quality, gene-based drug targets. With these targets, the pharmaceutical companies could then do what they did best: complete the process of drug development by taking drug candidates through the costly, prolonged, and arduous animal and human research process, seek and obtain Food and Drug Administration (FDA) marketing approval, and then manufacture, market, and sell the new drug. Included in the value proposition for Millennium was that, at least at first, the company would avoid all these downstream costs and uncertainties, but would share in the revenues when a Millennium-enabled drug was eventually marketed. Mark Levin put it this way: "We don't have to risk the company on clinical costs, which are an order of magnitude more expensive than what we do… Our approach is, instead of risking the company on a single drug or two drugs, to take the 5% to 15% cut of dozens, if not hundreds of [drug] leads."[8] Eventually, however, Millennium wanted to set itself up to become a fully fledged biopharmaceutical (drug) company—to discover, develop, and market its own drugs. Recognizing that it could take 10 years to market its first drug, partnering with large pharmaceutical companies was a way to obtain interim revenues that would allow it to grow into a drug company.[9] To achieve this goal, the company would first need to become successful in the discovery of drug targets and then structure its deals, enhance its infrastructure, and purchase or license expertise so as to position itself for a transition into a drug company. Keys to success in this endeavor included the ability to maximize large pharmaceutical company revenues while retaining significant assets: scientists, knowledge, and product rights.

Starting with publicly available information from the Human Genome Project, Millennium developed and acquired methods, automated equipment, and powerful computer programs to analyze gene sequences to identify genes involved in the biochemical pathways of certain diseases.[10] To produce a useful "deliverable" to a pharmaceutical company partner, Millennium identified disease genes in a specific category. Other corporate competitors were doing the same thing, so Millennium distinguished itself by also identifying the function of the genes and the specific proteins expressed by the genes that were implicated in particular disease processes.

---

[8]Prewitt, E. What You Can Learn from Managers in Biotech. *Harvard Management Update,* 1 May 1997.

[9]Drug discovery and development required contributions from bench scientists (such as chemists, biologists, and pharmacologists) to work on understanding the structure and function of new drugs in laboratory assays and animals. Clinical scientists (physicians and clinical pharmacists) were needed to conduct studies in humans to test for safety and efficacy. In addition, regulatory affairs, manufacturing, marketing, and sales groups were needed to get a drug approved by the FDA and on to the market.

[10]Millennium used a technique, called *positional cloning,* to identify disease genes. Positional cloning involves the study of the genomes of large families, many of whom are afflicted with a particular disease. Comparisons are made within the family between those with and without the disease to identify genetically linked molecular markers that allow scientists to discover the position of the sought-for gene on a particular chromosome.

Exhibit 4.1   **Millennium Pharmaceuticals, Inc. (Major Pharmaceutical Partnership Deals)**

Company	Date	Value	Purpose
Hoffmann-La Roche	3/94	$70M	Obesity, type II diabetes
Eli Lilly	10/95, 4/96	$90M	Atherosclerosis and cancer
Astra	12/95	$60M	Respiratory disease
Wyeth-Ayerst division, American Home Products	7/96	$100M	Central nervous system disorders (Alzheimer's, stroke, epilepsy, substance abuse, etc.)
Eli Lilly	5/97	$47M	Secreted proteins
Pfizer			
Beckton, Dickinson			
Monsanto	10/97	$343M	Agricultural
Bayer AG	9/98	$465M	225 new drug targets, including cancer, cardiac disease, pain, osteoporosis, and infectious and blood diseases
Bristol-Meyers Squibb	11/99	$32M	Cancer drug pharmacogenomics
Aventis	6/00	$450M	Inflammatory, allergy, and arthritis diseases
Abbott Laboratories	3/01	$250M	Metabolic diseases

These extra steps made it more likely that a useful drug could then be developed, and these drug targets were thus called *validated* or *druggable* in the industry jargon.[11]

Beginning in 1994, Millennium started on a path of creating more than 20 strategic alliances with leading technology, biotechnology, and pharmaceutical companies (see Exhibits). By the end of 2002, these alliances provided Millennium with close to $2 billion of committed funding used to conduct its work in discovering drug targets. The large pharmaceutical company deals were often combinations of equity investments, upfront and milestone payments for R&D that led to the identification of drug targets, license fees, and combinations of profit sharing and royalties when the pharma collaborators were successful in developing and commercializing products using Millennium's technologies and discoveries. In many later deals, Millennium also retained rights to co-develop and jointly commercialize products and/or rights to claw back those drug targets not developed by its partner. Millennium also entered into technology development and transfer arrangements with other companies to enhance its technology platform and to license its technology, in exchange for fees or royalties if products using the company's technology platform were commercialized.

During most of its history, Millennium was known for having the greatest amount of pharma alliance revenue of any genomics-based company. Millennium's first big hit occurred in Spring 1994, when it signed a $70 million, 5-year research collaboration with the large pharmaceutical company Hoffmann-La Roche, Inc.,

[11]Gura T. After the Gold Rush: Gene firms reinvent themselves. *Science.* 2002;297:1982–1984.

a deal typical of many for Millennium. Hoffmann-La Roche took a small equity position in Millennium, and the majority of the $70 million was devoted to milestone payments to support Millennium's research, which would focus on drug targets for obesity and type II diabetes. Both are thought to have genetic bases and, with more than 100 million clinically obese people in the world and 12.5 million Americans with type II diabetes, cures could prove quite profitable. The companies split the marketing rights to the drugs that would be developed through this collaboration. As with most of Millennium's large pharmaceutical company deals, the companies had to acknowledge the uncertainty about the possibility of identifying disease genes whose expressed proteins were amenable to drug treatment and, if it were possible, whether Roche could develop and get marketing approval for these drugs. Despite the fact that a gene can play a crucial role in the development of a disease, a drug that interacts in this pathway can still fail because of unacceptable side effects, or because other genes may override its effects. In any case, it would take years before it was known if Roche could capitalize on any targets identified by Millennium. This double-sided uncertainty was a feature of each of Millennium's large pharmaceutical company deals.

Over the next 2 years, Millennium added several deals with large pharmaceutical companies (see Exhibits). Considered very promising income-generating ventures, they gave Millennium sufficient credibility to proceed with a successful initial product offering (IPO) in May, 1996.[12] A string of subsequent such deals added revenue and employees to the company rolls and allowed Millennium to begin shopping for companies that would initiate expansion efforts into drug development. In 1997, Millennium merged with ChemGenics, a local biotech company that gave Millennium new abilities to validate drug targets and the ability to begin to discover drugs that would interact with Millennium's drugs targets. In 1998, after losing money in most preceding years, Millennium reported $10.3 million in net income from its large company deals, and was spending $200 million a year on research. By June 1999, Millennium's payroll had reached 800 employees and its research collaboration commitments totaled $1.25 billion.

The year 2000 was different from any other. Renewed interest in the biotechnology industry coincided with the imminent completion of the human genome-sequencing projects. Investors were betting that genomics offered unprecedented ability to discover the causes of disease and to shorten the time it took to produce vastly more effective and less toxic drugs. Despite the fact that many biotech companies were years away from market approval of any drug and were losing more money than ever, their stock prices were at all-time highs, and investors poured a record-breaking $31 billion of capital into the biotech industry. A huge run-up in Millennium's stock price that year gave it a September capitalization of $12 billion.[13] Millennium was able to complete a public offering that year that resulted in net

---

[12]4,500,000 common shares were offered for initial public sale at $12 each.
[13]See Exhibits.

EXHIBIT 4.2   **Financial Information**

Income Statement (in millions)

	Sales	EBIT	NA	Total Net Income	EPS
1996	31.8	(8.0)	3.9	(8.8)	(0.10)
1997	89.9	(83.2)	12.2	(81.2)[1]	(0.72)
1998	133.7	(1.4)	16.3	10.3	0.09
1999	183.7	(351.0)	21.0	(352.0)	(2.62)
2000	196.3	(182.1)	79.3	(309.6)	(1.84)
2001	246.2	(182.7)	100.1	(192.0)	(0.88)
2002	353.0[1]	n/a	n/a	(590.2)	(2.13)

[1] Revenue in 2002 included $193 million of strategic alliance revenue and $160 Million of copromotion revenue from worldwide sales of INTEGRILIN[R] (eptifibatide) Injection.

Balance Sheet (in millions)

	Current Assets	Current Liabilities	Long-Term Debt	Shares Outstanding
1996	72.1	11.8	9.3	95.7
1997	101.9	16.4	19.8	116.7
1998	202.7	24.3	24.8	139.7
1999	286.5	59.2	27.5	178.6
2000	1485.6	136.2	125.3	214.0
2001	1514.6	203.2	118.4	224.3
2002	n/a	n/a	n/a	287.0

Source: Financial reports.

proceeds of $767.4 million. This positive financial climate helped Millennium take big steps toward the goal of becoming a drug development company.[14] By the end of the year, the company had 1300 employees, had spent close to $270 million in R&D, and the company had 6 drugs in human clinical trials. It had also developed a fully functional medical diagnostics and pharmacogenomics unit, MPMx, that would produce tests to predict which patients would respond to specific drugs. By then, the value of Millennium's partnership deals totaled $1.8 billion, and the company was still leading the biotech industry in earned alliance revenues.[15]

By 2002, Millennium had succeeded in becoming a biopharmaceutical company. At the end of 2001, it had announced its intent to acquire COR Therapeutics,[16] which not only added to its ability to work in the area of cardiovascular research,

[14]LeukoSite was purchased in December 1999 for $550.4 million in a stock deal. Cambridge Discovery Chemistry was purchased in July 2000 for $51.8 million in cash. Millennium was also actively engaged in licensing rights for many technologies.

[15]Licking E. A Pharma Star is Born? *Business Week.* 2000 September 25:100.

[16]Millennium acquired this company through the issuance of about 55 million shares of common stock. The deal was valued at $2.2 billion.

but also gave Millennium its first drug, INTEGRILIN® for injection, an intravenous cardiovascular drug with annual sales, at the time, of approximately $230 million.[17] COR also brought with it a sales and marketing organization. In addition, Millennium had co-developed a leukemia drug, Campath®, with a subsidiary of ILEX Oncology, Inc.[18] The commercialization of a third product was pending. This was the diagnostic Melastatin™, a test used to determine the aggressiveness of melanoma tumors. Eleven other products were in various stages of human research. Millennium had close to 1900 employees, almost $2 billion in committed funding from its pharmaceutical and biotechnology alliances, and had begun to earn royalties on drugs. It planned to increase its drug pipeline and, even though it needed to spend heavily in clinical research, expected to turn profitable in the next several years. Millennium's future depended on being able to: manage its growth; maximize income from its alliances (both from the initial deals and from downstream royalties from marketed Millennium-enabled drugs) while diluting the company as little as possible; navigate the expensive and uncertain process of clinical research; and retain rights to sufficient technology, intellectual property, and markets so as to preserve the ability to grow into a company that, on its own, could discover, develop, and take new drugs to market. These were daunting tasks, but the company was highly optimistic about its chances for success.

INSTITUTIONALIZING CORE VALUES AND ETHICS

Millennium's focus on values and ethics was instilled in the company by its founder, Mark Levin, and then institutionalized in several ways:—by hiring a CBO who also focused on ethics; through educational programs; hiring ethics-minded employees; setting standards of best practice for internal and external relationships, learning from the mistakes of others; monitoring, assessment and feedback programs, corporate vision projects, corporate social responsibility programs, and repetition and reinforcement with the aid of artifacts.

FOUNDER'S ROLE

As part of its commitment to its core business, Millennium has been dedicated to the ideals of corporate values and ethics. This vision began with Mark Levin, who started his career as a chemical and biomedical engineer at Eli Lilly. Then, with some detours into unrelated fields (beer brewing and computer systems),

---

[17]Millennium copromotes this drug with Schering-Plough, Ltd. and Schering Corporation.

[18]At the end of 2001, ILEX acquired Millennium's equity interest in the partnership in exchange for which Millennium will receive a series of payments and obtain royalty payments on sales of the drug.

Exhibit 4.3   **Core values**

PATIENTS — We make a difference for our patients, their families, and healthcare providers.

PEOPLE — We hire, motivate, and develop outstanding people.

EXCELLENCE — We cultivate an environment where each of us can excel.

INNOVATION — We encourage innovation in the discovery, development, and commercialization of breakthrough products.

COLLABORATION — We seek results through collaboration — with each other, and with our customers.

RESPECT — We listen to and learn from everyone.

INTEGRITY — We insist on absolute integrity in everything we do.

RESPONSIBILITY — We are a socially and ethically responsible corporation.

VALUE — We are committed to creating sustainable value for our shareholders.

POSSIBILITIES — We believe that nothing is impossible!

Living Our Core Values

/W\ MILLENNIUM

ethics@mpi.com • helpline: 866.4MYMLNM

Levin went on to work for the biotechnology company Genentech. Having distinguished himself at Genentech as a consummate deal-maker, Levin then joined the San Francisco Bay area venture capital firm, the Mayfield Fund, where he founded 10 bioscience companies and served on their boards.[19] At each company, he collected perspectives that contributed to his vision for the corporate culture he brought to Millennium. According to Levin, Eli Lilly had a "tremendous sense of their place in the community"[20] and were highly cognizant of their history and what people meant to the organization. Lilly's culture included recognizing the value of employees and treating them well. Genentech believed in hiring the best, and tolerated and even encouraged eccentricities in employees so long as they were

---

[19]Rifkin G. (2003, January 12). The Boss: Always Avoid Boredom. *The New York Times*, p. 13.
[20]Quotes attributed to Mark Levin and not cited were obtained in an interview with Mr. Levin in March 2003 by Gary Cohen based on questions posed by the author.

passionate about their work, intelligent, and committed to excellence. "Passion, not fashion" was how Levin described it. Genentech also built a values-based company and, as a result, Levin saw employees respond positively to the company and adopt its values. Company values became part of employee consciousness, and Levin understood that this situation did not come about by accident. Later, when Levin operated as a venture capitalist and interim CEO of the start-ups he funded, he stressed the importance of values and vision, right from the beginning. He told Gary Cohen, Millennium's in-house ethics leader, "It's more than just vision, actually; it's about what each of these companies was going to be, about who the people at each of these companies were, and what they wanted to do with their lives."

According to those who have commented about him publicly, Levin has great energy and commitment to his work that attracts people to him. He has a different and sometimes offbeat way of thinking through problems that also allows him to spot opportunities faster than others. Levin was quick to recognize and respond to the commercial potential of the Human Genome Project, meaning that he understood that the genome sequencing would lead to the discovery of human disease genes that, in turn, would lead to the development of gene-based medical products of unprecedented value. At Mayfield, he became intrigued by the prospect of funding and founding a company with this as its business strategy. He enlisted his wife Becky to use her recruiter talents to find a CEO to run such a company. After a fruitless 9-month search, Becky became convinced that the company concept was so unusual, and the vision and understanding needed to run it so unique, that only her husband could do the job. So, in January 1993, Levin founded Millennium Pharmaceuticals, Inc. with $8.5 million in initial venture funds.[21] Although, in the past, Levin had served only as interim CEO for the companies he founded at Mayfield, in January 1995 he agreed to become Millennium's permanent CEO and has stayed on at Millennium full-time ever since.

Millennium was going to be about something very important, Levin decided. Dedicated to the discovery of genetically based drug targets, the company would be "transcending the limits of medicine" by revolutionizing the process of drug development. The disease genes the company would seek would be for complex and widespread diseases, for which the current treatments were inadequate or nonexistent. Its founders recognized early that this business plan would raise ethical issues, primarily because of misunderstandings about the link between genetics and disease: for instance, the prevalent belief that there is one clear genetic cause for a disease that is discovered to have a genetic connection, rather than that the disease may have multiple causes. Such a misconception could lead to stigmatizing individuals found to have a particular disease gene.[22] Any social misuse of genetic and genomic technology could create a public backlash that could

---

[21]Blanton K. (1999, June 13). Biotech's Pied Piper: Relentless in Recruiting, Generous in Research Budget, Levin Draws Experts, Money. *The Boston Globe*, Section F1.

[22]People found to have genes that predisposed them to certain diseases could be denied health insurance or employment, for instance.

threaten the firm. The human genome sequencing projects had already caused some public concern over the harm that could result from manipulating genetic knowledge. Millennium's founders viewed social and ethical issues such as this as an important force in the new genomics industry. Devoting attention to these issues was seen as essential to the success of the firm. With Levin's philosophy of corporate values as a guiding force, this view easily generalized into the determination that Millennium would operate with integrity in all fields of endeavor; the scientists would strive for excellence, and corporate practices would be socially and ethically responsible. According to Steven Holtzman, for all companies that produce medical products, the two goals are linked: being innovative and socially responsible both contribute to making a company competitive.

To implement this vision, Levin met with the top executives of the fledgling company to discuss his ideas about key corporate values—those he would abide by personally and expected everyone else in the company to follow. From this, a set of core values was developed that ended up on every employee's mouse pad:

Millennium Core Values.[23]

We hire only the best people.

We are creating an environment where our people can excel.

We listen to and learn from everyone.

We encourage innovation in the search for breakthrough products.

We insist on the absolute integrity of all our interactions.

We believe that both outstanding individuals and focused teams are critical for our success.

We have created and continue to build a socially and ethically responsible corporation.

When asked what prompted him to institute Millennium's core values, Levin said:

At Millennium, we were determined to think about, and talk about, and act on a set of core values, right from the beginning. So, I suppose my experiences at Lilly, and Genentech, and the start-ups influenced my thinking. We hired people who were interested in a values-based employer, people who would share our vision on Day One about the kind of place they wanted to build. We talked about our values a lot, at major company meetings and in smaller groups. We took actions based on our values, and there were real consequences for the few people over the years who weren't acting according to those values... It's not just about having a statement of core values, it's about executing those values.

[23]In November 2003, based on the company's year-long discussion of its values, Millennium adopted and reaffirmed the Core Values included in Exhibit 4.3.

By all accounts, Levin both talks the talk and walks the walk. Below is an example of what employees said about Levin's leadership:

Linda Pine, Senior Vice President of Human Resources and one of Levin's earliest nonscience hires:

> It is just the type of person Mark is. Passionately committed to being open, inclusive, listening to and learning from anyone. Someone who works in the mailroom can walk into Mark's office and share his ideas. There are not a lot of CEOs who start from the perspective of creating a great workplace. [He has a] grand vision of entrepreneurism, great science, creating great products that save lives. [He understood] early on that the area of mapping the human genome and doing genetic studies meant we should be on the front end of understanding ethical issues. He connects ethics to bottom-line business issues. [His mantra is still] 'We believe that nothing is impossible.'

Steven Holtzman, original (and former) CBO, and also responsible for ethics:

Holtzman knows that Levin did not give him the ethics assignment to make the company ethical. "All people in business want to deal with people of integrity. I was on a panel [with someone] where he said that if you need to hire a [corporate] bioethicist, it must be because you were in trouble. I agree and disagree. If you lack integrity, hiring someone to make you have integrity, I can tell you, it ain't going to happen." Holtzman added that Levin's sense of integrity was what made him believe that the CBO should also focus on maintaining corporate integrity. However, Holtzman maintains that employees will not behave in an ethical manner just because they are told to do so by the company ethicist. Rather, the company leadership creates an environment that supports ethical decisions, and provides avenues for participation and education about ethics, so that each employee is engaged and will then incorporate ethics as an aspect of his or her work. As a result, the corporate community as a whole will then function in an ethical manner.

Jack Douglas, Corporate Secretary, and former Senior Vice President and General Counsel:

Jack Douglas believes that Levin's leadership is critical to the company's core values.

> Early on [Levin] created a set of core values and he made commitments to integrity and corporate and social responsibility. When I interviewed with Levin for the position of general counsel, I asked him about an exit package given that this was a vulnerable position. He turned around and asked me what about my secretary, what would she get? I got so excited about this general approach, I gave up what I was asking for and came to work here, where no executive has more benefits than anyone else has.

Gary Cohen, Vice President, Ethics and Corporate Responsibility:

> In the business ethics arena, it is particularly important, notwithstanding any programs you do, to have an identifiable leader with charisma and a deep sense of commitment to the institution, and who models ethical behavior for his or her employees. Most stuff I've read suggests that a good business ethics program needs as its anchor an ethical and deeply committed CEO. We're fortunate to have a business ethics 'rock-star' in Mark Levin. He's a remarkable guy.

Throughout Millennium, any employee who talks with him or reads his periodic e-mails will see that he brings a deep respect for core values and connecting to them in meaningful ways. Our CEO sets a beautiful ethical standard.

Fabio Benedetti, Vice President of Medical Affairs, responsible for investigator-initiated studies and medical affairs activities:

Dr. Benedetti came to Millennium from a large pharmaceutical company, already imbued with a set of values that drives his work of implementing and supervising late-stage human research trials. He said:

My boss at Roche used to say, 'Good medicine is good business.' I'm a big believer in it. If I make decisions that help patients, this will ultimately translate into good medical outcomes. I think the opposite is also true. Short-sighted decisions will catch up to you in the long run. This seems obvious, but many don't do it. [What attracted him to Millennium is that] everything in this firm seems to reinforce behaving ethically. The patient-first emphasis comes from Mark [Levin]. It is part of the reason I joined. And I had a lot of other offers. When I was done with the one-on-one interview with Mark, that was when I knew I wanted to join. In my position, I can assure ethical reporting of outcomes for this firm in clinical trials. I can assure we won't put out anything that isn't appropriate since I head Medical Affairs. While I can't stop individual rogue behavior, it appeals to me that I can make these assurances.

Joanne Meyer, Director of Human Genetics:

Dr. Meyer believes that "genetics has always been a hot-button issue for ethics because of implications of profiting from DNA." When she was hired at Millennium, she was attracted to the company's ability to make a difference with the technology, which was greater than the impact that could be made in academia. In addition, she liked the corporate leadership. "When upper management displays an interest in ethics, it trickles back down to employees."

Chuck Gray, Director of Scientific Development:

Dr. Gray was hired to negotiate and monitor important academic research and technology agreements and budgets. He said that he was initially attracted by the financial package, but soon also became impressed by the senior people he met and their passion and excitement for the place. He especially liked Levin's style of leadership, which was "unusual, off beat, and he had integrity. He created this environment and, early on, it was evident that the Company was very entrepreneurial." The core values "start with the CEO and go all the way down the line. For instance, Levin flies coach, he doesn't go for expensive meals. He doesn't have the perks. He has a small office, and doesn't have an entourage. He's an honest guy. He meets with lots of employees throughout the year and answers e-mails and takes suggestions seriously. Other CEOs can be very different, they take what they can get."

Rob Kloppenburg, former Senior Director, Corporate Communications:

Kloppenburg said that Millennium has shown him that the ethics focus at a company starts at the top. He sees the management teams approach issues with "eyes wide open, keeping an ethics perspective, and this influences the culture of the company." He believes that the idiosyncratic effect of corporate leadership has more influence on how the company behaves than is often realized. And because of the leadership at Millennium, people feel empowered to make ethical decisions. He is especially impressed by the fact that Levin has committed to fund a full-time ethics position (Gary Cohen's) in such a young company.

Sue Nemetz, former Vice President, International Commercial Operations:
When the Company was filing new drug applications for a cancer product, discussions took place about which countries would be suitable to seek marketing approval. Obviously, there were various business and regulatory factors that needed to be considered. However, the business decision not to file in any particular country meant that citizens would not have access to the new drug, and some managers wanted that factor to be included in the decision making process. They wanted a discussion about whether the company considered all the ethical implications of this launch. According to Sue Nemetz, "Mark made it okay to look at such a question."

IN-HOUSE ETHICS LEADER

In the beginning, CBO Steven Holtzman was directed by CEO Levin to spend 10 to 15% of his time on corporate ethics issues. From the time he was hired in 1994, Holtzman implemented or advised on most of the programs described below. He was also available to advise anyone in the company on any ethical questions regarding company activities. Holtzman, therefore, played a vital early role in implementing and sustaining Millennium's values and ethics culture. In 2002, however, Holtzman decided to leave Millennium to start his own company. His challenge at that point, he said, was to "clone himself" to ensure that the ethics and values focus of the company would not leave with him. Eventually, Levin, Holtzman, Pine, and Douglas decided that Gary Cohen, the company's first corporate lawyer, would take on full-time responsibility for ethics. Cohen was chosen because of his interest in public policy and ethics, which he'd first cultivated at Genetics Institute, Inc., and later as a Vice President and General Counsel to Genzyme Transgenics (now GTC Biotherapeutics).

Cohen accepted the offer to become responsible for Millennium's ethics and corporate responsibility. He interviewed its top managers and learned that they were interested in seeing if he could expand on the bioethics endeavors. Cohen then set about expanding and enlarging on Holtzman's ethics work. According to Cohen, his program was "modest" in the first year, costing the company his salary and half of an assistant's, plus overhead and limited outside spending for consultant fees and conferences. Since then, Cohen, at Levin's request, has broadened the ethics programs to regularize business ethics practices firm-wide. Spending on ethics and values programs has remained stable, culminating in a year-long "Living Our Core Values" program to celebrate the Company's 10th anniversary (see below for a full description). Cohen has future plans to reissue an updated statement of the corporate core values and to conduct training to help employees connect the core values to business practice standards.

ETHICS EDUCATION PROGRAMS

IN-HOUSE ETHICS INSTRUCTION

Early on, Linda Pine identified, and Steven Holtzman hired, Dr. Phil Reilly to create an ethics education program for Millennium employees. Dr. Reilly is a physician specializing in medical genetics and a recognized expert in issues of public policy raised by advances in genetics.[24] Holtzman's goal with this eminently qualified teacher was to give employees an opportunity to become knowledgeable about biomedical and research ethics, and how they applied to company activity. In addition, ethical reasoning was taught so that employees would feel empowered to incorporate an ethical aspect into business decision making. According to Joanne Meyer, these seminars were started in 1994-95 when the whole company could fit in the conference room. She described the seminars as very interactive, with plenty of give-and-take discussion. Over the years, Dr. Reilly's sessions were well attended and popular, especially in the later years, by mid- to lower-level employees from the medical and science departments. Sue Nemetz found Dr. Reilly's class to be "fascinating; my brain would hurt with the challenges he posed."

FILM SERIES

Later, in September 2002, the in-house ethics education program was expanded to include a film series open to all employees. Employees who sign up for the program leave work at 4:00 PM and, with pizza and soft drinks provided, view and then discuss films that have an ethics subtext. These events last up to 3 hours and originally were moderated by Rahul Dhanda, an executive of Interleukin Genetics, Inc., who has training and an interest in biotech business ethics.[25] In advance, Dr. Dhanda provides employees with film reviews, background articles, and a list of possible questions for ethics discussions, and then moderates the group discussion after the film. The films presented to date include *Gattaca* (about a future where personal and professional destiny is determined by one's genes); *Jurassic Park* (about the wisdom of cloning dinosaurs); and *Inherit the Wind* (a fictionalized version of the Scopes "Monkey Trial" of 1925, in which a schoolteacher is prosecuted for teaching evolutionary theory). According to Dr. Dhanda, the post-film discussions are always wide-ranging, but someone always swings the discussion around to address the applicability of the issues to work at Millennium. Plans are underway to continue the film series in 2003.

---

[24]Dr. Reilly is also a lawyer who is past president of the American Society of Law, Medicine, and Ethics and was the executive director for 10 years of the non-profit Eunice Kennedy Shriver Center for Mental Retardation, Inc. He has also held numerous teaching positions at Tufts University School of Medicine, Harvard Medical School, and Brandeis University. Later in 1999, Dr. Reilly became the CEO of Interleukin Genetics in Concord, MA.

[25]Dr. Dhanda has written a book on the subject: *Guiding Icarus*. New York: John Wiley & Sons, Inc.; 2002.

DISCUSSION GROUPS

Additional ethics education occurs with in-house discussion groups focused on a particular topic. Gary Cohen has briefed upper management and various management teams (Safety Review, Marketing and Sales Committees, etc.) on ethical issues relevant to current corporate activity. He also initiated a series of management discussions about the book *Dispensing with the Truth* by Alicia Mundy. The book was about the corporate misbehavior behind the Fen-Phen debacle.[26] For these discussion meetings, Cohen purchased close to 100 copies of the book for managers to read, and he prepared the same number of detailed briefing books and discussion materials. The purpose of the discussions was to present the Fen-Phen case as a cautionary tale and to prepare the corporate managers to think about drug safety and how to manage a drug recall, should it ever be needed. Rob Kloppenburg, public-relations senior director, endorsed these discussion sessions, because "companies can potentially collapse over ethical issues, plus, as an individual you have to be able to sleep at night. Doing the wrong thing to get out of a short-term jam is ruinous."

CORPORATE GOVERNANCE PRACTICES

Governance issues were addressed early in the firm's history. Jack Douglas came to Millennium as General Counsel with an interest in reinforcing good corporate governance practices, not only because he and the chief financial officer (CFO) "were on the hook for it," but also because he wanted to have the company run according to the highest standards of integrity. Douglas saw the adoption of ethical corporate governance practices as one way that the legal department could participate in value creation for the company. He put into place a set of best practices for corporate governance dealing with auditing, conflicts of interest, compensation practices, disclosure policies that focused on clarity and transparency, and provisions for objective input from outside directors. In 2001-2002, when the corporate governance scandals involving energy-trading companies such as Enron were uncovered, Douglas took the opportunity to review and compare Millennium's practices with those companies that had come under fire. Douglas found that the company was already doing what auditors and the regulators were asking for, and he felt good about its position. Another aspect of the post-Enron review, according to Gary Cohen, involved asking the question of "why firms with impeccable codes of ethics, like Enron, didn't drive it into the behavior of employees." Cohen came to believe that, for Enron and other companies, developing a corporate code of ethics was an exercise that lacked carry-through; the ethics codes ended up "sitting on shelves collecting dust bunnies." Cohen understood that having an ethics-guidance manual is only the first step in

---

[26]Fen-Phen was a drug combination for weight loss that was withdrawn from the market after the regimen had been shown to cause life-threatening heart valve problems. The book highlighted various corporate failures in the drug recall process.

a process to motivate behavior change. Enron was not the first company that had struggled and failed to make that connection but, at Millennium, this was an ongoing effort.

PARTNERSHIP RELATION

Partnering with pharmaceutical companies was fundamental to Millennium's early success, and the approach to partnering included aspects of its core values. Steven Holtzman, who negotiated the company's earliest agreements with large pharmaceutical companies, believed that the company's focus on values and ethics was one reason for Millennium's attractiveness to these companies. He listed several values that contributed to this, among them being that Millennium hired the best people and then created an environment where "great people can do great things, both as individuals and as members of a team." He told one interviewer: "Partnering would not have happened but for the fact that there was an incredible group of scientists and business people working together to build something very special. The partners recognized this and therefore realized the need and desire to tap into it." In addition, Holtzman believed that Millennium's partners appreciated the integrity with which business relationships were maintained:

> Once you have established relationships, it is a small world, and people essentially reference you with your own existing partners. The most important things then are whether you have a reputation for integrity and that, when you enter into a relationship or partnership, you really strive to have open dialogue and live up to your commitments in so far as you can. The fact is that the nature of drug discovery is that you are going to fail more often that you succeed, so all one can expect from each other is your best effort and acknowledgment when things are working and when they are not.[27]

EMPHASIZING VALUES IN EXTERNAL RELATIONS

Steven Holtzman took Millennium's vision of corporate ethics outside the company: he served as a galvanizing force in organizing BIO's[28] Bioethics Committee in 1994; he was the only member from industry appointed to President Clinton's National Bioethics Advisory Committee; he also coauthored a paper published in a scientific journal about gene tests and patient privacy.[29] The paper was a collaboration with the Chief Patent Counsel and the company ethics-educator, Phil Reilly, and addressed the ethical aspects of genetic testing. The authors stressed the need to broaden the disclosures made to research subjects that would allow them to fully understand the risks of genetic research, including genetic discrimination and emotional distress flowing from detecting a gene predisposing to a disease.

[27]Lawrence RN. Holtzman Discusses Partnering. *Drug Discovery Today.* 2001;6:1042–1045.

[28]BIO is the Biotechnology Industry Organization, the main industry trade group.

[29]Reilly PR, Boshar MF, Holtzman SH. Ethical Issues in Genetic Research: Disclosure and Informed Consent. *Nature Genetics* 1997;15:16–20.

Millennium also commented on drafts of the new PhRMA[30] code of ethics regarding corporate interactions with healthcare professionals.

MILLENNIUM ORGANIZATIONAL SURVEY

Initially done twice, now once, a year, the employee survey is used as a way for employees to comment on such things as how satisfied they are with their work and whether their work aligns with their goals, how committed employees feel, etc. In 2002, Millennium was pleased to see that 81% of the employees responded to the survey, giving management a good sense of the workforce. A section of the survey is devoted to how effectively the company is living up to its core values. Following is the percentage of employees who "strongly agreed" or "agreed" that:

- Millennium is a socially and ethically responsible organization (85%)
- My manager provides a good example of how to live the Millennium core values (82%)
- My manager treats me with respect (87%)

Because the company uses a professional consultant to perform the survey, Millennium can also obtain comparisons with other companies who use the same consultant. Millennium found that their ratings were "off the chart," according to Linda Pine. The lowest-scoring statements (those questions about which employees were either "neutral" or "disagreed or "strongly disagreed") included:

- Millennium hires only the best people (50%)
- Employees listen to and learn from everyone(50%)

Senior and line managers are encouraged to use the results of the survey to improve on the lower-scoring items. Given the importance attached to treating employees well, even neutral ratings on some items are addressed as areas of concern by human resources. One such rating involved the core values of listening to and learning from people. The rating had slipped from the prior year and prompted the company to initiate a management communication and training program run by its Internal Communications Director, Mike Houldin.

EMPLOYEE PERFORMANCE REVIEW PROCESS

Ethics and the core values are also explicitly included in a rigorous employee performance-review process. The wording in this survey includes language that allows the appraiser to relate behaviors linked to core values. In the forms for "Self/Manager Assessment" and "360 Feedback Form" (which allows employees to evaluate all their peers), how colleagues are "living up to the core values" is included in the rating of an employee's effectiveness, and "teamwork, cooperation, information sharing, respect, and listening" are included in the rating for collaborating.

---

[30]PhRMA is the Pharmaceutical Research and Manufacturers of America, the industry trade group.

In the "Self/Manager Assessment," employees are also rated on how well they empower staff to "own part of Millennium and take action" and foster the company culture by asking, "Does this person set an example and take responsibility for building local culture?" and "Does this person model the Core Values?" Another section of the same form asks whether the person is recognized as an external ambassador for the company: does the person "transcend turf-orientation and hidden agendas?" "We're strongly committed to having people who live up to, and adhere to, core values," said Human Resources executive Linda Pine. "Even senior employees who are competent and smart will derail at Millennium if they do not treat people fairly and consistently or do not communicate honestly."

Chuck Gray, in Scientific Development, described his impressions of how the employee evaluation program works to reinforce the core values, such as creating an environment where all can excel.

> We have a 360 feedback program here that can show you how you may not have lived up to the core values... At my performance review with my boss, [the way I complained about employee performance] came back to bite me. I learned a lesson from it. All of an employee's good points and warts are exposed in the 360 process. They are the most complete reviews I've ever had—and not just from close buddies. The process is really designed to make you better—for instance, that I have to work on not being impatient. You get told that you need to work on some things. It is tough to change some things that are 'hard wired,' for instance, to go from being an introvert to an extrovert. But you may be able to change on the edges. You end up knowing what you need to work on. The 360-degree review process takes a lot of time and effort. They can cripple work for a couple of weeks. But the company cares that much about it, that they devote the time. The focus in the reviews is on how we are all behaving with the other people... If I'm not treating people with respect, I would get nailed on it. So everyone takes the core values seriously.

ETHICS HELPLINE

In 2002, Gary Cohen proposed to Linda Pine that the company set up an internal ethics helpline for employees. Pine agreed, seeing in this program a significant link with what Human Resources was trying to achieve for the employees by way of an existing Employee Assistance Program. Cohen also saw the helpline as an early warning system for ethics violations or employee problems. The system would allow employees to speak candidly and confidentially with a trained ethics advisor, without fear of reprisal. The company manager would then receive advance notice so that problems can be dealt with proactively when they are more amenable to resolution. In addition, while it was not the driving motive for the program, a helpline would meet the federal sentencing guidelines for a system that allows employees to report corporate misbehavior. Other options under consideration for the ethics helpline include a mechanism to advise employees on how to have conversations with other employees about ethical problems, with advanced training for all employees on how to respond to others who seek such conversations.

CORPORATE VISION PROJECT

Sue Nemetz believed that she was handed the "dream job" of proposing a commercial vision for Millennium. Nemetz, who had a long-term career in the pharmaceutical industry and had started a business unit within Dupont Pharmaceuticals, could not believe that Millennium would pay her for this work, something she loved to do. The reason for her enthusiasm was that not only was the company's appetite for innovation very high, but that she was also asked to design the "Nirvana view" of how a commercial organization would operate and interact with customers. "A key theme in this project was to consider trust and integrity," she said. She worked with key cross-functional groups at Millennium, including Gary Cohen, and she initiated a program of interviews with customers and intermediaries, clinical research organizations, government agencies, industry lobbyists, public relations firms, medical educators, and industry consultants. With each group, she asked what they would consider an ideal working relationship with a company like Millennium. How would they design an ideal commercial organization? At the end of the first series of interviews, Nemetz had six primary components, some of which were overlapping. People told her that the company should:

1. Be truly customer-centric—and achieving this would be novel in the pharmaceutical industry
2. Improve commercial productivity
3. Have a strong clinical-commercial interface
4. Develop a new economic model
5. Redefine the talent profile
6. Be committed to trust and integrity

Nemetz's next step is to use this feedback to design the specifics of a business plan. She remains amazed that Millennium cares enough about its business integrity to seek this feedback and input. Nemetz described her impressions from this project:

> What we learned from the interviews of the intermediaries is that few customers or constituents really trust corporations, particularly in the pharmaceutical industry. A lot of what people said about corporations is very concerning. It was clear that there was a lot of mistrust of corporations, that a corporation could not be trusted to do the right things, and it pretty much confirmed what we saw in the studies and surveys about public trust of corporations. Trust level is low. Biotech companies are viewed more favorably because it's perceived that they put science first. We talked a lot about how a story gets spun. For instance, data is usually not black and white, there are nuances. Depending on company philosophy, they can emphasize information differently. If the data was not all positive, if you were up-front about the negatives, people could trust you. Trust is lost when you only talk about the good results. We learned from these interviews that it was OK to sell your product and make a profit, but people want full disclosure.

CORPORATE SOCIAL RESPONSIBILITY PROGRAMS

According to Gary Cohen, corporate social responsibility addresses the following questions for Millennium:

1. What are the unique responsibilities of biotech or pharma firms in this world—for instance, in the areas of clinical trials, and the pricing and other decisions made that impact access to products by developing countries?
2. How does a company "think globally, but act locally"—for instance, how should the company engage with the local community where Millennium is one of the largest tenants besides MIT and Harvard?

Gary Cohen's group mission addresses the questions of how the company acts responsibly as a member of the global corporate community. One of the decisions made under the umbrella of the legal department concerns the sharing of materials discovered or made by Millennium scientists. A common gripe about biotech companies is the degree to which they obtain proprietary rights and do not share basic science materials, thus limiting the research of others who wish to advance the science. Mindful of this problem, Millennium has had an active practice of sharing those genes or other materials discovered by Millennium scientists, especially ones that do not fit within the company's commercial plans. The vehicle used to share such materials is called a material transfer agreement (MTA),[31] which is issued to outside academic scientists for their research. Even the simplest MTA costs Millennium several hundred dollars to complete. According to Cohen, there is no expectation of any subsequent benefit from these MTAs, except for the accompanying goodwill and satisfaction of knowing that the MTA will help to move the science forward. As he said, "There is only a slight chance that the academic will be able to make a commercially valuable discovery with Millennium's material, to which we have the right to negotiate a license, but the sole expectation of profit is not the reason we do this."

An older aspect of the company's corporate social responsibility program is a grass-roots community outreach program called Millennium Makes a Difference (MMAD). This program was initiated by Peter McLoughlin, a long-time Human Resources employee, in response to employees' requests for help in organizing community service programs. McLoughlin refrained from making this an executive-driven effort, and he maintained the low-key and employee-centric nature of MMAD. Consequently, MMAD reflects the company culture as defined by the employees; they choose which initiatives to undertake according to the connections

---

[31]A Material Transfer Agreement (MTA) is a contract that governs the transfer of materials from one organization to another and allows the recipient to use the material for research purposes. The MTA defines the rights and obligations of the provider and the recipient with respect to how the materials and any derivatives can be used, and usually specifies that no further transfer of the materials can be made to any third entity. MTAs also usually specify what intellectual property and licensing rights are retained by the materials provider for any discoveries resulting from the recipient's research.

they want to make with the community. MMAD programs, which are run by employee volunteers, include donation drives, environmental programs, blood drives, and tutoring in community schools. They also make suggestions about the manner in which other company programs are run: one MMAD suggestion involved the Biomedical Science Careers Program, where local bioscience firms sponsor scholarships for minority students pursuing an education in biology and related sciences. One Millennium employee noticed that the awards ceremony, a fancy black-tie event, typically consisted of Anglo, or Asian, male business executives presenting awards to African American or Latino students, a situation often reflective of the racial mix in the executive ranks of the local biotech companies. The next year, the MMAD initiative addressed the imbalance by having Millennium's award presented by the young African American woman who had been Millennium's past award recipient, and who had since become a Millennium employee. Gary Cohen, who attended this event, said that the difference of all white-male presenters followed by the young African American female was startling to see. Cohen believed that the corporate and community audience immediately understood that Millennium's employees were trying to make a difference, and that this gesture was "emblematic of the ethos, if not the ethics of place."[32]

ARTIFACTS: MOUSE PAD AND THE ETHICS QUICK TEST

In keeping with its ethics mandate, Millennium has deployed several tools for self-reflection and reinforcement, which have often led to a dialogue about how decisions should be made. The first was a computer mousepad on which was printed the list of core values. The mousepad is distributed to each employee during orientation, for placement on the employee's desk. Another such tool, developed and launched by Cohen in 2002, is an Ethics Quick Test, which encourages employees to ask the following questions before making important business decisions:

- How do I FEEL about this action?
- Is this action the RIGHT THING TO DO—does it align with our Core Values?
- How would I feel if my MOST RESPECTED PERSON knew of my action?
- Would I behave differently if I knew my action would be REPORTED in the newspaper?

These questions were made into posters displayed throughout the company as, one interviewee observed, "a reminder to every employee that they were encouraged and empowered to make decisions consistent with whether it feels right, is the right thing to do, and would not bring disrepute."

---

[32]Philanthropy is also part of Millennium's social responsibility effort; a "modest" program devoted to provide support to institutions and agencies that meet important healthcare needs of society.

IMPACT OF INSTITUTIONALIZATION OF VALUES AND ETHICS

To assess the impact of the values and ethics programs implemented by Millennium, information about the following was obtained: whether the core values influenced the hiring and retention of employees; whether employees integrated the company values and ethics in their work; how the company's values and ethics influenced company-sponsored human research. Various employees also gave examples of how they were motivated to make certain decisions based on the company's core values.

HIRING

One of the company's core values is the dedication to hire only the best people. In addition, according to Linda Pine, one way to maintain a focus on the other core values is to start during the hiring process. Pine admits that it can sound trite to say that Millennium "is all about our people [but] our assets are our intellectual capital devoted to pursuing the goals of the firm. [Our hiring is] all about attracting world-class people and creating an environment for them to be innovative and entrepreneurial." Millennium decided that it not only wanted its scientists and other employees to be excellent, it wanted those who could fit the corporate culture. Most employees tell tales of the series of interviews they experienced to determine how they would fit in. Ethics and values are explicitly included in the employee-selection criteria, and behavior-based interviews are used to learn about this aspect of a candidate. According to Pine:

> One of our core values is building an ethically and socially responsible firm. We want to recruit people who sign up and are passionate, not just about the work, but about being part of our core values. People who want to make a difference. We select for culture-fit as well as skills, education, and experience. We use extensive interviewing and multiple visits with many people. We make clear in the interview process what life at Millennium is all about. The flip side to the 'Nothing is Impossible' attitude is a fast-paced and demanding work climate where the bar is raised all the time. New people have to know what they are getting into.

This approach to hiring was evident from the time of Levin's earliest employment choices. Early on, for instance, he hired the veteran Linda Pine to run Human Resources, and gave her a senior position in the company. This was an unusual move, as most other start-up biotech companies tended to hire junior managers to provide only basic Human Resources support. Knowing this, Pine had to be convinced to take the position. She had had 20 years of corporate Human Resources experience and had left in-house work to concentrate on high-level consulting. She had worked for 25 to 30 CEOs as a consultant and had no intention of going back inside a firm, especially for mundane work. However, Levin convinced her that his company would be "all about the way you treat people" and his commitment to this value convinced her to reverse her career path. "There is no one else I'd have done it for," she said.

According to Linda Pine, "Steven [Holtzman] came with a philosopher's bent and an interest in ethical issues. He's a brilliant, very engaging person with a broad range of skills. He understands science and business. He had founded another biotech firm, so had business experience, but I'm sure commitment to people and ethical issues was one thing that attracted Mark to Steven." Holtzman readily admits his interests in public policy and bioethics, a trait that sets him apart from most corporate business officers. He told one interviewer that, by the end of his career, he wants to have "contributed to the creation of a bioscience industry that takes seriously our social responsibility to ensure that our technologies are used broadly for the benefit of as many people as possible."[33] Steven Holtzman later hired Gary Cohen as a corporate lawyer. Holtzman liked the fact that Cohen was thoughtful (something Holtzman attributed to Cohen's having studied literature) and because Cohen had interest and experience in corporate ethical issues. Cohen had been at Genetics Institute, an early biotech pioneer, and later at Genzyme Transgenics (now GTC Biotherapeutics), a company in the forefront of making human proteins in animal milk. Because of the nature of its products, GTC had been aware of ethical concerns surrounding the use of animals to produce human therapeutics, and then became immersed in the ethics of cloning after the birth of Dolly, the first cloned sheep. For these reasons, Cohen's Genetics Institute and GTC work had involved components of public policy and ethics. Cohen was prompted to join Millennium, inspired, he said, because Holtzman's "company work on bioethics really engaged me and my attention."

Holtzman also helped to hire Millennium's General Counsel, Jack Douglas, a highly experienced business lawyer. According to Holtzman, Millennium hired Douglas because we "wanted someone who was good as well as wanted him to be values-based, a person who fit the culture." Douglas fit the bill since he had had experience of, and an interest in, corporate social responsibility when, as General Counsel at Reebok, he started a human rights program that dealt with the ethical responsibilities of American manufacturers operating in developing countries. Fabio Benedetti had a similar affinity to the Millennium's focus on values. Dr. Benedetti was an oncologist at Sloan Kettering who had worked at large pharmaceutical companies before coming to Millennium. He accepted the job as Millennium's Vice President of Medical Affairs because Levin "told me about his vision and concept of putting the patient first. The company platform focused on finding the right drug for the right patient, which is a great underlying ethos. This was one of the things that inspired me. This fits with the ethics oath that all doctors take." Another prominent hire that came about because of a mutual interest in ethics was Rob Kloppenburg, to public relations. He had previously worked for Bayer Canada and had been responsible for the development of what was, at the time, the first corporate-financed, independent advisory bioethics council in

---

[33]Lawrence RN. Steven Holtzman Discusses Partnering. *Drug Discovery Today.* 2001;6:1042–1045.

the world.[34] Of all his work, he said that he is proudest of this, and he wanted to continue to find similar purpose and meaning when he accepted employment at Millennium.

In the hiring and interview process, prospective employees are introduced to the company's core values and learn about such things as the policy not to grant one employee more benefits than another. During employee orientation, the core values are explicitly covered and ethical issues are discussed, applying the core values to expected employee behavior. The sexual harassment policy is reviewed as well as the general expectation that employees will treat each other with dignity and respect. When Millennium grew to the point where it needed sales representatives, a separate policy was developed to govern interactions with health professionals. Future hiring initiatives include addressing the question of diversity in the workforce, especially in upper management where, typical of most biotechnology companies, most executives are male.

EMPLOYEE BUY-IN

In addition to the views of employees documented above, and the responses from the organizational survey, there is other evidence that the employees of Millennium have integrated company values and ethics into their work. The manner in which this is done seems to be a mixture of pragmatism, business necessity, and idealism. Steven Holtzman's approach typified this triad of influences. As he told one interviewer in 1999, "Many companies have become savvy enough to understand that the single greatest obstacle to utilizing new technologies is the potential for a public backlash. In 1995, it was more difficult to get companies involved in these issues, because they didn't see the relevance of them. That changed for the most part when Dolly came along."[35] According to Holtzman, a growing number of executives began to recognize that ethical issues could impact the companies that were commercializing biotechnologies. However, from what he has seen, the biotech companies that pay attention to bioethics tend to treat the issues as part of public relations or government relations, with responsibility assigned to a lower-level employee. These people, for instance, would be assigned to lobby against the anti-stem-cell legislation, part of which included addressing the moral and religious objections to stem-cell derivation. At Millennium, in contrast, ethics has its own standing, and those responsible for ethics leadership are in senior management. Holtzman verified that the CEO often displays the core values at company meetings, and that they are discussed repeatedly to emphasize what

[34]The Bayer Advisory Council on Bioethics was initially set up to address the ethical issues associated with the provision of blood-collection and transfusion services in Canada. At the time of Kloppenburg's involvement, Bayer was bidding on these services after the Red Cross had lost the business due to a failure to introduce screening and withdrawal procedures quickly enough to prevent infections and deaths from hepatitis C and AIDS.

[35]Brower V. Biotechs Embrace Bioethics. BioSpace.com, June 14, 1999. Available from http://www.biospace.com/articles/061499_ethics.cfm.

the company stands for. Holtzman endorses this approach and believes that if the message is good and right, people will "absorb it and believe."

When asked for examples of how he sees the core values and ethics being integrated into employee decision making, Holtzman listed a series of questions that would be asked about an issue and then addressed the same questions from an ethical perspective. Examples of key questions about whether to develop a research project would include, he said, "Why am I doing the research? How am I doing it? What will we do with the results?" The related ethics inquiries would go something like this: "Why am I doing the research? Is the goal of the research ethical? How am I doing it? Does the research method include obtaining informed consent or including people who can't afford the product, etc. What do I do with the research results? Are there social justice issues?" According to Holtzman, asking such questions is fundamental to making ethics operational. "What is important to ethics is the inquiry." From what he has seen, this inquiry approach is applied to all major company decisions.

Another Millennium executive, Chuck Gray, the Director of Scientific Development, also spoke about how the company's core values had influenced the way he worked and made decisions. He said that he was initially surprised about the impact that Millennium's core values had:

> A lot of firms have core values but not much is done about them. But here, they are. I've gone to 15 or 16 all-firm meetings and each time the CEO goes through the core values. I realize that the core values keep coming at you. The people here kind of live by them. For instance, that nothing is impossible—so you are encouraged to go out and try new things. I've generated lots of new initiatives, and they support it. Another value is that you don't discriminate—I'm an Orthodox Jew, and was surprised at how they go out of their way to accommodate me. The general counsel called me, for instance, and wanted a list of all Jewish holidays to avoid conflicts with board meetings. Another core value is to hire the best and brightest—and there are a lot of smart people here. So, the core values are taken very seriously here. They begin to influence how you behave and how you treat other people.

Gray, who is responsible for research collaborations with academia, listed some examples of how the core values were implemented in his case. Early on, he would attempt to obtain the "killer agreement." Now, instead of looking for maximal advantage in contracts with academic researchers, he views these relationships as:

> … a mix of practicality and ethics. If both sides can share views on what they need, we can end up with good decisions. There have been cases where I could squeeze the professors who need our funding, but what would I gain? We're in the center of an academic universe here and we have to worry about our reputation. We need the academics and they need us. I have to deal with partners with integrity and honesty. If I say I will share data with them but then I tell them they have to wait a year or more, then I have shot one of our core values with holes. My word is key to my reputation.

Irving Fox, a Millennium Scientific Fellow, also described how the company focus on core values had influenced him. Prior to Millennium, he'd had a successful career at Biogen, had done some teaching, and was consulting part-time for biotech companies. He was financially secure enough to pick and choose what companies to work for. His criteria for consulting with a company were: technology that could improve medical care; people and resources to make it happen; colleagues who would listen. He started at Millennium part-time, and quickly became an 80% employee.

> I learned about Millennium's core values at the employee orientation—I got the mouse—and it is what keeps me here. I got more excited as I learned more and believed it was possible. I decided that this is how I want to spend the rest of my career as an older person, because time is my most valuable commodity.

For Fox, living the core values was reinforcing and he wanted to be associated with a company that was making products that mattered.

For Rob Kloppenburg, there is a definite connection between the core values and his public relations work. "An ethics approach is encouraged here from the top down and there tends to be a consistent approach." Kloppenburg sees public relations and ethics as closely aligned. "If the company screws up, I'm the person who has to face the public and the press." His main philosophy is that "good corporate policy equals good PR. Defending the reputation of the company is a lot easier if the company behaves responsibly."

CLINICAL RESEARCH PROGRAM

Clinical research (meaning research in humans) is the most heavily regulated aspect of a bioscience business; it is also the activity most closely linked to ethical concerns. Modern concepts of ethics in human research arose from the medical research atrocities of Nazi Germany, have been promulgated though the Nuremberg Code and the Declaration of Helsinki, and incorporated into the United States regulations intended to protect the interests of human subjects.[36] Medical products companies must follow these regulations to obtain FDA permission to conduct human research, on which data the FDA bases its decisions to approve marketing. Thus, ethical practices in clinical research are essential, and any bioscience company dedicated to ethical practices would naturally focus on its clinical research activities.

Early on, Millennium made commitments not only to adhere to the regulations, but also to consider the broader ethical consequences of its research activities. For instance, academic scientists conducting Millennium-sponsored genetic research

---

[36]FDA regulations regarding human subjects rights are found generally at 21 CFR Part 50. Research ethics guidelines generally include the following provisions: a subject's well-being takes precedence over the interests of science; the voluntary and fully informed consent of the human subject is absolutely essential; the researcher should avoid all unnecessary physical, mental suffering, and injury; medical information about the subject should remain confidential; the degree of risk should never exceed that determined by the humanitarian importance of the problem to be solved; research should be justified by its benefit to society; and research should be conducted only by qualified persons using scientifically sound methods.

knew they were not obliged to disclose to subjects who provide tissue samples that the tissue will be used by a for-profit company. However, Holtzman decided that the research consent document would contain the disclosure, because this fact would matter to some people. This decision, though not legally required, is representative of Millennium's approach to consider the wider consequences of its research. Following Holtzman's lead, other research managers at the company routinely made decisions consistent with this approach.

Joanne Meyer, Millennium's Director of Human Genetics, described how Millennium's business and legal groups were early promoters of ensuring that human subjects recruited for Millennium-sponsored studies gave full informed consent so that they understood the consequences of deciding to donate tissue samples for genetic research. Also, responsible systems were put into place to protect the privacy interests of those who volunteer to be human subjects. Even though the DNA samples could reveal limited information at the time collected, in the future, advances in genetic testing were likely to reveal extensive amounts of medical information about the tissue donors, some of which would be unwelcome news to them. For instance, studies to identify disease genes often involve collection of tissue samples from families. According to Meyer, in any given genetic study of families, 10% of the samples will reveal genetic evidence of non-paternity. Anticipating this problem in advance, Steven Holtzman and Phil Reilly helped the group decide how they would handle this information and address questions such as whether to reveal the information to the tissue donor, and what to do if the tissue donor or the donor's spouse or child asked about paternity. Ultimately, Meyer's group decided that under no circumstance would they reveal paternity information to anyone: it was not the subject of their research and there were other means available for family members to obtain the information in ways that protected the interests of all involved. In addition, Millennium's informed consent document would warn subjects that paternity problems could be revealed to Millennium scientists, but that this information would be disclosed to no one, including the tissue donor.

This heads-up approach influenced the entire process of deciding what constituted acceptable informed consent practices for Millennium's initial genetic research. Rather than rely on the existing practices of its research partners, which could vary from site to site, Millennium was proactive in drafting informed consent templates for the company and its research partners. In this process, Meyer's group talked with 30 to 40 different academic groups to obtain input before the consent and protection practices were implemented. As a result, in Meyer's experience, Millennium was significantly in front of the existing standards to ensure informed consent for disease gene studies—that all subjects fully understood in advance the consequences of providing tissue samples for genetic study. Other systems were put into place to ensure that tissue samples and data were protected. Practices to ensure the privacy of subjects also included mechanisms to limit both access to DNA information and disconnects to protect against the discovery of the identity of the tissue donor. Meyer was dedicated to ensuring that "every step along the way, we implement policies that protect the patient."

Although it was not a common practice in the industry, Mark Levin decided that the company needed an internal board to monitor the safety of its clinical trials. Dr. Fox was chosen to run this clinical trials Safety Review Board and, according to him, with clinical trials, "ethical issues arrive everywhere." The Safety Review Board makes decisions about the human trials sponsored by the company, such as when it is appropriate to test a new chemical entity in humans, and when it is safe to increase the dose of an investigational drug. In addition, the Board addresses the emergencies or unanticipated events that occur during clinical trials, such as when a production glitch causes doubts about the safety of research materials, or whether the appearance of abnormal blood tests in human subjects presages an unacceptable risk of side effects such that the trial should be stopped.

In making calls such as this, Dr. Fox's motto is "safety first, no shortcuts." The Board evaluates the impact of the decision on the human subjects: how significant is the potential harm; do patients have other options for treatment; how sick or close to death are the human subjects; and what does the therapy contribute to subjects' health and longevity? Many of these clinical trial decisions must be reported to shareholders and can have significant financial impact on the company, such as when the company announces that it has abandoned the trial of an important drug. However, Dr. Fox views these decisions as ones that need to be made with "blinders on for business issues. We leave [the business ramifications of our decisions] to someone else. We make rulings that might not be in the best interests of firm, but are for the patient," he said. Many of the decisions the Board must make are of the "nail-biting" type, as neither the company scientists, nor the external research review boards, nor the FDA can reliably predict how a compound will behave when studied in humans. Unexpected toxicity is always a possibility. The board's job is to ensure as much as possible that everything has been done to minimize this possibility in clinical trials. Dr. Fox views this work as a trust that is given to him that he cannot fail to uphold.

Dr. Fox is assisted in his work on the Safety Review Board by some very senior managers. The Board is composed of the Vice Presidents of Regulatory Affairs, Preclinical Research, Medical Affairs, and Quality, and the Director of Drug Safety. Gary Cohen was a non-voting member of the board when he worked as the company lawyer. However, when Cohen's position in the legal office changed into being the "company ethicist," the Board invited him to be a voting member. Prior to that change, the Board considered itself as having a technical function (to analyze the preclinical data and decide if it was safe enough to put into humans, etc.). Even though the technical function in the hands of experienced and knowledgeable, professionals could have been enough, in Fox's opinion, to ensure that the board "did the right things" to protect subjects, with Cohen's presence and focus, an ethical dimension to such issues was made explicit. According to Dr. Fox, even though Cohen is still learning about the process of drug development, he raises questions that "make people think… He has an influence… I was skeptical initially when he came as a lawyer. But I am comfortable with Gary's role as an ethicist."

Dr. Benedetti, who is responsible for investigator-initiated studies (usually, later stage or post-marketing studies), took a similar approach and developed a research protocol review team for the studies that he supervises. He was motivated to do so because of his belief that not all questions about conducting clinical trials were answered by consulting the regulations. "We won't accept deviations from them," he said. "But they are not always super-detailed, so there is room for flexibility and differences in interpretation." For these reasons, research review boards need to work out the specifics of what constitutes a medically sound and ethical study. Dr. Benedetti described an example of such a dilemma. Should a clinical trial be stopped early when a data-monitoring board discovers that an anti-cancer drug has shown a potential clinical benefit (i.e., has reduced the rate of occurrence of a cancer) over a placebo, before the study has matured enough to reach its most important clinical endpoint? Stopping the trial early and offering the drug to the placebo group possibly benefits the subjects in the placebo group and future patients who receive the drug. But the early stop also prevents the researchers from studying whether the drug can actually prolong survival, or just delays the onset of cancer. Such a question had arisen in the famous Tamoxifen breast-cancer trial,[37] and no consensus had been reached about whether the decision to stop the trial early had been correct. In addition, the precedent created by the Tamoxifen trial decisions will make it difficult to conduct future survivability studies in similar circumstances. According to Dr. Benedetti, there is no clearly right decision in such a case, and reasonable people can disagree on what is the ethical choice. Do you focus mainly on the subjects at hand, or do you continue the research hoping to benefit the many other cancer patients? When asked how he would make such a decision, Dr. Benedetti said he would engage in careful reasoning about the dilemma, consider all consequences, and make a decision with the right intentions. Part of the careful analysis in a corporate setting, he said, is the necessity to distinguish between the commercial needs of the company and the medical needs of the subjects and future patients. Dr. Benedetti believes that it is his responsibility to know when these two interests are (or are not) aligned, and to ensure that the medical needs of patients always come first.

Another clinical trials question being explored at Millennium, by Rob Kloppenburg, involves the manner in which human trials are conducted overseas: can studies be done in other countries that cannot be done in the United States? Do the same values prevail, or should the company proceed under the rubric of cultural relativism, operating under the standards of the foreign country? The active discussion of these questions was triggered by an incident in 2000 that had occurred in a Chinese study. At the time, Millennium was one of few genomics

[37]The Breast Cancer Prevention Trial began in 1992 and was largely funded by the National Cancer Institute, a branch of the National Institutes of Health. The study enrolled more than 13,000 women at high risk for breast cancer, half of whom received tamoxifen and the other half placebo. The study was intended to last 5 years but was stopped early when data showed a 45% reduction in the incidence of breast cancer in the tamoxifen-treated group compared with the women on placebo.

companies to sponsor genetic studies involving the collection of tissues from people in remote regions of the world, where genetic diversity was expected to be at a minimum. As is in the nature of pioneering research, some unanticipated questions arose about Millennium's partial sponsorship of a Harvard researcher in China who was conducting population studies looking for disease-linked genes. Millennium's interest was in an obesity gene in a group of people who were obese despite low caloric intake. The study was being conducted in a remote province where there was a high degree of poverty, low educational levels, and a correspondingly low ability to comprehend the nature of the study. As detailed in a lengthy front page *Washington Post* article,[38] the Harvard researcher was accused of inadequate disclosure about the nature of the study and of duping the local people to enter the study by promising, but not delivering, free or low-cost medical treatments. In addition, the Harvard researcher unexpectedly encountered genes linked to schizophrenia in the same population of people and failed to consider the stigma attached to mental illness when disclosing this information. Once the *Washington Post* article was published, the study came under wide criticism, some of it quite vitriolic.[39] Each time an article was written about the study, Millennium was listed as one of the study's sponsors. This criticism was painful for a company that had prided itself on being a leader in ethics; managers worked hard to brief employees on what had occurred before the article appeared and to assist them in explaining Millennium's role to inquiring outsiders. With Phil Reilly's help, Millennium's management reviewed what had happened, and the company's obligations as a sponsor of an academic researcher who used Harvard University's research review board to approve the study, and whose field work Millennium could not fully control. As part of the review, Reilly and others consulted with many different research review boards and analyzed the newly emerging national rules about the research treatment of genomic material. As a result of this project, Millennium now structures its genomic studies differently to reflect the different requirements, sensibilities, and needs in each research site, and takes more care to consider in advance the impact of the gene hunting research it sponsors.

INVESTOR RELATIONS

Employees were asked to give an example of an investor relations decision that was influenced by the core values. Rob Kloppenburg recalled such an instance. A suggestion had come up at a high-level meeting that Levin attended that Millennium should disclose some positive interim data from a company clinical trial.

---

[38]Pomfret J, Nelson D. (2000, December 20). An Isolated Region's Genetic Mother Lode— Harvard-Led Study Mined DNA Riches; Some Doctors Say Promises were Broken. *The Washington Post,* Section A1.

[39]One UK group accused the researchers of being "gene vampires." See "Gene Vampires Strike in Rural China," published by the Institute of Science in Society. Available at http://www.i-sis.org.uk/isisnews/i-sisnews7-24.php.

Someone else at the meeting objected that Millennium did not have a practice of releasing interim study data. Further, in fairness to shareholders, if the company adopted a practice of disclosing interim data, it should be comprehensive, meaning that both good and bad data should be included. According to Gary Cohen, Levin then asked for more information ("he's a data hound"). Levin wanted to be consistent and follow the industry standard. A 2-month study commenced, headed by Gina Brazier (Investor Relations) and Ruju Srivastava (Business Development), to research industry practice about releasing interim study data. Scores of press releases were studied and information was compiled, leading to a recommendation that the company not engage in this practice. Employees who conducted this study explicitly invoked Millennium's core values. And when the press reported the problems with the timing of ImClone Systems Inc.'s announcements about an FDA filing for a drug approval, when ImClone was quick to announce the filing and slow to announce when the FDA rejected the filing, Millennium made this an opportunity to study FDA filing announcements, to guide the company in adopting responsible practices.

PERSONALIZED MEDICINE AND STAKEHOLDER EDUCATION

New business endeavors have also been developed, taking into consideration whether the business plan is consistent with core values. Many of Millennium's future drugs will be developed based on "pharmacogenomics"—drugs tailored to an individual patient's genetic predisposition to respond to the compound. As an example, if a gene is discovered that is responsible for causing drug toxicity, the drug would not be prescribed to those patients who test positive for the gene. This personalized medicine approach offers great promise to both enhance the efficacy and reduce the toxicity of future drugs. However, Millennium's molecular medicine group has recognized that personalized medicine (gene testing prior to drug prescribing) poses medical, business, and ethical dilemmas that need consideration prior to commercialization. Joanne Meyer has engaged in active discussions and debates, both within and outside her group, to develop personalized medicine strategies that will benefit patients, physicians, and the company. The strategy group meets weekly and deals with these broad questions as well as many more tactical issues.

PRESERVING A FOCUS ON CORE VALUES

10TH ANNIVERSARY—LIVING OUR CORE VALUES

To keep the vision of the company's core values alive, Millennium decided to celebrate its 10th anniversary by making the year 2003 dedicated to one theme, "Living Our Core Values." Over the course of the year, numerous discussions among employees across the company took place, guided by Gary Cohen and a discussion book containing questions to address, worksheets, and feedback forms. The discussions were intended to re-evaluate, renew, and reinforce the company's core values and seek input for enhancing their impact.[40] At the end of the year, Millennium wanted to know how to enhance and integrate their mission and value statements into all aspects of corporate life. It also wanted to fill in the gaps to reflect changes in the company business plans that have occurred over last decade. For instance, Gary Cohen predicts that the reaffirmed core values will address the company's responsibility to patients who receive its drugs. Gary Cohen views the anniversary exercise as an attempt to mirror and recreate with the current employees what Mark Levin did with the first 15 to 30 employees: a new corporate Constitutional Convention, as it were. The idea, Cohen said, is to "try to bond people to the values in the same way as the founders were bonded to our original core values and mission."

THE FUTURE

By the early winter of 2002, there was good news and bad news for Millennium. The company had continued to pursue joint development and merger and acquisition opportunities to provide itself with products on the market, products in later-stage development, and additional capabilities necessary to accelerate its own downstream efforts in drug discovery. It was succeeding in making the leap from genomics to being a fully integrated biopharmaceutical company. In addition to COR, it had purchased other companies with drug-making capabilities and ramped up infrastructure and in-house expertise. R&D growth was continuing, and the company expected to spend from $525 to $540 million on R&D in 2003. As a result, the company was steadily filling its product pipeline. Ten drugs had progressed into human clinical trial testing. Millennium was about to file a new drug application to market Velcade®, a treatment for relapsed and refractory multiple myeloma, a life-threatening condition for which there were few treatment options.[41] In addition, Millennium was receiving co-promotion revenues from sales on Integrilin®.

However, the year 2000 industry boom was over and the financial community was looking more skeptically at genomics companies that were attempting to

---

[40]The planned discussions were February: "Why do values make a difference?" March: "What does it mean to 'hire only the best people?'" April: "How can I help create an 'environment where our people can excel?'" May: "Do I really 'listen to and learn from everyone?'" June: "What does it take to 'encourage innovation' in a growing company?'" July: "What does 'absolute integrity' mean to me and my colleagues?" August: "How are 'outstanding individuals and focused teams' critical for our success?" September: "What's required to build a socially and ethically responsible corporation?" October: "Walking the talk." November: "10th Anniversary Statement of Core Values."

[41]On May 13, 2003, Millennium received approval from the FDA to market Velcade® for the treatment of multiple myeloma patients who have received at least two prior therapies and have demonstrated disease progression on the last therapy.

move into the pharmaceutical industry sector with gene-based drugs.[42] Because it was moving into this industry sector, Millennium would be facing much more formidable competition against other genomics-to-drugs companies (such as Human Genome Sciences and Celera) and against pharmaceutical companies with their greater resources and experience in drug development, manufacturing, and marketing. Millennium was also caught in an industry downswing that had started in late 2000 and had not recovered. From its high in November 2000 of $89, Millennium's stock price was bumping along in the $13 area. The company was still reporting losses that were expected to reach, in 2003, between $295 and $310 million on revenues that had been expected to be in the $450- to $475-million range. New, marketed drugs were also years away, and the company had to maintain its ability to weather the financial dry spell. However, the expected cash balance of more than $1 billion by the end of 2003 meant that Millennium's prospects were better than most.

Gary Cohen knows that Millennium will face other challenges as well; principally, it will be difficult, as Millennium grows, to keep employees focused on the set of core values that were fundamental to the founding employees. Once products are on the market and the employees are held accountable for sales results and growth, marketplace pressures could both divert attention from, and tend to undermine, the ethical aspects of business decision making. He uses this analogy: "People in large apartment buildings don't behave the same way to each other as those in a three-unit building. It is the same issue in firms as they grow. There is a danger of losing the sense of shared commitment that we had as a start up." In addition, employees were reporting in the annual survey that they felt increasingly under pressure, so Cohen knew that he could not "drop huge ethics training programs on the workforce at this stage. We have a preposterously overworked group of people. We want to infuse an awareness of ethics into the workforce without creating more [of a] burden on them."

Cohen was focusing on managing corporate values for his growing company at a time when business ethics was part of the national discourse. Egregious corporate ethics violations, involving venerable companies such as Enron, WorldCom, Adelphia, Tyco, and Global Crossing, and some of their accounting firms, including Arthur Andersen, had been widely reported throughout 2001 and 2002. While these companies were being prosecuted for self-dealing and fraud and were filing for bankruptcy, newspapers and websites printed the companies' values and ethics statements and detailed how each had been ignored and violated. These publications were devoted to highlighting the hypocrisy of, and disconnect between, corporate values and corporate behavior. Business values and ethics were definitely taking a beating in the public press and in Congressional investigation committees. Business consultants were also publishing articles about how companies could

---

[42]Carey J. Nice Job on the Genome. Now, Let's See Some Profits. *Business Week*. 2001 February 26:34.

both benefit from adhering to values and be harmed when they were ignored. According to one business consultant, not related to Millennium:

> Enron—although an extreme case—is hardly the only company with a hollow set of values. I've spent the last 10 years helping companies develop and refine their corporate values, and what I've seen isn't pretty. Most values statements are bland, toothless, or just plain dishonest. And far from being harmless, as some executives assume, they're often highly destructive. Values statements that are empty create cynical and dispirited employees, alienate customers, and undermine managerial credibility.

The consultant went on to say that strong commitments to core values can make a company stronger, both competitively and culturally. But the commitment is not pain-free:

> Indeed, an organization considering a values initiative must first come to terms with the fact that, when properly practiced, values inflict pain. They make some employees feel like outcasts. They limit an organization's strategic and operational freedom and constrain the behavior of its people. They leave executives open to heavy criticism for even minor violations. And they demand constant vigilance.[43]

Steven Holtzman and Gary Cohen believed that Millennium's core values had been successfully woven into most aspects of the company business. But it was not a time for complacency. Cognizant of the prevailing cynicism and the challenges for Millennium, Cohen is working on devising programs that fit the needs of his growing company. Steven Holtzman and he knew that many of the ethics programs had been so strongly identified with Holtzman that corporate ethics could easily have dissipated on Holtzman's departure. Therefore, Cohen and Holtzman worked closely together, over a transition period, to ensure that Holtzman's legacy would carry on after he left. Cohen was successful in increasing the number of employees engaged in ethics programs and initiatives. The ethics and business law activities were solidified within the legal department, and other programs expanded. So far, with respect to filling Holtzman's shoes, Cohen felt pretty good. "It's not possible to replace someone as unique as Steve, but with support from Mark [Levin] and Linda [Pine], the ethics program has grown, and is deeper and richer," he said. Cohen's next step is to determine for himself what differences and similarities exist between biomedical and business ethics so that he could build on the first to inform the second, to create complementary in-house programs and system approaches. He found this to be a difficult task, as the two disciplines did not coexist often enough in the same setting for him to learn from the experiences of others. While working on the question of how to integrate biomedical and business ethics within the company, Cohen was also addressing some management issues that were strictly of business ethics. His first step was to learn from the business ethics failures of other companies, such as Enron. Cohen was meeting with other corporate ethics officers at large companies in an attempt to

---

[43]Lencioni PM. Make Your Values Mean Something. *Harvard Business Review.* 2002 July:5–9.

learn why the company's statements of core values and ethics had become so disconnected from the way employees spent their work day. With these insights, Cohen hopes to raise the ethics perspective for such issues as corporate transparency, audit procedures, tax practices, and executive compensation, loan, and stock option plans.

Another challenge for Cohen was to address those who were skeptical of any large company that described itself as ethical or dedicated to set of core values. He had spoken to one such sometime skeptic, George Annas, Professor of Health Law, Bioethics, and Human Rights at Boston University School of Public Health.[44] According to Cohen, Professor Annas sometimes has seen corporate bioethics as an oxymoron, a view, Cohen knows, that is not uncommon. The suspicion is that companies use bioethics programs to enhance public relations and shareholder value, and that in-house ethics officers such as Cohen are simply mouthpieces to serve that effort. Rather than being discouraged by these views, Cohen finds it worth knowing they exist. "This is something I ask myself each day, and I conclude that it isn't just that," he says. "I can't concede the bioethics field to academia."

Other Millennium employees voiced concerns about the fate of company values and ethics similar to Gary Cohen's. Rob Kloppenburg's experience told him that "just having ethics statements doesn't matter if employees are continually asked to step over the line." Dr. Fox is interested to see if the new marketing groups blend with Millennium's values. Dr. Benedetti, having come from a large pharmaceutical company, has a perspective on the differences between large and adolescent bioscience companies and is counting on Millennium's flexibility and past history to prevent it from becoming a company that, for instance, decides what studies to fund based primarily on commercial needs. Sue Nemetz had also done quite a bit of thinking about how the core values would fare in Millennium's future. She wondered whether, as Millennium grew, it would be able to balance the competing demands of "making our numbers, ethics, and organization survival." She had seen and heard about managers at larger businesses bowing under the need to "make the numbers" by doing such things as "booking sales" by overloading the supply chain and then failing to inform the CEO for fear of losing bonuses tied to the sales figures. Another industry practice, driven by similar motives that she hoped Millennium would avoid, involved paying physicians to conduct small "studies" of product use in order to garner product loyalty. The practice was viewed by some as "buying new business," and it raised ethical questions in her mind. Having a strong ethical foundation would help to keep Millennium from falling into the same traps. In addition, the lessons from the energy trading industry scandals were on everyone's mind. Several managers who were committed to the company's values noted that Levin's strong moral leadership would have to be reflected throughout the company for Millennium to continue to pursue financial imperatives ethically.

[44]Beginning in 2003, Cohen began a Masters in Public Health (bioethics) course of study at Boston University. He asked Professor Annas to be his academic advisor.

Some employees worried that the values and ethics focus might slip if Levin's influence diminished, despite Cohen's involvement. Although most in-house ethics education classes dealt with biomedical ethical issues and not business ethics, some employees were discouraged that very few business managers attended. There were also reports of some employees voicing opinions contrary to the core values that, according to one interviewee, would trouble Levin if he learned of them. The merger with COR Therapeutics also had prompted introspection at the combined company about the "right" way to maintain sales growth for their one marketed product, according to Nemetz. As Millennium grew into a company that marketed products, the decisions that had ethical dimensions would be addressed not by one group, but by many, which would add complexity. For instance, in exercises to define the long-range products plan, some senior managers appeared more focused on business imperatives; less thought was given to questions such as constrained healthcare resources and affordability when deciding which products to develop. Issues that also might be difficult to manage because they were not specific to one group were product pricing and factoring in physicians' off-label prescribing practices,[45] both of which had ethical dimensions that might affect financial goals. As such pressures begin to accumulate, employees at Millennium are confident that their CEO will continue to focus on doing what is right. However, with strong external pressures and potential distance from Levin's vision and influence as the company grows, one interviewee wondered about whether "doing the right thing" would always prevail.

It was encouraging, nonetheless, that, as Millennium was revving up to fulfill its expanded business plans, many of the senior managers were cognizant of the challenges, asking the right questions, and dedicated to keeping Millennium grounded in its core values. All the managers interviewed for this case study were optimistic, as was Levin himself. As he wrote in his message to the company about the "Living Our Core Values" anniversary program:

> When I reflect on my experience over the last ten years, it's often the way our accomplishments have been achieved, rather than the accomplishments them-selves, that have provided me the most joy… We have a bright future ahead. We'll achieve our goals, and be proud of how we get there, if we continue to act with integrity.[46]

---

[45]Off-label drug prescribing occurred often in one of Millennium's main target businesses, oncology, where a drug is prescribed by physicians for diseases other than those approved by the FDA. The term comes from the fact that the FDA regulates what is contained in a drug's labeling (indications, side effects, etc.) and does not regulate the practice of medicine. This allows physicians to prescribe approved drugs for any indication deemed medically appropriate without fear of FDA sanction. Millennium does not engage in off-label drug promotion and trains its sales force to comply with laws and regulations.

[46]Living Our Core Values, Discussion Guide, Millennium Pharmaceuticals, Inc., 2003.

5

MAXIM

PHARMACEUTICALS (A):

INTERNAL AND

EXTERNAL DIALOGUES[1]

AN UNUSUAL BIOTECH LEADER

Larry Stambaugh is not your typical biotech chief executive officer (CEO). The silver-haired leader of San Diego, California-based Maxim Pharmaceuticals wears pinstriped suits, crisp dress-shirts and neckties, and looks like he would be more comfortable in a bank's boardroom than the more informal, lab-coated, tie-free environment of most biotech start-ups. Unlike most CEOs of early-stage biotech firms, Stambaugh spent a large part of his career in the financial services industry and does not have a scientific background.

Stambaugh's outsider status led him to adopt a different approach to building a biotech company:

> The general principles of business that I had can work in any industry if you surround yourself with people who have the right technical skills, whether [in] finance or electronics. You need leaders who can pick talent, drive for results, and motivate people. I like to think I brought good business practices to the firm... What I see in this industry a lot is scientists who try to run businesses without the necessary skills. They start their firms and try to gain good business knowledge and fall short—they don't know how to put good culture, processes and controls in place, or the ways to raise finance. Scientists are not given good education in these areas.

[1]This case study was written by David Finegold based on interviews he conducted with employees and a company consultant. Interviews were conducted with Beatrice Godard at Maxim and by phone in April 2003. Unless otherwise specified, all quotes were obtained during these interviews. This case study is one in a series produced under grants funded by The Seaver Institute and Genome Canada to study models for ensuring ethical decision making in bioscience firms. Grateful appreciation is extended to all those who participated and assisted in this case study, especially to Larry Stambaugh, without whose efforts this case study could not have been written.

Like all biotech firms, Maxim has faced a series of business, financial, and ethical challenges as it has navigated the long voyage from scientific discovery to trying to get an approved drug on the market. To anticipate and prepare the firm for these challenges, Stambaugh has sought to build a culture and management system that would actively solicit ideas from all employees, board members, and external stakeholders. "I've long believed that culture is critical to performance and productivity," said Stambaugh. "I want to make it a place where people can do their best work without politics, without distractions of excessive meetings and bureaucracy. I want a culture in the firm and boardroom that gives the best results."

THE IMPACT OF BUSINESS ETHICS IN THE BIOSCIENCE INDUSTRY

Culture and ethics are important in any industry. They have been particularly vital to the high-risk biotech industry, which has been built on faith: the faith of the founders and employees of biotech start-ups who have been willing to invest years of their lives and often to risk their personal financial well-being on the belief that this new technology can create cures for terrible diseases and wealth for those who can turn the technology into products; and the faith of the investors, who have put billions of dollars into early-stage biotech companies to finance the development of these products, despite the fact that more than 90% of them will never be approved.

But this faith is fragile, as the volatility of biotech company stock prices has demonstrated since the industry began in the late 1970s. In 2002, faith in the industry, already shaken from the collapse of the dot-coms and the high-tech stock market bubble, was shattered by a couple of high-profile scandals. First was ImClone. The company had negotiated a landmark $2 billion partnership with Bristol-Myers Squibb (BMS) in 2001 for the rights to its promising colon-cancer drug, Erbitux, but was denied the right to file for drug approval by the Food and Drug Administration (FDA) because of the poor design of its Phase III clinical trial. It is not uncommon for biotech firms to have difficulty navigating the FDA approval process. ImClone's case first became notorious—because of the very high expectations that had been created for the drug by statements made by its CEO, Sam Waksal, before and after the BMS deal—and then scandalous, because of the revelations that Waksal had shared insider information about the impending FDA decision with friends and family members and had committed bank fraud and check forgery in securing personal loans for $44 million.[2] Waksal admitted his guilt in October 2002 and was sentenced to more than 7 years in jail and a fine of $4.3 million in June 2003.[3]

[2]Burrill & Co. *Biotech 2003*, p. 374.
[3]CNNMoney. *Waksal Sentenced to Seven Years*. http://money.cnn.com/2003/06/10/news/companies/waksal_sentencing.

A month after the ImClone scandal broke, the Securities and Exchange Commission (SEC) announced that it was investigating the accounting practices at Elan Corporation. Elan had grown from a relatively small biotech company in the early 1990s into Ireland's most valuable corporation and, according to *Forbes Magazine,* was the world's top-rated "A+" healthcare company in 2001, with sales of more than $1 billion and a market cap of more than $20 billion.[4] Elan achieved this tremendous growth through a series of acquisitions, joint ventures, and aggressive accounting practices. Of the financial mechanisms it had used to enhance earnings, the SEC chose to investigate:

- establishing separate, off-balance-sheet companies in Bermuda, which then raised money from investors and paid Elan to do new-product development, enabling it to book this as revenue, rather than as expenses
- acquiring companies, such as the biotech firms Sano and Neurex, and then writing off the entire expense as "in-process R&D"
- investing in many early-stage firms that planned to use Elan's technology, who then paid the money back to Elan as a licensing fee.[5]

At the time, some of these practices may have been considered technically legal, but they became discredited after the collapse of Enron for similar off-balance-sheet deals in 2001.

The result of these two scandals was a decline of more than 90% in the value of these firms and a sharp drop in the financial community's faith in the biotech industry, with the market capitalization of the sector dropping 41% in 2002.[6] Industry expert, Steven Burrill, noted that although "it would be unfair to pin biotech's woes in the capital markets entirely upon ImClone…few events in biotech roiled investor confidence…more than ImClone and Elan." The greatest cost from these scandals, however, may not be measured in loss of dollars, but loss of life. As Dr. Robert Mayor, who led the clinical trials for Erbitux at the Dana-Farber Cancer Institute in Boston noted:

> There is no doubt that the compound works. We have had people who benefited quite dramatically… Yes, the stock is down. Yes, there are unhappy people on multiple levels. But my concern is that this is an effective drug—and it is not getting to patients.[7]

Erbitux was eventually approved by the FDA after the submission of new clinical trial data in late 2003, but a number of colon-cancer patients who had no alternative effective therapy are likely to have died waiting for that approval.

What these two examples suggest is that ethical issues in the biotechnology industry are not confined to traditional bioethical topics, such as informed consent, cloning, and stem cells. Business ethics problems, and the good corporate

[4]Koberstein W. Little Elan Becomes a Big Deal. *Pharmaceutical Executive.* October 1, 2001.
[5]McLean B. Just How Real are the Irish Drug Star's Earning? *Fortune.* March 29, 1999.
[6]Burrill, p. 372.
[7]Where ImClone Went Wrong. *Business Week.* February 18, 2002.

governance practices that can help avoid them, are also vital to the viability of inherently high-risk biotech start-ups and the new therapies they are seeking to develop. Under Stambaugh's leadership, Maxim has tried to build an innovative culture and set of governance practices that can help prevent such ethical problems.

COMPANY HISTORY

Maxim was founded in 1993, and Stambaugh was brought in following his prior position as Chairman and CEO of ABC Laboratories, a contract research organization, to be Chairman, CEO, and President. Maxim's core scientific platform was licensed from Sweden, where researchers discovered that histamine, a naturally occurring molecule most commonly associated with allergies, may also play an important part in the body's immune response to a variety of diseases by protecting critical immune cells.

For the first few years, Maxim operated as a virtual company. Stambaugh ran a small corporate headquarters in San Diego that focused on raising capital. He brought in $11 million that was used to contract with scientists and clinicians working in Sweden on research and development and early-stage clinical trials to explore the potential safety and effectiveness of histamine, before deciding whether to make a major further investment in the technology and building a company.

Based on promising early clinical results and an extensive patent portfolio covering the therapeutic use and manufacture of histamine that this research produced, the firm was able to go public in the United States in July 1996, raising $22 million. This was followed by a secondary public offering in October 1997, during which the Company listed on the Stockholm Stock Exchange and raised an additional $37 million. These funds were used to build Maxim's own organization. They also financed a worldwide set of clinical trials to test the effectiveness of using Ceplene™ (the name of Maxim's histamine drug) in combination with other therapies, to treat several forms of cancer (advanced malignant melanoma, acute melanoma, leukemia and kidney cancer), as well as hepatitis C and topical disorders.

Stambaugh described some of the advantages of being global from the outset of the firm:

> Not all the smart doctors are in the US. We needed to get the right talent and people who were motivated and wanted to work with us. The inventor and clinicians were in Sweden, so we went there. Moreover, we found that their socialized medicine system made trials much less expensive. With our particular [drug] we benefited in... Europe because they have a different approach to the treatment of cancer. In the US, doctors will use anything, no matter how toxic to the patient. Europeans tend to pay more attention to quality of life during treatment. Our drug was a good fit with Europe [because it had minimal side effects].

There have also been some challenges to this virtual model, because trials have been spread over many countries:

> It is more complex to manage work going on in Stockholm, Frankfurt, and Tel Aviv from San Diego. We have very limited infrastructure to manage 150 clinical

sites around the world...and have to use e-mail, phone, and some face time to manage from a distance.

Stambaugh estimates, for example, that he has made over a hundred trips to Sweden over the last decade.

In 2000, Maxim added to its internal research capabilities by purchasing Cytovia, another San Diego-based bioscience company with a strong research program and high-throughput screening platform for small-molecule drugs in the area of apoptosis, or cell death. This is being used to develop drugs for prostate, lung, and breast cancer.

EMBEDDING CORE VALUES

From early in Maxim's development, Stambaugh sought to build the firm around two core values:

> The choice of humble heart and open mind as key values was influenced more than any single thing by the greatest entrepreneur of the 20th Century, Matsushita, who built Panasonic around these values. Over my career I've found that these are the simplest form that results in all outcomes and behaviors you want.

He defined what is meant by these two values:

> Humble heart speaks to arrogance, a big problem in pharma, or any business. Success makes people feel like they are the best, that they are Teflon-coated and invincible, but it just isn't true. Arrogance gets in the way of being innovative and staying on top. It also interferes with listening to others. Keeping an open mind seems self-explanatory, but we tie it to the concept of collective wisdom. We believe collective wisdom ends up with the best answer. The best way to avoid unnecessary risks in a company is to be sure to get different perspectives.

One senior manager, who had recently moved from a pharmaceutical firm, could see the effects of these values in the daily operations: "The values trickle down into all of the meetings. When I worked at [ex-employer], there was no respect for each other; I would get yelled at in meetings. It was horrible. Here we don't have that; people respect each other and value each other's experience and their viewpoints."

The emphasis on respect for and inclusion of all employee perspectives does not mean that everyone has an equal say in the firm's direction or that the leader avoids hard choices that may prove unpopular with some parts of the workforce. "I hasten to add that while we want collective wisdom in a team or the board, while we want to listen to issues being discussed from many points of view, this is not a democracy," said Stambaugh. "We have to make a decision, and it is not going to be a vote or always a consensus. I listen to all viewpoints before deciding. I've learned that if I can truly get all views on the table, we usually get to the best decision."

While the CEO clearly articulated and believed in the values, they appear not to have fully penetrated into the organization until it reached a crisis: the decision by the FDA in 2000 not to approve the firm's first application for New Drug Approval (NDA) for Ceplene. One manager said, "While Larry always had it [the core values], before the FDA decision it wasn't embedded in the behavior of the organization... Failure was the best thing for the foundation of the organization. Had we been successful, I worry about the arrogance we might have had in the organization; failure later would have been much worse."

MECHANISMS FOR EMBEDDING CORE VALUES

In seeking to help the organization recover from the FDA setback, Maxim has worked hard to incorporate the values into its main human resource (HR) and other processes. When screening potential new hires, they probe hard through behavioral-based interviews for "the ability to do active listening" and whether the person is a team player. Pam Gleason, the Vice President of HR, described the process:

> Past behaviors are good predictors of future ones. It's easy to pick up on where they stand [relative to the values] through their examples. For instance, we'll ask: 'Give an example of a valuable lesson you've learned in life or the workplace.' People convey their values through their attitude or demeanor. If they act like 'I know everything'—I get an immediate sense of arrogance. If we ask about collaboration, and they respond with 'I,' we know it isn't a fit.

The values can help recruit some employees. "One of the things that attracted me was humility," said Elva Mazabel, a recently hired Clinical Program Manager. "I think that, in a way, the company is moving more toward that humility. We submitted the NDA and didn't get it. It wasn't a positive outcome, but we learned a lot. We are now realizing the true benefits of being humble." Some of the employees remained skeptical, however. Said one, "I don't know how much of a difference they make. I see them more as window dressing."

When new employees arrive, the core values are a key part of their introduction to Maxim. "We have a write-up on the wall in the training room describing the values (see Exhibit 5.1)," said Stambaugh. "Everyone gets it as part of orientation. These are the two principles we really mean." The CEO then holds monthly company-wide meetings and regular informal breakfasts with a cross-section of employees to give each person a chance to share their views on how to make Maxim more effective. Said one recent participant, "Larry left it up to us to discuss what we wanted... One of the things that came out of the meeting was in talking about team spirit, there were instances where some weren't as busy as they could be and they were willing to volunteer to help others."

The values are then embedded in the performance management process, with leaders assessed not just on whether they are getting results, but also on whether they are doing it in accord with the core values. Recalled the HR Vice President, "We had one executive who didn't appear arrogant...but was ignoring others' viewpoints. We got him a coach, but after three months of coaching [there was still no change]. We had to transition that person out."

One HR process that is not used as much to reinforce the core values is training. "It's hard to train on values," said Gleason. "It's an over-used remedy. You can train

EXHIBIT 5.1   **Maxim's Culture and Shared Values**

Culture

The Maxim Shared Values are the foundation of Maxim's programs, practices, and guidelines. We understand the complex blend of work/life balance and the need for today's professionals to work in an environment that supports that balance as well as accomplishment of results. That is why Maxim strives to find unique solutions for individual situations and to create a work environment that values and promotes each associate's talents.

Maxim shared values
Humble Heart
Characterized by modesty in behavior, attitude or spirit. Unpretentious.
Open Mind
Receptive to new ideas.
Collective Wisdom
Draws on the strength of the team, allowing each individual to make a contribution, rather than drawing on the strength of one individual.
Respect
Demonstrates consideration or appreciation of others and the company.
Trust
Total confidence in the integrity, ability, and good character of one another and the company.
Truth
Committed to determining fact or actuality.

Source: http://www.maxim.com/company/career_culture.htm

on the behaviors you'd like to see, and try to reinforce this message through positive recognition, rather than punishment. You can educate and make employees aware of the values, but each person brings their own [when they join the firm]. We're not likely to change their values; they have to want to change for themselves."

BUILDING A SUCCESSFUL GLOBAL BOARD

In addition to trying to build "collective wisdom" through frank and open sharing of opinions and different perspectives within the firm, Maxim has also sought to engage in an open dialogue with another key stakeholder: the firm's Board and the wider shareholder and community interests that its members represent. Said Stambaugh:

> The Board should be made into a team just like any other one in the company. To do that, you need a culture and working relationship where directors like each other and have personal relationships with each other. We keep them informed and get them information well ahead of meetings. We have all the standard good governance practices you would normally want. I try to get discussion in the meeting focused on key issues, not taking up time reporting on everything.

Although biotech is a very global industry, a number of companies have run into ethical or business difficulties, or both, by failing to appreciate the different attitudes and regulations relating to biotechnology in different countries. Maxim has been a global company from the outset, with a technology originating in Sweden, a corporate headquarters in the United States, and clinical trials in 20 different countries. To help anticipate any ethical and business issues associated with its global scope, it has constructed a very international board, with individuals from Sweden and Germany among its first five directors. In 1997, Maxim became only the second United States firm to list on the Swedish stock exchange. Said Maxim's then Chief Financial Officer (CFO), Dale Sanders:

> Having several directors based there gave us comfort. [Being listed in both the US and Sweden] has served us well over the years when the markets weren't moving in parallel. It allowed us to raise money in Europe when it was not possible in the US, and vice versa. The other thing that helped us is that, though the US is the [world's] largest market, there are very good markets and doctors around the world. Accessing them has been very important to our development program. If you limit yourself to one population [for choosing directors], you do not get the best guidance.

In order to cope with the wide geographic dispersal of its directors, Maxim holds four face-to-face board meetings per year, one in Europe, one on the East Coast of the United States, one in San Diego, and the fourth at varying sites for the strategic retreat. Stambaugh described the other elements of how this transatlantic board operates:

> We follow US rules; the Europeans acquiesce to this since US rules are as tough as, or tougher, than Europe's. We have to deal with or address some different rules than a typical US firm, however. For example, shareholder-right plans are the norm or required in many European countries and these are not typical or required for a US-based company. European directors are sensitive to differences in the two continents and help sensitize us to how investors there will react. Europe requires the separation of CEO and Chairman, but we don't do it, and we explain why not. There are pharma issues that they have which we don't. For instance, we get one approval [from the FDA] for all of the US, but that is not true in all of Europe. You face potential intra-Europe trading and ethical problems if you price differently in different markets. Our European directors are very aware of, and helpful on, these differences. US directors might not pick up on these subtle issues.

He continued:

> There are cultural differences that also matter...In the US, if I make a presentation to analysts, I expect blunt questions. In Europe, you don't get the same thing; there you invite the media, the public and analysts to annual meetings—we would never do that here, but there if you don't do it, you get a bad perception. In Europe they prefer understated claims; here they expect more promotion and higher energy. The presentation style is just different. Our European directors will coach us on how to best meet needs of their audience.

Dr. Andreas Bremer, a PhD of Biochemistry and biotech investor, summed up the benefits of having a diverse, global board: "Generally speaking, I would advise companies to have an international group of advisors and board members. These firms are competing globally for products and finance. Having different cultural

and geographic input is valuable and important." Along with the special challenges of building an effective global board, Maxim has tried to put in place the
best practices recommended by corporate governance experts, including:

• Appointing Duwaine Townsen, one of the firm's original venture capital
  investors, to serve as lead director, who can help shape meeting agendas by
  gathering the views of the other outside directors and provide leadership to
  the board in the event of a crisis. Stambaugh observed, "Having a lead director
  creates openness and enhances the ability of the Board to raise issues."
• Having all of the Board committees, such as those responsible for setting executive pay, appointing new directors, and auditing the company's financial performance and behavior, consist solely of outside directors, who are empowered
  to hire outside consultants if needed. As Stambaugh noted, "Committee work
  is something that is often not valued enough by boards for ethical oversight."
• Holding executive sessions in which only external directors are present, as part
  of each Board meeting. This was the most difficult practice to institute,
  Stambaugh recalled. "I tried to get the outside directors to meet regularly on
  their own, but it was a hard sell at first. I finally convinced them 2 or 3 years
  ago to start doing it. I prefer this to having them meet just when there is a
  problem, which would get me worried. Better to do it at every meeting."

ANNUAL STRATEGIC RETREAT

The most unusual feature of Maxim's approach to corporate governance is
having an annual, 2-day strategic Board retreat to focus on the key issues facing
the organization. The participants include not only key employees and the Board,
but also a few key industry experts who have no formal ties with Maxim and can
thus bring a fresh perspective. The CEO described the process:

> It's something I did in my two prior companies. We get all directors and manage-
> ment together, but also some people from outside, like a top investment analyst or
> a pharma executive. We get them to sign a CDA [Confidentiality Disclosure
> Agreement]. They typically don't know that much in detail about the company
> beforehand, and they ask tremendous questions. Some of these may seem way
> off-base initially, but then they lead to discussion of things that are very germane.

He continued:

> There are usually at least two, sometimes three, outsiders at each retreat. We send
> them a several-inch-thick business plan ahead of time. We tell them to read it
> carefully, but we're not going to talk about it directly. If they have questions, they
> should bring them with them. Then the two days are set aside to discuss five key
> questions. These are the strategic underlying principles upon which the plan is
> written: ie, How should we fund the firm? We have not partnered with big pharma
> until now, is that the right choice going forward? We want to discuss key under-
> lying assumptions, since if we need to change those it would force us to rewrite
> our whole plan.

BENEFITS OF THE PROCESS

According to participants, Maxim gets a number of benefits from the strategic retreat, ranging from new strategic ideas and identification of issues that need to be addressed, to directors who are more active and informed about the company and have better relationships with its leaders. "We have very lively conversations," said Stambaugh:

> And when we're done we always have some changes to the plan. Putting those changes into operation is up to management. We don't always take every suggestion, but discussing the options is what's important. When we leave the room the directors are glad to be on board and feel they have an impact on the firm's success; they are involved. Developing that attitude and belief is really important to having them engaged. If they're not engaged, then they're just picking up a check, and you're not getting the value you need from them. When they are engaged, then they are thinking of you between meetings; it open doors for the firm and they will pick up the phone to call you [when they learn a new piece of information they feel is relevant to the company].

Cynthia Tsai, founder and CEO of Health Expo, one of the largest conveners of conferences for the biomedical industry, took part in the 2000 and 2001 strategic retreats. She noted that the outside experts "don't come with any politics of the management team... You don't feel like you're offending anyone or taking away anyone's power. For example, we discussed [whether the company] should...hire its own sales force to market their own drug or outsource it. You know what insiders would naturally want."

Another issue on which this outside perspective was important was that of clinical trials. Recalled the CEO:

> A couple of years ago we had a guy from Merck who was a clinical expert. We hadn't talked a lot in the Board about clinical trials, since it seemed to be more of an operating detail. He started asking lots of questions about our trials—ie, how integrated safety information would be gathered? How we would report the results? Directors sat back and said, "We aren't asking about these things and we should." It raised a whole genre of issues we hadn't talked about on how to ensure proper and compliant behavior.

These discussions eventually led to some changes in the management of Maxim's clinical trials and a redesign of this process.

In addition to being free from internal politics, the outsiders also bring a fresh viewpoint. "I've felt for a long time that CEOs and Boards need to reflect their community of shareholders, but they don't," observed Tsai. "Typically the boards are made up of CEOs and their friends. This needs to change. In Maxim's case, there aren't any women on the Board. The head of HR is a woman who was participating very actively in the retreat. Don't undersell the importance of women's perspective in healthcare where women make 80% of the decisions."

Dr. Bremer, another outside participant who brought both a scientific and investor perspective, confirmed the benefits:

> I saw my role as articulating my outside view to the members of the Board. It is my experience that after someone serves on a Board for an extended period of

time, he becomes too much of an insider; you don't question as much as you should because you think you know the answers. You think you were involved in developing the strategy and that may make you myopic. One of my roles was to provide food for thought in my area of expertise—financing of companies and strategic direction... The director's role is twofold: one represents interests of shareholders, making sure the company is run in an effective manner. The second one provides strategic advice irrespective of a defined shareholder interest. To achieve that, you need to be able to step back to look at the situation a company is facing from a distance; therefore bringing outsiders in can be helpful.

Maxim's managers concurred on the value of these new ideas: said Connie Crowley, Maxim's Senior Director of Clinical Affairs, "It was very beneficial to have people who didn't have history with the company come to give unbiased opinions and a fresh perspective. They chose people who'd taken a small firm and made it a big one," which was the same transition that Maxim is hoping to achieve. This independence also gives the outsiders freedom to ask the hard questions. Tsai, who took part in a retreat right before the FDA was due to rule on Maxim's first NDA application, recalled, "The atmosphere was about as high as you get in excitement; everyone felt it would be approved. I was the only one who asked, 'What will you do if it is not approved?' That wasn't easy to do."

The outside experts could also help the firm make tough priority-setting decisions. "Maxim has a broad pool of product opportunities, but could not afford to pursue all of them, given budget constraints," said Bremer. "We discussed what programs to focus on and what strategy to follow in ongoing clinical trials given resource limits. I don't want to claim outsiders made great contributions to this decision, but discussions developed constructively in part because of outsiders' participation."

By building plenty of free time into the retreat schedule, Maxim provides opportunities for the members of the Board and management team to interact informally, which can help build additional lines of communication and assist the Board in its succession-planning process. Crowley said, "I never met the Board at other firms I've worked with before. It's unusual to me to have that kind of interaction with directors, but it was very beneficial."

In addition, by inviting the Board members' spouses, the CEO helped to further strengthen the directors' relationship with the firm. "They like Maxim and they get something out of it—a nice break," said Stambaugh. "They start to appreciate how interesting the firm is, which helps to engage the directors as well."

STRATEGIC RETREATS: WHY AREN'T THEY MORE COMMON?

Although annual strategic retreats appear to have produced a number of benefits for Maxim and would seem to be a useful practice for any Board charged with overseeing a public company, it is still not common among United States corporations. A 2003 survey of directors at Fortune 1000 companies found that only 54% had an annual strategic retreat, and the vast majority of these are unlikely to have included outside experts or stakeholders.[8] The individuals who participated in Maxim's retreats suggested a number of reasons why this may still be a relatively rare practice.

First, it is a risk. Bremer observed, "As a public firm, you need to be concerned regarding confidentiality; a very important issue. You need to make sure whoever you invite is experienced enough and not conflicted to be able to provide good advice and counsel." A second, related, reason is that it may be hard to find the right individuals to take part. Said Bremer, "I'm thinking of doing a similar thing for my own firms, though, as I said, it is hard to find people who will really add value. Few people have the breadth, depth and ability to stand up to a Board as an outsider." Sanders agreed, noting that the individuals invited may not always add value. "Use of outsiders in the strategic retreat is a risk. It doesn't always work; it is hard to pre-qualify [the experts we invite]." A third, more sobering reason is that many companies' Boards, particularly once they become public, may not be actively engaged in the key strategic decisions. "Public firms tend to have Boards that are more distant than private firms," said Bremer, who has sat on both large- and small-company Boards. This lack of board engagement may be due in part to many chief executives who do not seek to have their Board or outside experts regularly challenge their strategies. "From my perspective it is very difficult to make Boards effective. Typically they are suffocated by management and not empowered," said Tsai. "I think Larry is actively seeking advice and only a CEO who does seek it, and will consider it, can run effective Board meetings. Those who don't seek it don't see the need for effective Board meetings." Tony Altig, Sander's replacement as CFO, agreed, observing that in Maxim's case the Board "functions like it is and isn't Larry's board. You don't want a board that is a rubber stamp, but you also don't want them doing the CEO's job. In this case, the Board gets on well with Larry. He talks regularly to them and values their input."

DEALING WITH ETHICAL ISSUES

For a public biotech company that is now over a decade old, Maxim has had to deal with surprisingly few of the ethical issues commonly faced by firms in this industry. This is in part due to the nature of its drug development program, which is based around a drug that is known to be safe in humans and is targeted initially at a deadly form of cancer, melanoma, for which there is no effective treatment. And with no drugs yet on the market, Maxim has not yet faced the ethical issues associated with drug pricing, access, or promotion policies.

The main ethical issue the firm has dealt with, according to the scientific staff, concerns requests for compassionate use of the drug by patients not in its trials. Crowley said,

> We have to weigh the benefits for the patient versus the impact on the study. We try to look at the whole patient population in making the decision... We have

[8] 2003 USC/Mercer Delta Corporate Board Survey. Results, Mercer Delta, January 12, 2004, p. 15.

a compassionate use program, but we don't get a lot of requests. We don't advertise it, because it is not easy for a small firm to manage. You have to monitor each person, and the drugs they are sent, to make sure no adverse events occur. We're very resource-constrained. I've run huge compassionate-use programs in big pharma, and they take a lot of time and effort. If we had resources, we'd like to do it. But we have to weigh the benefits versus time and resources required, which may slow us from getting a trial completed and getting the drug to market.

Most of the compassionate use and other ethical issues associated with clinical trials come through the contract research organizations (CROs) that Maxim hires to run its trials. Said Crowley, "We work with large CROs because we don't have all of the on-ground resources to do the trials ourselves." She noted that Maxim's values help in building a good relationship with CROs that can help trials be conducted effectively and ethically. "Our site relations are better than in some companies because we're not saying, 'We're the customer and we know everything.' We want to be partners and help patients."

Stambaugh explained how he believes putting into practice the firm's core values has helped encourage an ethical approach to clinical development:

> In clinical trials safety is always an issue. You need to make sure your drug is not doing any harm and that your people do not take any shortcuts. Because of our agreement on collective wisdom, we've had employees question whether our reporting is being done right, whether there were decisions made that weren't according to GCP [Good Clinical Practice]. The team looked at it, came to me with the issue, and we went to find out what had happened. In this case, openness to straight talk really worked. We identified the issue and put in time to address it.

According to Stambaugh, this same openness has been important in helping deal with ethical issues at the board-level. "Part of using collective wisdom is having seasoned folks with lots of life experience as directors. In the boardroom, I'll allow them to set [agenda] items and ask about anything; if they feel any conversations are truncated, I want them to be able to keep going, and talk about anything." He noted that in addition to the Board's primary role, "to hire and fire the CEO," it also focuses on "helping to set strategy, putting in place general policies and guidelines…assessing and developing the executive team, and making sure that we have the right culture that will promote ethical behavior." More recently, the Board has focused attention on providing an assessment of risks. "This is a big area for us now," said Stambaugh. "We're going back through all of our operations, looking at each type of risk we take and how we mitigate them, through insurance, QA/QC [Quality Assurance/Quality Control], etc. We're looking at it in a very detailed way, examining risks of taxes, regulatory risks, the danger of business interruption. The Board is very interested in it."

The biggest ethical issue discussed by the Board concerned how to handle reductions in the workforce, as occurred after the FDA's rejection of Maxim's first NDA. Stambaugh observed:

> When those conversations come up, we have to look at what people are involved, where we invest our resources, and what are the consequences of closing

something down, including what impact it will have on people. When we put together a marketing group, then had to disassemble it, we were very fair with them. We wanted some of them to come back when we do go to market.

Other ethics-related issues the Board has dealt with concern shareholder lawsuits and the firm's communication with the public and investors. "Like all biotech firms, we got lawsuits when our stock dropped," said Stambaugh. "The Board gets actively involved with counsel to look at the merits of each case. They have an outside lawyer who reports directly to the Board. Both cases we've had were dismissed." He continued:

> Another area that is very sensitive is press releases. The Board sees all of them before they go out. We try to give them a reasonable length of time to review them. [The directors] have raised very good points in helping us communicate more effectively… One example: we would be asked by financial analysts when a trial or partnership would be done. We're inclined to try to come up with some reasonable dates; we might say, 'March,' when a director will suggest we change it to 'Spring' to give a little wiggle room, so we don't fail to meet expectations. We don't put things in that aren't accurate, but there may be things that could be misinterpreted or where we're erring so much on the side of full disclosure that it makes [the press release] hard to read—like if we put in too much detailed data and statistics, instead of just stating the main findings of a study.

Stambaugh summarized how he hopes the firm's core values can help the firm avoid ethical problems:

> My biggest ethical challenges as a leader are worrying that someone knows something isn't right and is not telling me. We're operating under GCP, GMP, and I worry that someone may get overwhelmed and fake data, or may just not want to talk about a problem and cover it up. The only way to deal with it is to talk about openness, to deal with an issue in a constructive way when it does come up and put in place good internal controls… If I'm not responsive, I want them to be able to go right to the Board, to show by example that we've put in place the right kind of checks and balances. You can't always know if someone is doing something covert, but the smell test will catch them eventually.

MAXIM PHARMACEUTICALS (B)

A GOVERNANCE CHALLENGE

Note to the Reader: The following five-part case is intended to be used after the background and context provided by Maxim Pharmaceuticals (A) case. This case illustrates a chain of decisions facing the firm's Board and Chief Executive. It is designed to encourage reflection and discussion at each of the key decision points before proceeding to the next segment of the case.

Act I

Despite the careful attention to core values and creating an effective Board, Maxim Pharmaceuticals was not immune from governance issues. In 1998, Maxim was in danger of running out of the cash needed to maintain the firm's operations and to finance its extensive global clinical trials. Stambaugh went looking for funding again, and after extensive searching he identified one group of investors who were willing to make a private placement or PIPE infusion of capital, but only under this condition: "They wanted me, the CEO, to have more skin in the game. I had plenty of options, but no stock," he recalled. This proposal left Stambaugh and his Board with a difficult choice: What should they do?

Act II

To secure the financing needed to enable Maxim to continue operating, the CEO exercised a number of his options and purchased more than 200,000 shares of his firm's stock in 1998-1999. With this additional capital, the firm was able to proceed with its development program. The promising results from these trials, plus the general biotech boom, led to a significant rise in the price of Maxim's stock to $78 per share, with the CEO's stake appreciating in value to over $15 million.

Although now wealthy on paper, the CEO had limited cash, and he needed to pay the taxes associated with exercising his options, as well as wanting to purchase a house in the expensive San Diego real estate market. He was concerned, however, about the ethics and perception of selling a significant number of shares just prior to the time when the firm was due to hear back from the FDA on its first NDA. He discussed the situation with his Board. What should the Board and the CEO do?

Act III

The CEO and Board agreed that it would not be appropriate for the CEO to sell his shares prior to the FDA decision. Instead, the CEO took out a personal loan, secured by his stock in the company, to pay the taxes and make the down payment on a house. Recalled Maxim's Lead Director, Townsen, "We didn't want him to sell his shares when they were high, before the FDA ruling. We asked him to take a loan, rather than be seen as a selling-CEO at that vital stage. That was absolutely the right thing to do."

In December 2000, the FDA turned down Maxim's NDA application, indicating that the firm needed to conduct a second Phase III trial before it would consider approval. The stock price plummeted, and the CEO was faced with the bank making a call on his loan. Stambaugh shared the situation with Maxim's Board. Faced with the prospect that the firm's Chairman and CEO would be forced to declare personal bankruptcy, what should the Board do?

Act IV

The Board made a $3 million loan to Stambaugh to enable him to pay off his bank debt. The loan was technically due at the end of 2002 but, said Stambaugh, "it was anticipated by himself and the Board that it would be renewed for 4 to 5 years until the stock price got up to where the stock price was expected to be."

In the wake of the Enron and WorldCom scandals, however, Congress passed the Sarbanes-Oxley Act in July 2002. It included a provision banning loans from a company to its CEO. According to Townsen, "The original loan was done because we felt it was of benefit to the shareholders to do so. Then out of the blue we get Sarbanes-Oxley, which was absolute [on the issue of loans]; there was no wiggle room."

Although the stock price had recovered somewhat from its post-FDA crash, even if Stambaugh sold all his shares, he was far short of being able to repay the loan. What should the Board do?

Act V

The Board formed a committee, composed solely of outside directors, who met in emergency session to examine the problem. Townsen described the process:

> We engaged outside counsel in addition to the firm's own to review the issue in great detail and look at all possible scenarios at great length. We wound up with four sets of lawyers, who all came up with the same conclusion in the end: we should just allow the note from the Board to Larry to go into default and report it in our financial statements, but under the Act there is no requirement for the Board to act on it.

Stambaugh summarized his view of the difficult choice the Board was faced with:

> Do you want to put the CEO in bankruptcy? Is that in the best interest of shareholders? The Board looked at how we got into this situation and how we could get out of it. Where we've wound up is: they intend to collect and I intend to pay, but the Board can't renew the loan, and they don't want to bankrupt me. I could have sold stock when it was high, but that wouldn't have been the right thing to do. The Board responded well after it looked at all the alternatives. I'll get the loan paid back as soon as I can.

Postscript: In Fall 2004, Maxim was surprised to learn that the confirmatory Phase III trial for use of Ceplene in advanced malignant melanoma had failed. The firm was compelled to restructure, laying off 50% of its workforce, including the CFO and CSO. Stambaugh took a voluntary pay cut for the 2004-2005 fiscal year. Maxim was concentrating its resources on seeking regulatory approval for Ceplene to treat acute myeloid leukemia (AML), where clinical trial results looked more promising.[9]

---

[9]Maxim Pharmaceuticals Announces Restructuring Plan to Support Commercialization and Development Efforts. Press Release, October 18, 2004, www.Maxim.com.

6

DIVERSA INC.:
ETHICAL ISSUES IN
BIOPROSPECTING
PARTNERSHIPS[1]

FROM BIOPIRACY TO BIOPROSPECTING

Thermus aquaticus polymerase chain reaction (TAQ PCR), a method used in most applications of biotechnology for greatly amplifying DNA, uses an enzyme found in a geyser in Yellowstone National Park (YNP) that improves the efficiency of PCR reactions. The commercialization of TAQ PCR has generated billions of dollars in profits for Roche Pharmaceuticals. (The Swiss multinational corporation acquired the technology from Cetus Corporation, the biotech company that developed PCR.) TAQ PCR has generated hundreds of millions of dollars for Roche and for Promega, a biotech company that also had claims on this invention. But neither of these companies has shared any of this wealth with Yellowstone.

This scenario is common in the life-sciences industry. Pharmaceutical companies, for example, have traditionally collected natural compounds, and the genetic resources they contain, and turned them into drugs without seeking permission or legal rights to collect them. Nor have they offered to share the benefits derived from discoveries made from these samples. To redress such inequities, the Convention on Biological Diversity (CBD) was created in 1992. It established obligations for the nations that ratified this convention to develop legal frameworks that would facilitate access to biodiversity and promote related benefit-sharing arrangements.

[1]The case study was written by Cécile M. Bensimon and David Finegold based on interviews they conducted with Béatrice Godard between February 2003 and November 2004. This case study is one in a series produced under grants funded by The Seaver Institute and Genome Canada to study models for ensuring ethical decision making in bioscience firms. Grateful appreciation is extended to all those who participated and assisted in this case study. Special thanks are extended to Leif Christoffersen for invaluable his assistance in this case.

Diversa Corporation, a biotech company founded in San Diego, California in 1994, set out to create a new approach to bioprospecting that would support the CBD principles. In 1997, Diversa established a ground-breaking, benefit-sharing agreement known as a CRADA (Cooperative Research and Development Agreement) with YNP under which Diversa would voluntarily pay the Park for the right to collect small samples of earth and water and guarantee the Park royalties for any products or commercial applications derived from the microorganisms collected there. But as a pioneer in public-private biodiversity collaborations, the Diversa-Yellowstone partnership quickly became a lightning rod for protests by those opposed to any commercial activity involving public resources. Soon after the Park announced its agreement with Diversa, opponents filed a lawsuit against the United States Department of the Interior (DoI) and the National Park Service (NPS), which oversees Yellowstone. The defendants won on all counts but one: the judge ordered the NPS to conduct a National Environmental Protection Act (NEPA) review of the impact of such public-private partnerships throughout the National Park system. The judge further ordered the suspension of the agreement until the completion of this review. While the review takes place, other companies continue to conduct research within the Park's boundaries under scientific research permits, and the Park continues to receive no benefits.

What is most difficult to accept for Dr. John Varley, Director of Yellowstone's Center for Resources (who negotiated the agreement with Diversa), is that the Park has had to return to the pre-CRADA system for providing researchers with access to Yellowstone. Summing up what the lawsuit has meant for Yellowstone and any potential discoveries originating from the Park, he noted the irony that "those who've brought the suit have contributed to what little gold-rush there is in bioprospecting." Mostly, though, he is dismayed that Diversa was "penalized for volunteering to be a guinea pig. They played fair, yet they were pummeled as evildoers. *That* was unfair."

After the court decision, Diversa was legally entitled to collect samples in Yellowstone by obtaining a scientific research permit, as other firms were doing. Varley was concerned that Diversa would lose valuable research time if it chose to suspend collection until the environmental review was completed and encouraged the company to "be here with their competitors." Diversa's co-founders and leadership team—CEO, President, and Chief Technology Officer Dr. Jay M. Short and Eric Mathur, Vice President of Molecular Diversity and Scientific Affairs— were faced with a difficult business and ethical decision: should Diversa follow Varley's guidance and continue collecting samples in one of the world's great biodiversity hot spots,[2] or should it stay true to its stated principles of only collecting samples from areas where benefit-sharing agreements were in place with local partners?

---

[2]Yellowstone is not listed as one of Conservation International's 25 global hot spots, although it is one of the greatest sources of diversity for microorganisms.

This case explores how Diversa developed its innovative business model and approach to the ethical issues associated with bioprospecting and presents three examples of Diversa's biodiversity collaborations in Yellowstone (Case B), Russia (Case C), and Kenya (Case D).

THE COMPANY

Diversa Corporation[3] was founded by a team of scientists who were among the world's leading experts in molecular biology and extremophiles—microorganisms that thrive in extreme conditions, such as terrestrial hot springs exceeding 100° C, polar ice caps, high-salt marine environments, high-pressure environments such as the bottom of the ocean, extremely acidic or alkaline fields, and waste from inside nuclear plants. Short and Mathur had come up with the idea for a company like Diversa when developing novel cloning systems and conducting research on extremophiles at Stratagene, a tools company for the growing bioscience industry based in San Diego. Recalled Short, "Stratagene built tools to be sold to other companies. Our ideas were not seen as part of their core business, yet the promise to extend our original research was compelling to us."

A team of scientists interested in products derived from extremophiles, Professor Melvin Simon of Cal Tech, Professor Jeffrey Miller of UCLA, and Professor Karl Stetter, formerly with Germany's University of Regensburg, together with Healthcare Investment Corporation, provided backing to establish Short and Mathur as management co-founders in the company. Although all the scientists involved were well known in their own right, Stetter became famous for his discoveries of hyperthermophiles and other extremophiles in 1982, isolating microbes from aquatic geothermal vents and terrestrial hot springs. In 2003, Stetter won the prestigious Leeuwenhoek Medal, awarded once a decade by the Royal Netherlands Academy of Arts and Sciences, for the most significant achievement in microbiology.[4]

Short believed that products from previously inaccessible uncultured microorganisms might be accessed as a mixed population of microbes and genomes through molecular tools and that the products would be ideally suited for industrial, medical, and other processes that require conditions similar to the environment in

---

[3]Diversa was launched as a sister company to Human Genome Sciences, the Maryland-based firm founded by William Heseltine, which was a leader in the early patenting of thousands of genes discovered during the sequencing of the human genome. Diversa was originally known as Industrial Genome Sciences, Inc. and then Recombinant Biocatalysis until the company changed its name to Diversa Corporation in 1997.

[4]The Leeuwenhoek Medal was established in 1875 in honor of scientist Antonie van Leeuwenhoek for inventing the microscope. Twenty years later, microbiology pioneer Louis Pasteur won the prestigious medal for solving the mysteries of rabies, anthrax, chicken cholera, and silkworm diseases and for his contributions to the development of the first vaccines. http://www.biospace.com/news_story.cfm?StoryID=14486320&full=1.

which the organisms are found. Diversa relies on bioprospecting to search for unknown biochemical forms from which to derive commercially valuable enzymes and other biomolecules. Using Short's inventions, company researchers isolate interesting genetic material from environmental samples for discovery. They then use a process of "accelerated evolution" to identify and create new products and improve old ones for the pharmaceutical, agricultural, chemical, and industrial markets. Bioprospecting, as noted, has always been a core part of pharmaceutical discovery; until recently, however, less than 1% of all the bacteria swarming in soil and water has been tested for medical and other useful applications. This is because conventional methods of discovering useful enzymes in nature involve culturing organisms from samples collected from the environment. Natural products discovery was limited by microbiology's inability to culture (grow) almost all microorganisms found in the environment. Diversa patented Short's proprietary technologies that allow its researchers to discover and evolve novel genes derived from the genetic material—DNA and RNA—in the microbes they find in a cup of water or soil (more than 70 patents issued to date).[5] "With Diversa's techniques," explained Mathur, "we can capture the other [unexplored] 99% [of microorganisms] on Earth."[6] According to Dr. Mark Burk, Senior Vice President for Chemical and Industrial R&D at Diversa, "in any given environmental sample there are probably 5,000–10,000 different organisms, generally bacteria and fungi, most of which have never been observed or characterized before."[7] In 2001, Robert Eilerman, Givaudan's Senior Vice President for Research and Development (R&D), expressed the interest his company had in Diversa: "[Diversa] maintains large gene libraries that represent significant biodiversity. This, together with its high-throughput-screening capability, makes the firm an ideal partner for identifying the new enzymes we require."[8]

To find potentially interesting new organisms, Diversa goes to biodiversity hot spots—areas with high concentrations of diverse living organisms—and forms partnerships with local institutes to collect small amounts of air, soil, water, insects, sponges, and other environmental samples. These samples are processed by these biodiversity collaborators and the genetic material is then sent back to company laboratories, where they are inserted into bacterial hosts that multiply quickly in laboratory Petri dishes. This generates gene libraries, called DiverseLibraries®, which express biochemicals normally found only in the original

---

[5]Alternatively, another method involved collecting large amounts of an organism harboring the molecule of interest (which potentially destroyed natural habitats) and spoiled organisms (which experience a very large loss of biodiversity the instant they are extracted from their habitat).

[6]Biotech Firm Diversa Turns to Russian Scientists. *Forbes.com*, November 17, 2000, http://www.forbes.com/2000/11/17/1117diversa_print.html. Accessed June 2003.

[7]Bush L. Biocatalysis for Efficient Manufacturing, *Pharmaceutical Technology*. In the Field—Pharmaceutical Science and Technology News, March 2004.

[8]Rouhi AM. Biocatalysis Buzz: Deals Underscore Interest in Biotechnology-Based Methods to Improve Chemical Processes. *Chemical and Engineering News*. 2001;80(7): ©2002 by the American Chemical Society, February 18, pp. 86-87.

soil or water. Thousands of these more than 1 billion member libraries are screened rapidly by Diversa's high-throughput systems to identify DNA fragments encoding molecules of interest—organisms with the greatest number of optimal properties that could potentially translate into tens of thousands of commercially viable products.

After identifying promising enzymes, Diversa scientists can then use the company's proprietary DirectEvolution® genetic-modification technologies, including Gene Site Saturation Mutagenesis™ and GeneReassembly™ systems, to try to improve on genes of interest. Diversa is unique in the strategy of combining discovery from nature with directed evolution and has been an awarded a United States patent for Short's novel approach. With these technologies, Diversa generates hundreds of thousands of gene mutations known as *complex recombinant libraries* and then tests them for biological activity to identify potential benefits of newly created genetic variations. This unique directed-evolution approach imitates Darwin's theory of natural selection: enzymes are altered, ineffective ones are discarded, and new ones are altered again. Only the most useful ones survive and enter Diversa's product offerings. It might appear that accelerating evolution would raise ethical issues about a company playing God, but as Dr. Joel Kreps, a Principal Scientist in Cell Engineering at Diversa, explains:

> Evolving different proteins is not the same as creating life, per se. From an ethical standpoint, we're not doing anything that is different from what other molecular biology labs do—making changes in the DNA sequence one protein at a time. We could come up with an enzyme no one has seen before, but not a form of life that no one has seen before. The advantage of our technology is that we can do it far more rapidly.

By September 1997, 3 years after it was founded, Diversa had discovered, characterized, and produced several hundred proprietary enzymes. By 2003, Diversa's collection contained millions of genomes representing billions of genes and included thousands of characterized enzymes. Speaking of the potential of these discoveries, Short described how Diversa's proprietary enzymes are:

> …next-generation biomolecules that enable new processes and significant efficiency gains for applications ranging from production of bio-fuels to synthesis of pharmaceuticals. The potential of these molecules is driving the dawn of a new business sector—Industrial Biotechnology. This industry bears little relation to the oil-based industry sectors of the past, but will more closely resemble the driving power of the pharmaceutical, software, Internet industries in terms of protecting the environment, driving economic growth, and improving our lives for generations to come.[9]

Diversa is using the versatility of its technologies to develop applications that range from industrial products to drug development, with strategic partners in each segment. Executive Vice President of Bioscience Products at Diversa, William Baum,

---

[9]*The Wall Street Transcript:* "Jay M. Short, Diversa Corporation." Company interview excerpted from UBS Warburg Healthcare Special, October 2001.

explained, "We can offer a fine chemicals company, for example, the enzyme, the reaction, and the process. And then they would take it forward, licensing the enzyme from us to make a product, and we would share in milestones and royalties."[10] Early investors saw the establishment of several important collaboration agreements as "a testament to the strength of Diversa's technology."[11] Mathur points out that this multi-pronged business model is also vital for showing tangible results to Diversa's biodiversity partners:

> When most people talk about biodiversity they're thinking about small molecules that can become drugs. These are potentially lucrative applications, but they take 10–15 years to develop. We're pursuing drugs, but also developing enzymes for industrial and environmental uses that have the advantage of yielding much quicker revenues to our partners, in some cases taking as little as 3 to 5 years from discovery to reach the market. This is important because it is a fast way to show the value of biodiversity partnerships to companies and countries.

Diversa was able to raise more than $40 million in venture-capital financing by February 1998. Two years later, in February 2000, Diversa completed an initial public offering (IPO), raising over $200 million. By then, investors were drawn to the company's promising and substantial portfolio of intellectual property that included more than 50 issued patents and 200 patent applications, along with the start of Diversa's own drug-development pipeline. By 2004 the company had 188 issued patents worldwide and more than 540 pending patent applications.

In January 1999, the company announced the sale of its first commercial product, a thermostable, or heat-tolerant, enzyme discovered in a hydrothermal vent in Yellowstone National Park, to Halliburton Energy Services (a business unit of Halliburton Company) for oil field applications. Today, Diversa has more than 50 products in development and several products on the market, including Cottonase, an enzyme that is used in the textile processing industry to treat cotton, and Luminase, an enzyme used in the pulp and paper industry to replace $500 million of harsh chemicals used in the bleaching phase. Among the wide variety of products in development are specialty enzymes, enhanced genes, and small molecules for use in crop protection, animal feed, animal health, and drug development; enzymes that act as biological catalysts in the production of polymers and specialty chemicals, such as amino acids, antioxidants, vitamins, and pigments; and enzymes used in consumer products, starch processing, oil and gas extraction, pulp and paper processing, production of modified oils, water treatment, and research reagents and diagnostics.[12]

---

[10]Thayer AM. Diversa Promises Products, Profits. *Chemical and Engineering News*. 2003;81(14): April 7, pp. 15–17, ©2003 by the American Chemical Society.

[11]Bear Stearns. *Diversa Corporation*. Equity Research Report, November 27, 2000, p. 6.

[12]Life Science Analytics. *Diversa Corporation*. Detailed Product Pipeline Report, August 4, 2003, www.lifescienceanalytics.com. Accessed August 2003.

By 2004, to access regions high in microbial biodiversity, Diversa had entered into 18 benefit-sharing agreements with entities providing access in 10 countries across 6 continents and to all international waters around the world.

THE CONTEXT FOR BIOPROSPECTING PARTNERSHIPS

More than 150 countries signed the CBD at the Rio Earth Summit in June 1992 that gave birth to bioprospecting partnerships. The CBD was ratified in the context of a global effort to promote sustainable development that would enable economic growth to coexist with the conservation of natural resources. However, the United States, home to the world's largest biotech and pharmaceutical companies, was one of the few advanced industrial countries not to have ratified the CBD (to date seven in total have not ratified the CBD, including the United States, Iraq, and Somalia).

Under the CBD, countries or entities providing genetic resources receive a set of benefits from those commercializing the resources, including a fair share of the profits generated, as well as non-monetary benefits such as technology transfer and the opportunity to participate in the research. In exchange, providers facilitate access to their genetic resources while preserving the environment that provides the samples. To this end, the Convention affirms that custodians of biological resources have sovereign rights over their resources (article 3) and promotes the preservation of the knowledge, innovation, and practices of indigenous and local communities embodying traditional lifestyles (article 8j). According to Leif Christoffersen, Diversa Biodiversity Manager,"Traditionally, there were no informed-consent or benefit-sharing arrangements with respect to, for example, the collection of genetic samples. Scientists would go on holidays with their kits and get whatever they could." Mathur confirmed that even after the CBD was signed by most nations, few United States-based drug companies took it seriously: "Many in big pharma used to make fun of us for setting up these benefit-sharing agreements."

DIVERSA'S BIODIVERSITY COLLABORATIONS

The CBD provided a framework for Diversa's biodiversity collaborations from the very outset of its operations. Said Short:

> Within the first days of setting up the company we determined that we would only access materials if we had legal rights to engage in commercial research with all samples collected by Diversa directly, and through Diversa's biodiversity collaborations network. That's true whether they came from our homes, or friends, or anywhere in the world.

To ensure these rights and access, the firm established benefit-sharing partnerships wherever it collected samples. As Christoffersen explained, "At Diversa, bioprospecting projects are the foundation of the company; close to 100% of all our research starts with biodiversity collaborations."

However, some members of Diversa's Board of Directors, shareholders, and others in the industry questioned whether it made sense for a company still seeking to reach profitability, and based in a country that had not ratified the CBD, to become a leader in promoting sustainable development. Short conceded that, at that time, biodiversity partnerships were a hard sell: "No one else was paying money; they were just putting samples in suitcases like the old days. It was very rare then to find people or industrial partners who valued biodiversity." Despite this skepticism, Short explained Diversa's rationale for the benefit-sharing partnerships:

> First, it allowed us to be sure that any investments we made in developing products would not backfire down the road [if someone contested their rights to the original genetic material]. Second, it allowed us to sleep at night, knowing we were doing the right thing. Finally, we believed if you can show an additional economic value for nature, then people are more willing to protect it.

In addition, the agreements provided the firm with access to the scientists and conservationists who had the most intimate knowledge of these unique habitats.

The hiring of Christoffersen in December 2000 highlighted the importance Diversa attached to biodiversity partnerships. His first contact with Diversa had been through his position at the World Foundation for Environment and Development (WFED) and Diversa's work with the National Institute of Biodiversity in Costa Rica, Yellowstone National Park, and the National Park Service. He had participated in the negotiation of the ground-breaking bioprospecting partnership between Diversa and Yellowstone. He had previously worked for the humanitarian organization CARE in Costa Rica and Kenya on community development and environmental projects. Through his interactions with Diversa, said Christoffersen,

> ...it became apparent to me that Diversa did more for the environment than most environmental non-profit organizations. It further supported my belief that the best way to protect our environment was by facilitating support and participation from the private sector. Furthermore, I realized that companies could frequently engage in these types of biodiversity collaborations directly with local actors without the need for intermediaries like NGOs (non-governmental organizations).

For Mathur it is a sign of the firm's commitment to biodiversity "that someone like Leif, who has committed his career to supporting biodiversity conservation, thought he could do more here in this company than in a non-profit to help the environment." Christoffersen's international experience brought an important new set of skills to the company. Recalled Mathur, "I told Leif when we hired him that part of his job was to make sure that we had legal rights to all of our samples."

PROCESS FOR ESTABLISHING AND IMPLEMENTING COLLABORATIVE AGREEMENTS

Diversa uses three criteria to identify potential new partners. First, the company looks for gaps in its existing sample collection, seeking to add habitats and ecosystems that are not well represented in their existing gene libraries. Christoffersen explained:

> Since we have a partnership in Costa Rica, it would be hard to justify establishing partnerships with Nicaragua or Panama, given that the biodiversity differences are unknown. In Africa, we have collaborations in Kenya, South Africa and Ghana because there isn't much overlap in the biodiversity found in these three countries.

The second criterion is to access biodiversity hot spots, especially the 25 global hot spots identified by Conservation International, a non-profit environmental organization.[13] Diversa prefers these areas because of the high diversity of habitats available in a single location, which they believe directly translates into a greater diversity of microorganisms and, therefore, genes as well. The third criterion is based on the company's industrial collaborations and the needs of its customers. For example, one of Diversa's industry partners asked the company to help find an enzyme that would work in specific pH and temperature ranges. To locate such an enzyme, Diversa searches for natural habitats that most closely replicate those conditions, such as the ultrahot geysers in Yellowstone National Park or in Kamchatka's Geyser Valley. However, Diversa does not limit its search to these closely matched environments because, as Christoffersen noted, "there have been many discoveries that come from other habitats [than the targeted conditions where it will operate commercially]." The company thus uses "a mix of libraries that represent targeted micro-environmental conditions with those that represent novel ecosystems."

Diversa's processes have evolved over the years and today, after identifying a potential region, Christoffersen and Mathur explore suitable local partners with which to form Diversa's biodiversity collaborations. Partners can be a government agency, private firm, academic research institute, or a non-profit organization, depending on their various local circumstances and capacities. A key criterion for identifying the right partner is finding an organization with the legal authority or support from the government to sign an international biodiversity agreement. To achieve that goal, Diversa starts by contacting the National Focal Point for the CBD in each country where it plans to establish a biodiversity collaboration. This contact in turn provides them with guidance on relevant laws, regulations, and recommendations on

---

[13]Conservation International identified such hot spots on the basis of diversity, the number of endemic species, and the number of threatened species. There is an implicit correlation between the diversity of plants and animals and the diversity of microorganisms. In other words, the larger the diversity of the first, the larger, it is presumed, will be the diversity of the other.

possible partners. In an area or country where national and international law is still often unclear, Diversa, at the very least, seeks an official seal of approval. Ghana, for example, has no laws governing bioprospecting. At Diversa's request, the University of Ghana approached the Ministry of Environment, Science, and Technology to seek support and authorization for a partnership with Diversa. Within 1 month, the Ministry had supplied Diversa with a letter from E.O. Nsenkyire, the Chief Director of the Ministry, authorizing the collaboration and confirming their support and approval of the agreement and activities related to the collection of samples for commercial research purposes from any habitat within Ghana, and the sharing of any benefits deriving from these discoveries with the University. In 2004 the first product derived from Ghana samples was licensed from Diversa by Givaudan, and first royalties will be paid to Ghana in 2005. With each partnership Diversa has signed, Christoffersen explained, "our system has been evolving," with the firm now requiring written proof that its partner has legal authority before finalizing any agreement. He feels that the partnerships that work best are those with local, scientifically strong partners endorsed by the responsible government body, such as the Costa Rican National Biodiversity Institute or the Kenya Wildlife Service.

The starting point for each partnership is a benefit-sharing agreement, which was developed by Diversa's Intellectual Property (IP) department, specifically the Vice President for IP, Carolyn Erickson, and subsequently coordinated by Associate Director for Intellectual Assets, Monica Sullivan. The company is well aware that if these are not designed to meet the needs of each participant, then there is little incentive for cooperation. Diversa seeks to strike a balance between meeting its objectives and guidelines while respecting its partners' decision-making processes and differences in capabilities. Christoffersen explained, "We try to be flexible. We explain our interests and allow them to figure out the best way to satisfy them." The level of benefit sharing, for example, will depend on whether Diversa's partners have significant scientific expertise and are able to participate in multiple stages of the research process, or whether their capabilities are restricted to sample collection. Diversa is also flexible in the financial structure of its partnership agreements: "We provide annual payments or per-sample payments, depending on the interests or needs of partners. We leave it up to the partner to determine which one they prefer," said Christoffersen. To avoid receiving too many "mediocre" samples, however, the company tries to reward quality and uniqueness of samples over quantity. Christoffersen elaborated on how incentives in the agreements are structured and evolve over time:

> We have milestones and royalties. With the royalties, we offer the same rates to all of our biodiversity collaborators. Differences can occur, such as when certain biodiversity collaborators prefer to stick with old benefit-sharing terms when renewing biodiversity collaboration agreements. For the milestones, we struggled to assess how much (of the value) is due to our technology versus the actual sample. To avoid that problem, we've created a different milestone structure that is based on the performance of the collaborator. The milestones we provide are based on meeting targets we set, so we're getting the productivity we want in

terms of the number and quality of samples, and adherence to, or improvements on existing research protocols. We also have the flexibility to come up with modified terms when we renew agreements with existing partners. This way, we can introduce changes slowly, which gives us the ability to evolve in an organized way.

Mathur noted that determining the value of samples collected, particularly when working with partners who have not been involved in the long and risky process of developing commercial products, may be a difficult process:

> It can be hard to explain what these samples are worth to our partners. If you have a drug that is a novel molecule, and it is safe and effective, it can be worth billions. If you have a microbe with some activity, it's worth less, and if you have some mud that may contain one of these microbes, it's worth even less. The further down the value chain you get, the less something is worth. We've spent over $200 million to develop technology that enables us to develop products from mud.

Because of the length of time and uncertainty inherent in new-product development and the problems of determining what constitutes a fair share of any eventual revenues, Diversa places a large emphasis on the non-monetary benefits of its partnerships. As Short commented:

> What is the right amount for a royalty? I came to the conclusion that there is no correct answer that would ever satisfy everyone. So the key, in addition to royalties, is to focus on capacity building. We train our partners to be able to do their own science and build their own companies. Not training in a proprietary technology, that will make them dependent on us, but giving them the basic microbiology and molecular biology tools to do their own research. It's like the old adage that it is better to teach a person to fish than to just give them a fish. This is fundamental to any real biodiversity collaboration.

This capacity building can include scientific training, both in the partner country and Diversa's labs, and transfer of technologies, as well as support to strengthen local conservation activities. In addition to building capacity, the company feels that it is very important for the success of the agreements to build trust and foster long-term relationships with its partners. As part of its effort to develop strong relationships, the company will often take part in special projects with its partners, such as supporting Yellowstone's efforts to reintroduce wolves into the Park (see Case B, Partnership with Yellowstone National Park).

ETHICAL DECISION MAKING

Diversa's approach to biodiversity and to ethical issues associated with their business begins with the top leadership. "Jay [Short] is very important," said Mathur. "He's a very ethical man. Basically most of the issues are dealt with by Jay, myself, and Leif…If we didn't care about diversity, we wouldn't be here." In addition to these three key individuals, Diversa involves a number of other departments when ethical issues arise: the IP group, finance and accounting, the R&D team, and public relations (PR).

Diversa "tries to be proactive and not reactive" in dealing with ethical issues, according to Christoffersen. He believes that the company has not faced any significant ethical problems to date because it continues to find ways "to make sure that we are not reacting to problems or criticisms, but are ahead of the curve." Despite this proactive stance, Diversa cites three types of issues it has had to confront. The first involves general ethical objections to bioprospecting and genetic engineering; the second focuses on opposition to a for-profit company that is seen to be exploiting natural resources that are public, and the confidentiality of the benefit-sharing terms; the third concerns publication of the results of its research and what information should be in the public domain.

The general objections to bioprospecting are part of a wider ethical debate on whether it should be permissible to patent any life forms. This debate reached a crescendo in the late 1990s with the sequencing of the human genome, when some companies, such as Human Genome Sciences and Celera, filed patents covering thousands of segments of the genome that were thought to have potential medical uses. In the case of patenting genes that encode proteins discovered through bioprospecting, Christoffersen describes the characteristics of the typical critics:

> There are some environmental organizations that are against genetic research and associated patents. If they can prevent a company from collecting samples, then they believe that they can prevent them from engaging in genetic research and securing patents for their work. No one has approached us about such issues, but they may tackle bioprospecting to get at it... Others are just totally against genetic research. No matter how great the package is and how happy the collaborator is, anti-genetic-research environmental organizations won't be happy.

The counter-argument that Diversa and the many countries and environmental groups that supported the CBD make is that the best way to preserve biodiversity is to show the direct health and economic benefits to humans that can be derived from these natural habitats without disturbing the ecosystem. By operating under the principles of the CBD, even in countries that have not ratified it, Diversa feels it is on safe legal and ethical grounds. In addition, Diversa notes that it is not patenting natural life-forms, but rather genes that encode proteins from the artificially generated, enhanced modifications of these microorganisms produced through its Directed Evolution technology, and that along with patenting comes the obligation to publicly disclose the results of this research. Because its biodiversity agreements are all non-exclusive, other firms or scientists who want to negotiate similar access to these natural resources can do so. "I've been offered exclusive agreements a few times, including one place just a few months ago," said Short. "I do not want us to be in that position and I believe it is a questionable stance for any company to take."

The other, related criticism of Diversa and other companies engaged in bioprospecting is that they are exploiting public resources for private gain. In part, this may be a result of public misunderstanding associated with the term *bioprospecting* and its association with the Gold Rush and other pioneers who were not sensitive to the environment, according to Yellowstone's Varley:

Everything is an education problem. Partly this is because of when you use the name 'bioprospecting' in the Western US, people see mining and logging. They see big stacks of spoiled land and toxic water. I didn't pick the term. There's a lot of concern out there that this will turn into a case like the Pacific yew tree, where the drug Taxol was found in its bark, and every tree was seen to be at risk. None of that would be allowable here [in Yellowstone].

In Diversa's case, because the size of the samples required for microorganisms is so small and there is no need to harvest particular microorganisms on an ongoing basis, there is no measurable negative impact on the environment. Even without a risk of physical exploitation, some critics are concerned that the company may be financially exploiting its non-profit counterparts by not fairly sharing benefits. "The other problem is not trusting that our partners know enough to effectively negotiate a deal," observed Christoffersen, adding that NGOs have "concerns about companies cutting corners and taking advantage of the situation." Christoffersen argues that it would not be in Diversa's own interests to negotiate a one-sided deal, even if they were in a position to take advantage of a partner less experienced in such negotiations:

We feel that to have a good and productive collaboration, we have to be fair. In order to better conserve the biodiversity of interest to us, we want to empower our partners as much as possible to reinforce their knowledge of their resources so they can more effectively manage and protect the natural resources under their care. The success of our company is based on the biodiversity, so we have an interest in maintaining or amplifying it by empowering our partners... For example, we have at times offered milestone payments, even though some of our partners didn't ask for them. We focus in each agreement on their needs and capacities to collaborate with us. Those that can only provide samples get different packages from those who process the DNA and save us a step.

Because of these variations among its partnerships and the central role that these agreements play in its business model, Diversa has kept the specific financial terms of each agreement confidential. This decision, however, has raised concerns among the critics of bioprospecting about the lack of transparency. Christoffersen explained, "Ethics also comes up through outside parties who try to evaluate and assess the collaboration. Some interest groups, such as journalists searching for a story, or environmental groups in need of controversy to help boost fundraising efforts, may find the mere fact that these benefit-sharing terms are confidential is unethical." He noted that the bulk of the agreement is made public, with only the precise level of payments for services, milestones, and royalties kept confidential, and described Diversa's rationale:

We're proud of being the world leader in this area. Other competing companies may be working with biodiversity, but for them it is only a small percentage of their activity [while it is core for Diversa]... You can look at it in several different ways: from our point of view it's important to keep some things confidential because it [the agreement] is based on our determination jointly with our partner of the value of the services provided and their financial needs. We don't want one biodiversity collaboration to bias another one. We could justify and explain how we came to each arrangement. But it is easier if we can avoid getting into that discussion. We would rather focus on what they need and the value of what they can provide. In cases where you're talking about a small area, for example, the Meadowlands where we had a small partnership with Rutgers, it is not the same as dealing with all of Alaska or Kenya... For us it is important not to have outside parties who have no direct stake in the matter assess the terms of our agreements. We don't want to be caught up in discussions of why these terms vary from partner to partner. You could spend the rest of your life debating what is equitable. Ultimately, what is most important is that the parties to the agreement are pleased with the terms.

Though not wanting to involve outside parties in these negotiations, Diversa has sought to explain its approach to biodiversity access and benefit sharing to stakeholder groups, but has had some difficulty engaging in a constructive dialogue. Christoffersen cited the example of Diversa's efforts to join the World Conservation Union:

They are a powerful and important environmental organization, whose members are governments and NGOs. They have no private firms as members. I've approached them to ask if they would allow corporate members. I was told that there is no way they can do it; the non-profits would refuse. How do they ever hope to have major impact if they shut out companies? The point of mentioning this example is to reinforce the problem we face with a significant portion of the environmental community that believes that companies should not be involved in any aspect of the conservation of nature. What many of these environmental organizations fail to recognize is that biotech companies are made up of scientists primarily from the field of biology. Hence, these are biologists that love life, the diversity of life, and nature. Their aim is not to destroy the environment, but rather more often than not to figure out ways to protect it.

In addition to biodiversity issues, the other ethical question that Diversa faces, like most life-science firms, is how to balance the need to protect its intellectual property and their shareholders' interests with the scientific norms around public disclosure of research findings. Mathur admitted:

As a company, you don't want to publish. But as a scientist, you do, particularly if you're going to have a scientific paper, and readers are going to want to have access to the underlying sequence. We decided to abide by the guidelines set down in the Bermuda meetings on genome policy, which allow for a 6-month window [to conduct research and file patents] prior to publication.

TOWARD THE FUTURE

Bioprospecting remains a work in progress. Despite the Convention on Biological Diversity, bioprospecting is still an unevenly regulated activity across the world. Many companies continue to collect samples without sharing the benefits they derive from them. This practice is still legal in many places, including the United States, and preventing it is almost impossible even where laws exist. Companies continuing to collect in this way are not behaving "in an ethical manner," according to Christoffersen. "That is the difference between what is legal and what is right." As a world leader in bioprospecting, Diversa will probably face controversy for some time to come and recognizes that it is in an environment of ongoing change. "The company's philosophy," stated Christoffersen, "is constantly evolving through new collaborations and the renewal of old ones."

One of the largest potential challenges facing bioprospecting is the emergence of a new entrant in the biodiversity space: J. Craig Venter. The controversial founder of Celera was looking for a new adventure after tying for first place in the race to sequence the human genome. Following the example of Charles Darwin, he has embarked on a new project: a 2-year, round-the-world cruise on his high-tech, 95-foot yacht, *Sorcerer II,* to systematically gather samples of microorganisms from all of the world's oceans, sequence their DNA, and then make the genetic code publicly available in the GenBank database. "The goal," says Venter, "is to create the mother of all gene databases," or, less technically, "just trying to figure out who f^*@#* lives out there."[14] In a pilot trip for the project, he collected samples in the Sargasso Sea near Bermuda, where he discovered 1,800 new species. He hopes to identify 100 million new genes by the time the project is completed, which would still be well below the billions Diversa has collected and uses in its screening process.

In contrast with Diversa, Venter has sailed into controversy not for patenting the results of his research, but for giving them away for free, and for the potential impact of this on the ability of countries, in particular developing countries, to share in any subsequent benefits resulting from their biodiversity. As a consequence, some governments, like France's, have denied him permission to collect samples in French Polynesia. Drawing a comparison with the controversy over the decoding and patenting of human genes, Venter noted: "The irony is just too great. I'm getting attacked for putting data in the public domain."[15]

One of the criticisms of Venter's approach is that he has been diluting the quality of the information in GenBank, by putting huge amounts of raw sequence into the database. Many would prefer to have a publicly available metagenomic database that is curated. More broadly, however, Venter's activities could ultimately threaten bioprospecting partnerships and the CBD's goal of equitable benefit sharing. To give just one example, Diversa has spent the last 6 years building a biodiversity partnership with the Bermuda Biological Station for Research (BBSR), providing annual payments for collection and research that have resulted in the identification of 100 to 500 interesting genes out of the millions analyzed. During Venter's quick trip through Bermuda's Sargasso Sea, he gathered samples that enabled him to identify and publish 1.2 million gene fragments.[16] Even if only 0.1% are useful full-length genes, that is 1200 potentially useful genes. By placing these unprotected and uncharacterized genes in the public domain, Venter

---

[14]Shreeve J. Craig Venter's Epic Voyage to Redefine the Origin of Species. *Wired.* 2004;Aug:104–151.

[15]Ibid, p. 146. As it has with universities and government labs, Diversa offered Venter a $1000 license to use its technology for research (noncommercial purposes), but as of November 2004 he had refused to sign an agreement.

[16]Ibid.

potentially damages the host country's ability to provide incentives for companies to engage in commercial research, and thus undermines their ability to share in economic benefits and meet their development needs. As a result Diversa and companies like it may now find it harder to justify to their shareholders that they should continue to pay BBSR for something that they can now initiate for free from a public database. According to Christoffersen, "This really lowers the value of genomic diversity and it harms that country, since Bermuda loses its ability to derive commercial benefits from those genes."

As the public debate on these issues evolves, Diversa must decide how to proceed with its biodiversity collaborations. Should it continue to seek access to areas ripe for bioprospecting through these scientific research agreements, as it has in the past? Should it wait until there is a more standard and broadly accepted approach? Or should it simply focus on evolution technologies applied to random gene fragment sequence data available on public databases, thereby escaping all royalty obligations, lest it become uncompetitive against its competitors? Without a clear regulatory framework in place, can it be certain that its partnerships will not later be deemed invalid? If it waits, however, does it risk losing its first-mover advantage in creating biodiversity partnerships and having fewer resources generated to protect threatened ecosystems? As Mathur summed it up, "This area is just a quagmire in the US, which has not ratified CBD, and where is often unclear who stewards the land. Few bureaucracies have the regulations in place, so it is often unclear how to get the rights, even if you want to do it the right way." For the moment, however, Short is optimistic that most companies, however reluctantly, are beginning to recognize the need for a benefit-sharing approach:

> The agreements are not onerous; they can afford the royalties. Furthermore, the parties to the CBD can seek some form of reprisal with any firm they feel has gathered samples without permission. There are plenty of attorneys out there willing to provide pro bono service to help in such cases. In some instances, companies may get away with [not disclosing where a sample came from], but there is always a risk that some disgruntled employee will disclose it. I can't imagine any reasonably sized company trying to build a business on hidden material (see Case D for an example of the risks of not adopting this approach).

In the specific case of Yellowstone, Diversa must decide whether to follow Varley's advice and collect in the Park without a biodiversity agreement on the same terms as other companies. If so, are there ways it could collect samples there and remain true to its ethical principles? Or can it maintain its competitive advantage by ignoring Yellowstone and focusing on its other biodiversity collaborations around the world?

DIVERSA INC. (B)

PARTNERSHIP WITH YELLOWSTONE NATIONAL PARK

Diversa's leadership ultimately decided that the company would not collect in Yellowstone until the court case was resolved. Eric J. Mathur explained

Diversa's decision: "The judge ruled that while the environmental review was occurring we could collect samples, but that we didn't need to pay since there is no mechanism in place yet. We decided to do the reverse—we continued paying under our agreement, but refused to collect until we had a legal agreement in place. We tried to take the high ground." The following cases provide more details on the history of the Yellowstone agreement and two other ongoing partnerships Diversa has established in Russia and Kenya.

Yellowstone National Park was the world's first National Park and an obvious candidate for one of Diversa's first partners. The CRADA[17] was reached in August 1997, 3 years after Diversa had begun conducting research in the Park under scientific research permits. In the 3 years preceding the CRADA, Dr. John Varley, Director of the Yellowstone Center for Resources, had been leading an effort to define the Park's policy on the commercialization of Yellowstone's genetic resources. A scientist who began his career as a fisheries researcher, Varley had devoted his life to research, public education, and the conservation of natural resources. Understandably, he was committed to ensuring the sustainable use of Yellowstone's rich biodiversity and thought it was imperative to seek the views of other stakeholders on how best to achieve this.

He began exploring the idea of benefit-sharing partnerships for bioprospecting after a *New York Times* reporter pointed out that the enzyme that had improved PCR had been discovered in Yellowstone, but that the Park had not shared in any of the revenue it had generated. Upon further investigation, Varley found "about a dozen patents from organisms originating from Yellowstone." As he raised the issue with officials in the Federal Government, it became clear that "no one agreed that we should proceed with the status quo, including the Secretary of the Interior, and Republicans and Democrats alike." As part of the process of defining the Park's policy, he also looked at statutes that would authorize the Park to manage bioprospecting at YNP. Varley's team found that Yellowstone would be legally able to enter into benefit-sharing agreements under the Federal Technology Transfer Act (FTTA) of 1986 and the general framework for CRADAs. It was thus decided that Yellowstone would enter into biodiversity collaborations with research and commercial partners that would enable the Park to gain a share of any profits derived from Yellowstone's microorganisms.

After this decision, YNP invited potential bioprospectors to a conference held at the Park's Old Faithful Inn to outline YNP's new policy on bioprospecting. This meeting became known as the Old Faithful Symposium. At the end of the conference, Diversa's Eric Mathur approached Varley to tell him that his company wanted to be Yellowstone's first partner, as per guidelines established by Short as head of Diversa research. Varley saw that it was important to Diversa to access Yellowstone's biodiversity. "I asked him why," he recalled, "and [Mathur] said, 'Where else can we be in the middle of half the unknown genes in the world in half a day from our corporate headquarters?'" In actual fact, it wasn't hard for

---

[17]Cooperative Research and Development Agreements were designed for national laboratories, such as Los Alamos and Lawrence Livermore. Yellowstone was already designated as such in recognition of its long history of research.

Varley to understand why YNP would be a strategic partner for Diversa. "I see all bioprospecting in nature as looking for a needle in a haystack. But we have many needles here." Perhaps foreshadowing the controversy that the Yellowstone-Diversa partnership would spur, Varley also asked Mathur if "he knew what it meant to be a guinea pig, and though he said he did, in retrospect, I don't know if he really knew what he was getting himself into."

Neither did Varley. At the start of the negotiations, he conceded that he did not know the right questions to ask. The National Park Service had no history of working with life-science corporations, and there was virtually no cultural precedent for this type of partnership. But Varley had decided to take this route and, to pursue it effectively, he knew that he needed "a test case" that would serve as a model for future collaborations between the Park and other commercial research partners. "At first, we tiptoed, but Diversa was very outgoing," he recalled.

To prepare for the negotiations, Varley spent a lot of time in Costa Rica studying the Costa Rica–Diversa biodiversity collaboration. To design Yellowstone's benefit-sharing agreement with Diversa, he adopted the Costa Rican model, adapting it to his own context. The Costa Rican model, for example, relies on a central authority based in San Jose, the capital city. A consequence of this centralization is that payments do not necessarily reach the local areas where samples were taken, which means that these areas do not receive any direct benefits to improve the management of conservation efforts. Varley chose to decentralize the model, similar to American universities being able to set their own terms for commercializing federally funded research following the Bayh-Dole Act. Varley explained:

> The way we did it, if someone wants to bioprospect in Yellowstone, they need to come here to make a deal. Our apparatus was a small cadre of experts on IP to advise the Park on making the deal as they go through negotiation. Everyone really likes that: small is beautiful. The reason we were able to pull it off was the FTTA, which requires a decentralized system.

Another difference is that, in contrast to Costa Rica, where government and Park scientists collect samples that are then sent to a central Costa Rican facility for processing, samples collected in Yellowstone are sent directly to Diversa to process, as the issue of job creation for Yellowstone is less of a priority than for Costa Rica.

In August 1997, coinciding with the Park's 125th birthday, YNP and Diversa signed an agreement that provided the Park with an annual payment, training, and equipment in exchange for access to Yellowstone's extremophiles and other biological resources. In addition to training YNP staff in modern molecular biology techniques, transferring high-tech equipment to the Park, and receiving payments, Yellowstone's collaboration with Diversa proved to be advantageous in ways that Varley had not anticipated. For several years, he had been "a fence sitter" on the issue of collaborating with corporate interests. During the process of defining Yellowstone's policy on bioprospecting, and long after the Park's policy shift had been announced, he had experienced a great deal of inner turmoil over whether he had compromised his personal convictions and violated the philosophy of

National Parks: that they should remain in a natural state and not be exploited for commercial purposes. The turning point was when he saw first hand that "what Diversa could provide us in terms of knowledge was *vital.*" He explained that 10 years ago, no one in his business would have conceived of going to a company to help in describing biodiversity and identifying species; today most firms have more and better information on biodiversity and cutting-edge technology, and they can go further and faster than any government agency or university.

To exemplify how the Diversa collaboration has assisted science in the Park, Varley cited two projects that involved extracting DNA, which Yellowstone could not afford to do on its own and which Diversa did pro bono. The first project was to map the genetic diversity of a wolf pack that had been reintroduced in the Park. Diversa put together a team of researchers who provided a detailed analysis that ensured there was sufficient heterogeneity in the wolf pack for healthy breeding, "a huge benefit" to the Park, according to Varley. The other example is of a study of the bottom of Yellowstone Lake, which only a few years ago was discovered to be a geyser basin under 400 feet of water. When YNP sent down submersibles to study it, they found coral-like spires on the bottom of the lake, including one that had been knocked off. Varley gave a midwestern university permission to study it, but was very angry when he discovered that, after 18 months, it had done nothing with the specimen. So, as he recalled: "I sent it to Diversa, and asked them, as a favor, to do the DNA extraction. They agreed and came back with 17 new species of organisms never seen by science before. That kind of thing is just stunning to us. It's the sort of thing we can't afford to do routinely."

Varley views this work, which went beyond the terms of the benefit-sharing agreement, as evidence of how Diversa differs from other corporations. "I see them as having an altruistic dimension, an ethical dimension that other companies don't have," he said. He has talked to many Diversa scientists who have told him that they would leave the company before doing something they deemed unethical. He is regularly invited to Diversa's corporate headquarters to give seminars about YNP's activities, which at least 60 to 80 company employees attend. This demonstrates, in his view, their "genuine enthusiasm for what we're trying to accomplish." On the whole, Varley felt the YNP–Diversa partnership was working well and provided many benefits to the Park.

But the CRADA was met with opposition almost immediately. The day Diversa and Yellowstone issued a press release announcing their formal agreement, environmental groups began to protest. The first objection centered on the fact that the Agreement's royalty formula would remain confidential. Varley explained that they had agreed to release the range of royalty payments (from 0.5–10%), but not the details because Diversa viewed this as proprietary information. Because of this decision, Yellowstone lost the support of many of the environmental groups whose backing it had fully expected for an agreement that would provide extra resources for conservation and that was designed in the spirit of the CBD.

Varley's team countered that every other aspect of the agreement had been made public, with the exception of the exact percentage of royalties the Park would receive. Varley strongly believed that it was the right decision to keep this information confidential because, in his view, disclosing royalty rates would set the "high rate," taking away YNP's future ability to negotiate more generous deals. He described his rationale:

> Our experience in oil and gas, where we [the US Government] have [publicly disclosed] royalties, has shown us that we end up with rules to follow. But we continue to think we can improve on these agreements, but never will as long as companies' competitors know what they are up to and where they are going. If competitors know, companies lose their competitive advantage and we lose our ability to negotiate competitively.

But this was just the beginning of the controversy. In March 1998, not long after the CRADA was signed, three public-interest groups, the Edmonds Institute, the International Center for Technology Assessment, and the Alliance for the Wild Rockies, filed a lawsuit against the United States Department of the Interior (DoI) and the NPS, which oversees Yellowstone. The lawsuit charged that the Park did not go through public review to analyze its environmental impact of the agreement with Diversa on Yellowstone that, according to the Plaintiffs, is legally required for proposed projects within YNP. It also charged that the FTTA, which states that such agreements are designed solely for companies to work with government laboratories, had been breached. It further argued that Diversa's extraction of microorganisms would erode Yellowstone's biodiversity. Finally, it contended that, in addition to the FTTA, a number of other laws had been violated, including the National Park Service Organic Act, the Yellowstone National Park Organic Act, the National Environmental Policy Act, and Public Trust Doctrine. These specific legal challenges were part of a broader opposition by these groups to the concept of bioprospecting in general, patenting of life, and more specifically, any private use of public natural resources. Other opponents denigrated genetic research and genetic modification altogether (see above, "Main Ethical Issues").

In March 1999, 1 year after the lawsuit was filed, a Federal judge ruled in favor of Yellowstone on all issues with the exception of the NEPA review (Environmental Assessment [EA] or Environmental Impact Statement [EIS]). He ordered the DoI to conduct an EA or an EIS and suspended the Yellowstone–Diversa CRADA until its completion. The judge had discounted YNP's argument that its scientists had already determined, during the process of defining YNP's policy on bioprospecting, that it would not have any environmental effect. Putting the environmental impact of bioprospecting in perspective, Varley commented that at Yellowstone, "we allow trout fishing and we allow you to keep and eat what you catch. The average fisher takes out more biomass than all of the bioprospectors put together." He further explained, "We consciously chose not to use the EPA and go into a formal Environmental Policy Act process because we didn't see any environmental effect. The courts disagreed that a NEPA review was not required—not because an effect was identified but because the issue remains controversial." Opponents of the CRADA saw this ruling as an important step toward ensuring that National Parks did not become "outdoor laboratories for commercial research" and because it "prevents the exploitation of our National Parks for commercial gain."

The judge further questioned whether Yellowstone had the legal authority to negotiate benefit-sharing agreements with corporate interests. In his ruling, he underlined the importance of the case:

> The Yellowstone–Diversa CRADA marks the first time in history that an American national park would stand to gain financially from scientific discoveries made within its borders. To understand the significance of this shift in policy, it is necessary to briefly examine the emerging field of 'bioprospecting' and how it relates to the Yellowstone National Park… As far as the court is aware, the defendants have not conducted a rulemaking procedure for this change in policy, nor have defendants solicited public comment informally.

In April 2002, 1 year after this preliminary ruling, the same Federal judge dismissed the lawsuit with prejudice filed against DoI and NPS, concluding that the NPS properly determined that the CRADA was consistent with the governing statutes, because it would produce direct concrete benefits to the Park's conservation efforts by affording greater scientific understanding of Yellowstone's wildlife, as well as monetary support for Park programs. The judge further declared that the CRADA did not violate any laws and that the plaintiffs' allegations revealed "fundamental flaws in their challenge." This was a major victory for Yellowstone and the Park Service. Upon hearing the judge's decision, Varley commented, "We couldn't agree more with Judge Lamberth's statement that the arrangement is a 'thoughtful and rational approach to research conducted on park resources.' It's good for science, good for parks, and good for the citizens of our country."

The Judge did not, however, revoke his first ruling ordering either an EA or an EIS that he deemed was required under the NEPA. In June 2001, NPS set in motion a NEPA study to determine the impacts of bioprospecting at all 387 of its Parks. NPS is also exploring alternative policy directions on bioprospecting to formulate a consistent national policy on future benefit-sharing agreements: whether to prohibit them, manage them, or return to a pre-CRADA laissez-faire approach. By fall 2004, the EIA still had not been completed, but it was scheduled to be released for public comment in 2005. Its release will probably set the context for the future of biodiversity collaborations in the United States. Whatever the outcome of the Assessment, Jay Short is convinced the firm made the right decision. "In this space, you need to adopt the highest ethical standards. My gut feeling was that ours would be the best approach. In hindsight, [the decision to pay Yellowstone while abstaining from collecting] has been incredible in elevating how we are viewed by other partners."

DIVERSA INC. (C)

PARTNERSHIP IN RUSSIA

Diversa's partnership with the Center for Ecological Research and BioResources Development (CERBRD) in Russia was set up under the auspices of the United States Department of Energy's (DOE) Initiatives for Proliferation Prevention (IPP). The IPP seeks to employ Russian scientists, engineers, and others who were formerly engaged in the development of weapons of mass destruction (WMD). The CERBRD, established in 2000 as a non-profit partnership of Russian research institutes under the umbrella of the IPP program, brings together former Russian WMD scientists with scientists from other Russian institutes to form the first ecological center in Russia and the first institutional mechanism to facilitate Russia's efforts to harness its own biodiversity.[18]

The Center's Executive Director, Dr. Vera Dmitrieva, met Diversa's Senior Director for Molecular Diversity, Eric Mathur, at a NATO workshop held in Novosibirsk in September 1999. She had been invited to make a presentation about the non-governmental organization (NGO) devoted to the study of man and the biosphere that she had founded and run for 14 years, to help commercialize Russian research. In partnership with academic laboratories, her NGO has implemented more than 70 projects in biology and the environment. Prior to that, Dmitrieva had been a university lecturer and researcher at the Institute of Biochemistry and Physiology of Microorganisms of the Russian Academy of Sciences. She first learned about what other countries were doing in biodiversity conservation and collaboration from two former Institute colleagues, Drs. Boronin and Volkov, who had themselves learned about it at the international BioTrade Conference held in France in November 1998, where they also had lengthy meetings with Jim Wolfram, Senior Scientist at the Civilian Research and Development Foundation, Christoffersen, who was working with the World Foundation for Environment and Development (WFED) at that time, and Preston Scott, WFED's Executive Director. By the time she attended the NATO workshop the following year, she had became acquainted with the emerging concepts of biodiversity and bioprospecting. At this workshop, she talked to Mathur, Wolfram, and Scott about the idea of setting up a biodiversity project in Russia. It was Wolfram's idea that such a project could be established in cooperation with the DOE's IPP program, and he subsequently facilitated DOE discussions with Dmitrieva and other Russian scientists that eventually led to the creation of the CERBRD based in Pushchino, Moscow.

"Pushchino is a good ground for establishing such a centre," said Dmitrieva. "There are eight institutes of the Russian Academy of Sciences here. I can always find a local expert." The six institutes that form the CERBRD are:

- Institute of Biochemistry and Physiology of Microorganisms of the Russian Academy of Sciences
- State Research Center for Applied Microbiology
- Research Center of Toxicology and Hygiene Regulation of Biopreparations
- All-Russian Research Institute of Phytopathology
- All-Russian Research Institute of Biological Plant Protection
- State Research Center for Virology and Biotechnology

---

[18]CERBRD Document describing the Center's biodiversity collaboration with Diversa, given to the authors August 2003.

The CERBRD, explained Dmitrieva, acts as a broker between the various institutes and research and commercial partners who are seeking access to Russian technology, talent, and natural resources. It coordinates negotiations and contractual relationships, and provides financial support to its member institutes who are eager to enter into collaborations. According to Dmitrieva, biodiversity partnerships, such as the one with Diversa, are mutually advantageous arrangements: they generate economic and scientific benefits for Russia's scientists in exchange for providing the company access to Russia's biological resources and highly skilled scientists. One of the benefits to the CERBRD of the Diversa partnership is that the company has provided training in, and state-of-the-art equipment for, the process of discovery and isolation of microorganisms.

An interesting feature of the CERBRD–Diversa partnership is that it had identified a potential commercial and social benefit from Russia's Cold War legacy: samples are being collected from the contaminated and radioactive sites to see if they contain valuable bioactive compounds for product development. Just like other extreme environments, contaminated sites have been discovered to support life and to host microorganisms that are thought to be useful in industrial processes requiring bioactivity in similar harsh conditions. In addition, the partnership provides Diversa with the opportunity to work with Russia's well-educated cadre of scientists and well-developed scientific institutional infrastructure, which allows the company to strengthen its scientific and resource capacity.

The CERBRD and its member institutes derive four value-added benefits from the Diversa partnership. First, in addition to promoting the fledging efforts at conservation of Russia's biodiversity, Russian scientists have the opportunity to implement an environmental restoration effort, using Diversa's environmental bioremediation and phytoremediation technologies. Diversa has volunteered to assist Russian scientists in cleaning some of the country's contaminated sites. Second, in addition to attracting investments to commercialize Russia's biodiversity, the CERBRD and its members have the opportunity to develop capacity in product development and entrepreneurship, new skills that Russia needs to succeed in the global economy. As Dmitrieva commented:

> We've only begun investigating [product] commercialization. In Russia, we're not experienced in bringing products to market. Sometimes we've already obtained a microorganism that is a good producer of bioactive compound, but we don't have the skills to take it further. Diversa can guide us through the development process and will help us bring it to the level of commercialization.

Third, Diversa and CERBRD member scientists work on joint publications and are looking at patenting proprietary discoveries together. A fourth direct benefit is remuneration for Russian scientists, allowing them "to improve their situation, since salaries for scientists in Russia are extremely low," according to Dmitrieva.

All this is possible, said Dmitrieva, because there is a great deal of trust between the Russian and Diversa scientists. "We believe in each other; we never deceive each other. Diversa is always very helpful and efficient. It is an *unusual* company." What she admires most about Diversa is the respect that company researchers have shown for both Russia's biological resources and the country's legal and regulatory restrictions. Speaking of the company's ethical standards, she said, "Diversa always requests that we work in a way that respects Russia's laws and that we never violate them. In addition, one of the company's main prerequisites is that we have an access agreement from anywhere we collect samples, including farmland, industrial sites, and parks."

Perhaps one of the most unexpected outcomes of the CERBRD–Diversa partnership is that "it helped four of our institutes run by four separate ministries become partners." Dmitrieva added, "We have become real partners through our collaboration with an American firm. This is really something." She hopes that her Center's collaboration with Diversa will go on for many years to come because she strongly believes that everyone will continue to derive many benefits from this unusual partnership.

In July 2004, Diversa announced the most tangible benefit thus far from the partnership with the launch of Luminase, "a product that allows pulp and paper manufacturers to improve the cost and quality of pulp processing" by reducing the need for bleaches by as much as 28%. The enzyme, explained Mathur, "comes from a microbe discovered in a thermal feature in Kamchatka that breaks down xylanase. [The 100% biodegradable enzyme is] an environmentally friendly solution that greatly reduces cost. It's a win-win. The royalties from it will eclipse anything we've paid to Russia until now." Diversa estimates the potential market for Luminase at over $200 million and is hoping that this first major success will help confirm and sustain the virtues of benefit-sharing arrangements. "What we'd like to have Russia do is then flow some of the money back to us to fund continuing activity in the partnership," said Mathur. "That's a way to create a model that is sustainable; we can't rely on DOE's IPP funding forever."

DIVERSA INC. (D)

PARTNERSHIP IN KENYA

In 2001, Diversa partnered with the Kenya Wildlife Service (KWS) and the International Centre for Insect Physiology and Ecology (ICIPE) in Kenya. KWS is the Kenyan government body that is mandated, under the Wildlife (Conservation and Management) Act (Chapter 376 of the Laws of Kenya) to conserve and manage all forms of wildlife in Kenya. ICIPE is a non-profit international organization with headquarters in Nairobi, Kenya, with a mission to help alleviate poverty, ensure food security, and improve the overall health status of the people and environment of the tropics through research, development, and capacity building. The Leader of the Applied Bioprospecting Programme at ICIPE, Dr. Wilber Lwande, a chemist who has been with the ICIPE for more than 20 years, began exploring bioprospecting opportunities in 1995. He was looking for ways to diversify the Institute's research focus in natural products discovery for insect pest control. Lwande was well aware that international researchers had been conducting research in Kenya, collecting materials that at times had led to new products and not sharing the benefits that had accrued from these discoveries. He also knew that some independent local scientists had identified products and then sold them directly to companies who were, in turn, reaping profits from the sale of these products.

So in 1995 Lwande began looking at existing bioprospecting programs in Costa Rica and Yellowstone that could serve as models for ICIPE. He had learned about Merck's agreement with INBio in Costa Rica at a conference he had attended in the United States. When he returned to Kenya, he contacted the Washington office of the International Organization for Chemical Sciences in Development (IOCD), a non-governmental and non-profit organization that helps developing countries build capacity in chemistry, to talk further about bioprospecting. In 1998, IOCD and ICIPE hosted a conference in Nairobi to introduce the concept of bioprospecting to Kenyans, during which Lwande met Christoffersen, who was participating as part of the faculty for that conference while working for WFED. During that conference, Lwande spent a lot of time highlighting the benefits of bioprospecting, especially as a tool to promote conservation and as a collaborative effort to build capacity. He initiated discussions with Dr. Richard Bagine, the Deputy Director for Research and Planning at KWS, to establish a collaboration model for bioprospecting projects that would address issues of equitable sharing of benefits that are geared toward contributing to conservation of biodiversity. This initiative led to the signing of a memorandum of agreement in 2000 between ICIPE and KWS for partnership in discovery and development of products identified from Kenyan arthropods, microorganisms, and plants. The partnership ensured that, through KWS, the Kenyan government participated in the discovery and development activities, and potential benefits would accrue to Kenya and, more specifically, would support conservation of Kenya's biodiversity.

By 2000, bioprospecting was firmly established within the strategic research focus for ICIPE. That year, Lwande attended a workshop held in Botswana where he met the Director of Diversa's Diversity and Discovery Group, Martin Keller. At that workshop during the summer of 1999, Keller made a presentation on Diversa's global bioprospecting activities, after which Lwande approached him to talk about the possibility of establishing a partnership together. Within a few months, ICIPE, KWS, and Diversa started formal negotiations. In October 2001, KWS, ICIPE, and Diversa officially entered into a partnership that effectively met the needs of both the company and Kenya's conservation efforts and benefit-sharing requirements.

Under the agreement, ICIPE provides the resources for the technical aspects of the program, KWS arranges permits, and ICIPE and KWS carry out sample collection and processing before shipping them to Diversa. In turn, Diversa funds the collection and processing of samples, pays for personnel, including some overhead costs for both ICIPE and KWS, and provides training and equipment to researchers involved in the project. In addition, doctoral candidates trained at ICIPE use Diversa's cutting-edge technology to do their work on other projects to advance their formal education. The agreement is periodically renewed and renegotiated, and can be terminated within 60 days by any party.

The Diversa–KWS/ICIPE partnership was the first such biodiversity initiative in East Africa. As such, Lwande "wants to develop a model that can be copied by others." Part of this model, which ICIPE has since replicated in other projects, is to generate income for local communities. What really motivates Lwande are the potential benefits arising from discoveries made in the region. "If we make a discovery," he said, "people will see that the value of biodiversity is real. This will have a major benefit for conservation." Another corollary of making a successful discovery is that "the government will see the real benefits [of bioprospecting] and be ready to crack down on those collecting samples without agreements," which Lwande is sure is still going on, though to a lesser extent than before the CBD.

A clear sign of the government's new approach to biodiversity came in September 2004 when KWS announced that it was pursuing a multimillion dollar claim with the help of a newly formed,United States-based non-profit group, Public Interest Intellectual Property Advisors, against Genencor and Proctor and Gamble for biopiracy.[19] The claim alleges that the firms are using microorganisms, collected without a proper license from Kenya's highly alkaline "soda lakes," in detergents and to soften stone-washed jeans. The microorganisms were discovered by Dr. William Grant, a Leicester University microbiologist, who sold them to a Dutch biotech firm that was then purchased by Genencor. Genencor claims the samples were collected with a permit from Kenya's National Council for Science and Technology. Mathur will understand if KWS seeks punitive damages: "I told Wilber that I understand if they have to pay an order of magnitude more in royalties (for collecting without the rights to commercialize), because otherwise it will threaten your ability to form these benefit-sharing agreements upfront."

One of the things Lwande appreciates most about Diversa is its commitment to not collecting samples without an agreement in place. Speaking of Diversa's ethical decision making, he noted that "they could just have tourists scoop things up. But they have determined that they should not do that." In general, Lwande feels that Diversa is a company that has a great deal to offer. He noted that the researchers who were trained by Diversa "had rich interactions" with company scientists and that the company's CEO had recently spent 2 weeks with ICIPE. Moreover, Diversa is "focused and fast in how it delivers things." Perhaps the greatest satisfaction for Lwande is the strongly applied nature of the work: "We're not doing research for the sake of it or just to publish. It feels very different to be working on a product that will benefit our country and human society. It feels more real."

Although Lwande is very positive about his Institute's partnership with Diversa, ICIPE and KWS have not yet publicized the relationship on their website or in other ways. ICIPE and KWS have chosen this low-key approach because the partnership is only a few years old and has not yet produced any discoveries leading to commercial products. In Lwande's view, "Until people see something coming out of this, they won't believe that these things are possible."

[19]Barnett A. (September 5, 2004). Multi-Million Bio-Piracy Lawsuit Over Faded Jeans and African Lakes. *The Observer.* http://observer.guardian.co.uk/uk_news/story/0,,1297590,00.html.

7

PIPELINE BIOTECH A/S: COMPETING REGULATORY REGIMES FOR LABORATORY ANIMAL EXPERIMENTS[1]

INTRODUCTION

Pipeline Biotech, a Danish contract research organization (CRO) specializing in performing animal experiments for life science companies, was seeking to expand into Europe's largest biotech cluster in Cambridge, England. When Pipeline's leaders began talking with potential British customers in the fall of 2003, they were surprised by what they found.

Huntingdon Life Sciences (HLS), just outside Cambridge, is one of Europe's largest animal research organizations. It had become a focal point for animal rights protestors. They had set up regular pickets outside the gates, physically assaulted several Huntingdon executives who ended up in the hospital, and threatened many other employees and customers.[2] These protests were then extended internationally to any firm with a remote connection to HLS.[3] Dr. Klaus Kristensen, Pipeline's

[1]This case study was written by David Finegold based on interviews he conducted with Pipeline employees and a company consultant. Interviews were conducted by phone or at the Keck Graduate Institute in Claremont, California in April to June 2004. Unless otherwise specified, all quotations were obtained during these interviews. This case study is one in a series produced under grants funded by The Seaver Institute and Genome Canada to study models for ensuring ethical decision making in bioscience firms. Grateful appreciation is extended to all those who participated and assisted in this case study, especially to Rasmus Nelund, without whose efforts this case study could not have been written.

[2]Cowell A. (January 17, 2002). Company in Animal Rights Battle, *New York Times,* reprinted from http:www.amprogress.org/News/News.cfm?ID=208&c=63.

[3]As an example, Mitchell Lardner, who works for Sumitomo Corporation of America finding investors for high-tech start-ups, had animal rights protestors marching in front of his home because the Japanese parent company was thought to have a connection with HLS. Lardner was targeted because he owns an historic house once occupied by the novelist Upton Sinclair. (See Chavez S. [June 28, 2004]. Raucous Protests Aimed at Family's Historic Home. *Los Angeles Times,* Section B1.)

founder and Chief Business Officer, recalled his reaction: "It was the first time I'd experienced this fear of animal experiments. I have the feeling people there are scared of using HLS. People go there in taxis so the protestors won't trace their cars. It is a terrible situation."

Through the head of the region's local biotech employer organization, Pipeline was able to identify an experienced local consultant, Lia McLean, to represent the firm and provide them access to local networks. Before McLean would accept the assignment, however, she visited Pipeline in Denmark to assess their operation. She explained:

> I went to see them and do my own audit of them. To feel comfortable offering their services, I had to know if they were any good. I spent a full day there interviewing people and inspecting the facilities and their GLP [good laboratory practice] status. It is extremely important that what a company claims is GLP really is. If it is not, the company [and their clients] can get into terrible trouble down the road when seeking regulatory approval. I thought they were excellent; they were shipshape on GLP. I asked them for a specific study and they were able to find all the files. Every time a technician does anything to a cage, they need to a fill in a form, and all those forms were there.

McLean was able to make introductions to a number of biotech and pharmaceutical companies and helped Pipeline secure several new customers. Pipeline had several advantages over United Kingdom providers and global CROs, according to McLean:

> In the beginning I stressed that they can do it more cheaply (there is about a 30% price advantage) and more quickly. More quickly, because in a small firm you don't have to go through layers of management; you can just get it done straightaway. [And more cheaply because] in the UK, costs are higher for extra security. [Huntingdon] is not just out in an open field like Klaus is. In the UK they need fences, guard dogs and 24-hour security.

Pipeline's newly appointed chief executive, Rasmus Nelund, explained another reason for the firm's rapid success:

> We found our ethical approach more sensitive and the public handling of the regulatory demands much more smooth and less bureaucratic; it was a much more emotional and militant issue in the UK than in Denmark. The dilemma is the same in both countries: a potential drug needs to be tested for efficacy and side effects before the industry starts testing in humans. In discussions with Lia, she suggested we needed to add a slide to our business presentation that said we'd take care of the whole regulatory and ethical issues. The CROs in the UK also take care of all the regulatory and ethical applications. This is just a far more lengthy and complicated process than in Denmark, where the process is more sensible and less time consuming. We were not aware of the fact that it would be a competitive advantage for us because these issues were taken for granted in Denmark.

Pipeline's major edge was the speed with which it could conduct a new experiment for a customer because of the differences in the two countries' regulatory regimes for animal testing. Pipeline was often able to begin a new experiment in 2 to 3 weeks, compared with the 6 months often required to get a license in the United Kingdom. Although Denmark and the United Kingdom both subscribe to the same European Union regulations governing animal experiments, the difference, according to Kristensen, is how they are applied: "They are so strict there (in the UK), because of the activists." In the United Kingdom, a firm not only must submit a very detailed protocol to the national government for each new set of experiments, it also must obtain approval from a local ethics committee. Said McLean:

> The UK is unprecedented in terms of what's required. The Home Office guidelines are extremely strict; they don't even compare with what's required in Denmark or the US. I don't think it makes animals any better off; it just means piles of paperwork. Things have to go through a local ethics committee, which every community must have... It can be quite hard to communicate with them when filled with non-scientists, for instance the local vicar. One problem I had (not with Pipeline) was with a woman who couldn't get the concept of a control group. The experiment was for a meningitis vaccine, and I tried to explain why we have to have a control group, to provide different doses to some animals and just a placebo to the others, and then expose all of them to the disease and hopefully the control would become ill and the vaccinated ones won't. She didn't get it; she thought it would be great if all were vaccinated and none got ill. I tried to explain that it wouldn't be a valid experiment, but I couldn't get through to her.

Nelund elaborated:

> Working according to the Danish regulations is more flexible than in the UK, even though both countries apply the same ethical standards. We apply for permission to do a certain type of study and are then permitted to use a specific quantity of animals. Once a year we turn in how many animals we're using in each area and submit the data electronically. If we'd like to do a new kind of study, we need to justify the use of animals... In England, you do get a license for each type of experiment the same as in Denmark. However, the description of the type of experiment is so precise, that in practice you often have to get an amendment approved once you do something a little bit different: if you want to give 4 injections instead of 3, or increase the number of bleeds from 2 to 3. If this was not specified in the original license, you need an amendment, which has to go though ethical approval as well as the Home Office and this can take 3–6 months. A full new license takes an average of about 9 months. In fact, the Danish authorities' administration of the regulations is a competitive advantage for Pipeline.

These regulations apply to where the work is being conducted, rather than where companies are based. "So," explains Nelund, "if English firms work with us, they are free of the practical obstacles of English law and at the same time fulfilling the high ethical standards of the Danish law."

COMPANY HISTORY

Kristensen was the head of the laboratories at M&B, a Danish animal breeding firm, when he got the idea for forming Pipeline. "Every day we got calls from customers asking us to do small experiments for them," he recalled, "but we couldn't because the firm was focused on breeding and the management saw this as a high-risk enterprise." So in the spring of 1999, with lab facilities rented from M&B and financial backing from one of its owners, Palle Schjødtz, he launched a new firm, Pipeline, to focus on performing pharmacology and toxicology preclinical studies for drug-discovery firms.

Unlike most bioscience start-ups, Pipeline was able to leverage the M&B connection to obtain several customers from the outset and to secure loans from a bank to cover its start-up costs, rather than relying on financing from venture capitalists or angel investors. "We were able to convince them [the normally conservative bank lenders]," said Kristensen, "because we had three clients and business from day one. We already had a place to do the work and didn't need lots of capital for building labs."

Pipeline grew quickly, more than doubling in size in each of its first 3 years. By 2001, it had grown to over 20 employees, spurred by its largest ever contract that came through its partnership with Aarhus University, where Kristensen had obtained his PhD in cancer research. Some scientists at the University had developed a new gene-tracking technology that enabled them to generate genetic mutations in mice that could then be traced to particular forms of cancer. When a United States pharmaceutical firm contacted the University about sponsoring some of this research, they suggested that Pipeline conduct the animal studies.

To cope with the firm's growth, Pipeline decided to move to its own facilities, purchasing a new farmhouse 10 miles outside of Aarhus. Kristensen also decided the firm needed to bring on someone with more business expertise. At the start of 2003, he hired Nelund, who had a degree from Copenhagen Business School, and had helped launch ScanBelt, a regional networking organization for the 11 countries surrounding the Baltic Sea that was striving to create what he describes as "a mega biotech-region that could attract E.U. funding and international finance" for new enterprises in the sector. Kristensen described his reasoning:

> We always knew we needed someone with a business education. We're all biologists; we don't know how to run a company. Rasmus doesn't either, but at least he's studied it. I'm very good at making business, getting new orders, starting things, but the other part is running the business. I discovered I needed another person to do this; it is not a skill of mine... We tried to circumvent the "founder effect" and I think we've succeeded in doing that.

When Pipeline made him an offer, Nelund contemplated for the first time the personal ethical issues involved with working for a firm conducting animal experiments:

> It was a new issue for me; I hadn't worried about it when I was doing my research on the biotech industry. I couldn't work there if I didn't believe in it... My integrity would be completely lost. I quickly made my mind up that I didn't have anything against animal testing in this industry. I don't see any alternatives to testing proof-of-concept before testing in humans. May be in the future we can grow organs to use for tests, but for the time being there is no good alternative.

Just after Nelund started as Chief Operating Officer (COO), the firm faced its first crisis. Coming off a very profitable year in 2002, Pipeline expanded to cope with anticipated further growth, only to find a sharp drop in orders caused by the sharp downturn in the biotech industry. Nelund was forced to lay off 6 of Pipeline's 22 employees to enable the firm to cut costs. He said:

> Firing two people on maternity leave was the hardest business ethical issue I've had to deal with. It was legal, but still very hard to do. When you're that close to being bankrupt, your ethical limits are different. Things are different today because we have money. At that time we were also slow to pay suppliers and creditors. We made employee wages the first priority... I don't like to pay 2 months late, but that's the way business is.

"A lot of people felt uncomfortable last year when we had financial problems," said Gitte Laursen, a laboratory technician. "They were unsure about whether they would leave or stay. Would the firm survive? It caused lots of headaches." The firm had meetings to discuss the issues with employees, but as Laursen observed, "You can never get all the information you want. The layoffs were necessary. It is never a comfortable situation, but I can see why they had to do it." Kristensen noted that the firm made sure to keep a veterinarian and the lab personnel needed to maintain ethical treatment of the animals, but that it was still a difficult situation. "I'm a nice guy, but at that moment, the firm was in deep #@%. You start to think differently; you're trying to survive, you can't afford to be a great philanthropist any more. It is simply a matter of deciding who is going to stay and who is going to go, unpleasant for a few months, but necessary and then you are able to move on."

With the help of the layoffs and restructuring, the firm was able to recover, returning to profitability in June 2003.

BUSINESS MODEL AND ETHICAL APPROACH

Pipeline has a high ethical standard but handles it in a relaxed manner. At the outset, Pipeline seemed to be surprisingly unconcerned with ethical issues for a firm that would be focusing solely on animal experiments, an area attracting growing protests in many parts of Europe and the United States. Kristensen explained:

> It was not a particular issue for me, or the people working with me, to do animal research. We were used to working with mice and rats. We started with the idea that we wouldn't go into areas completely new to us. We stayed away from dogs, pigs and primates, because we weren't used to them, and also because of ethical issues. We also thought we wouldn't do pain or wound-healing studies for the same reasons.

In addition to Kristensen and his team's long experience with animal testing from M&B, the other main reason for the firm's initial relative lack of concern with ethical issues was the difference in the surrounding environment in Denmark compared with places like Cambridge. As Kristensen remarked:

> Our company is situated in the countryside, where the people around us are farmers who are used to animals being raised for a purpose. They (Huntingdon Life Sciences) are in an academic area, with lots of young academics looking

for a purpose in life. Another reason may be that Danes are a more peaceful nation. We're a small country; in all aspects of our society, there are not so many extremists. The Danish people have been compared to a tribe of individuals, who accept that if a trusted member of the tribe is doing animal experiments, then it is probably ok.

In contrast, England has had a strong antivivisection movement for more than a century, which McLean, a Dutch native who has lived in Britain her whole working life, sometimes finds difficult to understand:

I'm puzzled by some people's attitudes to animals; they seem to view them as far more important than people... I totally agree that we should never cause unnecessary suffering to animals, but they seem to ignore the reverse side. Even the very violent activists, if their little girl had cancer, I'm sure would happily use the drugs [that were developed with animal testing]. Some are convinced that animal testing is never necessary, that it all can be done on cells and cell culture or simulated on computers. But anything that we can do like that is already being done, because it is far cheaper and easier. The costs pile up as soon as you start doing animal studies, so companies only do it if it is needed.

Jon Kold, a study director and Pipeline's veterinarian, pointed to a more recent explanation for British attitudes—an undercover video exposé of mistreatment of animals at Huntingdon, entitled *It's a Dog's Life*, shown on Channel 4 in March 1997 as part of its *Countryside Undercover* series. The video depicted beagles being punched by two HLS employees, who were later convicted under the Protection of Animals Act of 1911.[4] The video resulted in two government investigations of HLS and, according to Kold:

That specific incident damaged the image of the whole industry. It is the same thing as Iraq currently [referring to pictures of abuse of Iraqi prisoners by US troops]. Horrifying pictures can have a tremendous effect. When people's home animals are being treated wrong, it doesn't get out in the same way. The only people who know about that are the vets.

Once this incident occurred, it then had a snowball effect, according to Kold:

When you start to close in and get into bigger secrecy, it creates an environment where protestors can thrive. Secrecy is like putting gas on the fire for them. They think the industry must have something to hide. This contrasts with the more open and pragmatic approach of government and people in Denmark.

He argues that many of the protestors are hypocritical for focusing on laboratory animals and ignoring the much wider use of animals in agriculture:

In Denmark we kill 400 million pieces of poultry and 22 million pigs each year compared to 300,000 lab animals. Where's the balance in that? Something is wrong here. Groups like those in the UK don't get through in Denmark on lab animals, so they attack the fur industry instead. They depend on public opinion, and they don't have it here.

---

[4]Huntingdon Life Sciences Exposed. Report by Stop Huntingdon Animal Cruelty (SHAC). http://www.shac.net/HLS/exposed/xeno/xeno_report.pdf.

Once the firm began operating, however, Pipeline recognized that it would need to begin to address ethical issues. At a minimum, it needed to ensure it was complying with all of the Danish and European Union regulations governing animal experiments with lab animals. Although the cooperation between the Danish authorities and the companies about these regulations are interpreted more flexibly than in the United Kingdom, Pipeline had to be prepared to engage in dialogue with the industry's critics. Said Kristensen:

> There is an animal ethics board that consists of equal numbers of experimentalists and opponents. Many times a year we have meetings at the Parliament where opponents discuss the issues and alternatives to animal experiments in a very open, democratic way... When we start doing different kinds of experiments, for each disease we need to get a new permit. We've not had major issues, but we don't always get permission at once. First time we do it, the board might come and watch the experiment or call you in to explain and justify it. We've experienced this with a number of disease models like arthritis, MS, and the first time we did dog experiments.

The cooperative relationship between the firm and the animal ethics board is aided by the fact that, in this small country, many board members have worked in the past with Pipeline employees. Although the board has not prevented Pipeline from doing any proposed animal experiments, there is a powerful potential threat to ensure that the firm treats its animals in an ethical manner: "If you do anything unethical, you can get up to two years in prison," noted Kristensen. "I haven't heard of this occurring, but it is a way for authorities to show they mean it."

In addition to regulatory compliance, there was pressure from key customers to adopt an ethical approach, according to Louise Stab Bryndum, Pipeline's original veterinarian:

> Some clients are very concerned regarding ethics. Novo ([Nordisk], Denmark's largest bioscience company) is sending a team to inspect how we handle animal welfare. Novo is famous for its concern in this area; other Danish firms are not that concerned. Some Swedish firms have asked as well.

EMPOWERING THE WORKFORCE

The primary driver for a greater focus on ethical issues was internal to Pipeline. Bryndum, who had major exposure to animal rights issues during her prior work as a veterinarian at both Huntingdon Life Sciences and Novo Nordisk, was enrolled in a part-time Master's course at Copenhagen's Royal Veterinary College that included modules on ethics and law, genetics, and GLP. "After going on this Master's course," she recalled, "I said to the CEO (then Klaus) that we had to write a statement of how we were going to approach animal ethics. We discussed it and then I drew one up. We reviewed it and now have it approved to give to all clients and stakeholders." This statement includes a discussion of the 3 Rs—reduction, refinement, and replacement—that have become the industry standard for humane

EXHIBIT 7.1   **Pipeline's Ethics and Animal Welfare Policy**

Ethics and Animal Welfare

Pipeline Biotech is a biomedical contract research organization; hence, the majority of our activities involve the use of laboratory animals. Laboratory animals play an essential role in the research process that is necessary to ensure the highest standard of medicinal product efficacy and safety in man. We at Pipeline Biotech acknowledge that we have an ethical and scientific obligation to ensure the appropriate treatment of animals used in research.

We have defined the following policy on the care and use of animals:

Pipeline Biotech is committed to maintaining high standards for the humane care and treatment of all laboratory animals, and it is our policy that all due measures are taken to prevent or minimize pain and distress before, during, and after experimental procedures.

We are committed to implementing the three Rs—**R**educing the number of animals used for research, **R**eplacement by non-animal methods whenever possible, and **R**efinement of the techniques used to eliminate or reduce suffering and improve animal welfare.

Pipeline Biotech ensures that all studies are conducted in compliance with national law and recommendations.

All animal studies are planned and carried out with great care, taking into consideration the welfare of the animals.

Experienced technicians who are under the guidance and supervision of the dedicated veterinarians care for the animals.

Personnel who care for animals or who conduct animal studies will be appropriately qualified and receive ongoing education and training on the appropriate care and use of animals.

---

treatment of animals, along with Pipeline's formal policies for protecting animal welfare (see Exhibit 7.1).

What may be most unusual about Pipeline's approach to ethics is the decision to empower the employees who are working directly with the animals to end any experiment they view as unethical. Said Bryndum:

> The laboratory and animal technicians are very concerned about ethics. They will always come to me or the other vet if they have any concerns regarding an animal's welfare. If they see any suffering, they are authorized to put the animal down. They can do this on their own over a weekend if we're not available. It's quite a good system.

This system was put into place after an incident involving a customer who was conducting experiments in Pipeline's lab. "We had one client who was doing painful experiments on rats and the staff didn't like it," recalled Kristensen. "No one was there with the authority to stop it. After that we put the technicians' right to stop an experiment into our Animal Welfare Statement. This is sort of an emergency procedure. Normally techs have time to ask the vet or study-director whether a study should end." Laursen described how she and her fellow laboratory technicians reacted to this incident:

> I only recall one time when we talked about a specific experiment that we didn't want to do again. It was one where the scientist wasn't very experienced. He was doing surgery where he was inserting a small tube into the blood vessels.

If a more experienced person had been doing it, it might not have affected us so much. We wanted to have specific guidelines where we could say that an experiment wasn't good and should not continue.

"In my world it is necessary to distinguish between different definitions of ethics," Nelund said. "There is *usability ethics*—for example, what we discuss with NGOs and customers about what the animal will be used for; *regulatory ethics*, which is separated from the actions of performing the study; and *action ethics*, which concerns the decisions by technicians on when to end a study." He noted:

> It is easy to have high standards when you're not performing the study; you can be holier than thou. Then you have action ethics, where lab assistants are performing the work. They are the ones who decide whether to go ahead with studies or not. They are the ones injecting animals or placing a wound on mice. Not me, you or some animal rights organization, because we're not part of the process. We're not seeing mice suffer. Our lab assistants come to ask for permission to stop a study if they feel it is unethical. It is not management who decides—they are the ones who have doubts. They have carte blanche to stop any study, and we will support them. The frightening thing about ethics is you can't always know what is right beforehand; you don't know how an animal will respond to treatment. Lab techs have so much more experience with handling animals; their limits are much better and professional than ours.

Kold, the vet, agreed, arguing that technicians are not only better qualified to make these decisions, but that allowing them to act quickly when an animal is in distress is the only ethical thing to do:

> It all comes down to the skill of the technician. They are highly skilled and see more animals than I do. They have a better view of whether an animal is approaching death. We can't have them feeling uncomfortable with what they're doing… What if they were not allowed [to terminate animals]? Should the animal just be left there until I can make it into the office to make a decision? If you are being ethical, then you need a swift response; you can't have long chain of command.

How have customers reacted in the few cases where a technician has decided to end an experiment they have contracted for? To date, none of Pipeline's clients has objected to these decisions, in part because the firm tries to make the samples for the studies large enough so that removing animals who are too sick to continue will not jeopardize the results. Kristensen adds, "For everyone in the industry it is vital not to get a bad reputation. And this can happen so easily. They would rather redo an experiment than do something nasty to any animal." Kold argues that this was not just the right thing ethically, but also good science:

> An unhealthy or very pain-influenced animal is not a good research animal. The argument isn't just that it is bad for the animal, but because it is, this makes for not very good science; you won't get good results out of the study… If you define the endpoints correctly, it doesn't matter if you have to terminate the animal. The endpoint should not be death or terminal illness, but a step before that.

Pipeline's situation-specific approach to ethical decision making also raises the issues of how to apply a consistent standard across the firm and the best way to determine what constitutes unethical treatment of animals in a context where, for example, mice are being bred specifically to have diseases such as cancer. Government regulations provide some guidance, as Bryndum observed: "There are some rules in our Danish legislation about how big a tumor can get. This is how we monitor whether it has gone too far. We try to avoid pain as much as possible." Kristensen acknowledged, however, that it can be difficult to know just where to draw the line:

> It is always important to try to define humane endpoints... One way to do it is to look at all cancer animals every day to see how they are doing and if any are suffering. Another is to try to define some early endpoints. We're mostly involved in lymphomas, cancers that you can feel on the lymph nodes. The animals hardly ever discover they have cancer. Like when people develop the same disease, they usually don't know they have cancer until they go to the doctor.

Kold gave a more specific description of how workers are able to make a determination of the appropriate time to end an experiment:

> These techs look at 1000 animals each day. They will spot anything abnormal even before the animal gets sick. For example, in a tetanus toxin experiment, an animal will ultimately be paralyzed. At the point where technicians spot the influence of the toxin, they can feel its effects in the muscle hours before real paralysis sets in. It's thus not really a matter of monitoring for abnormal behavior or distress, but of identifying signs before these occur... It may be with mice that you see their hair standing on end, or them getting watery around the eyes. The key judgment is not really spotting these signs, but deciding, based on these, when the animal is in real distress.

Rather than rely on formal policies to try to dictate a common approach to ethics, Pipeline tries to encourage an ongoing dialogue on the issues among employees, to reach a consensus on how to operate. "We don't have any specific guidelines on what we do or we don't do other than Danish law," said Nelund. Instead he tries to foster "a self-organizing construction in the lab," by clipping relevant articles from the press and then distributing them to laboratory supervisors and technicians to discuss. Informal communication is fostered by Pipeline's small size and informal atmosphere. "We're a very small company," said Laursen. "Employees can talk about things at lunch, when we're all eating together; nothing is taboo." Kristensen agreed. "The key is to be able to discuss ethical issues openly and have the boss not get angry if they say 'We don't want to do it that way.' We're very much an attitude, not a regulation, firm. If you put in right attitudes, then you get the right regulations."

The employees confirmed that the leaders have created an environment where professionals can be free to follow the spirit as well as letter of the law. "Danish legislation is very strict and we very much agree with it and stand up for it," said Laursen. Kold acknowledged, however, that the law is not always a clear guide as to what is the right thing to do and, therefore, it is important that he be given professional autonomy to make these decisions without regard to the business consequences:

> We have wording in permits that no major distress is allowed. What is 'major' is the gray area. We need to think about how long a study should go, what is the amount of distress that is tolerable, then we have to make a decision. I'm lucky that I don't have a big, bad boss breathing down my neck, saying, 'This will cost us $2 million dollars.' I design and monitor the studies myself and take the decisions. Where the problems come in are when people get to the point when they are afraid to make a decision. If you end up there, you should not be the one making that choice. The one making that decision shouldn't have any other concerns. It's the way you build up a chain of command. You need the vet to be independent to take the decision. The law ensures it.

The other key to fostering an ethical approach to animal experiments that all employees are comfortable with is the hiring process. "It starts with the interview," said Nelund. "We explain what we're doing, make clear what the business is and that we have the permission to do it." Laursen agreed. "When we hire people we try to find ones with the same attitudes. It is very important that people be responsible, and so we end up following very similar kinds of ethical behavior." Kristensen argued that the firm has developed its ethical approach "by hiring professionals. Everyone here has at least four years' education related to animals, even the technicians."

When less-experienced lab technicians are hired, Pipeline offers a lot of on-the-job education to teach them Pipeline's ethical guidelines. "We have worked with animals for many years, so we can teach new hires a lot," said Laursen who has 15 years of experience in the industry. "That's where the discussions of ethics come in. Without covering walls with ethical standards, we can talk in plain language about how to treat the animals and make them feel comfortable." To stay current with the latest ethical issues and approaches, Pipeline's veterinarians have also been active participants in a small professional organization that brings together vets from Danish bioscience companies, as well as courses run by Denmark's leading animal ethics expert, Professor Peter Sandoe, Director of the Center for Bioethics and Risk Assessment at the Royal Veterinary and Agricultural University of Copenhagen. Among the ethical issues most frequently discussed at these meetings, according to Kold, are the design of endpoints in trials, the creation and use of transgenic animals, creating appropriate environments for animals, and what to do when there is a tension among the 3 Rs: "Should we use fewer animals who feel worse or more with a better life?"

ETHICAL ISSUES AT PIPELINE

The biggest ethical challenge that Kristensen feels he has faced since founding Pipeline is not what one might expect:

> Giving interviews to newspapers. I was always told this would be a catastrophe, that it could ruin the firm because we'd have all the animal rights people

standing outside our door. But it didn't happen. This may not be an ethical issue, but it has to do with ethics. Should I be silent or public about what we're doing? It was a major decision.

In addition to its rural location and Denmark's general acceptance of animal testing, Pipeline may have avoided the protestors that have plagued Huntingdon because, according to Kristensen, "They are all in prison in Denmark. The activists took action on some mink farms, and the police caught them. We have heard nothing from them for the last 2 years."

Pipeline may also not have attracted protestors because it has thus far kept a relatively low profile. Although customers, academic partners, and regulators have been frequent visitors to its labs, Pipeline has not yet initiated a dialogue with the wider community. "We haven't brought stakeholders in," said Nelund. "We're not engaged yet with NGOs... It's because we're so small. We don't have the resources to go seek wider discussion, and we're also not as visible." He conceded, "We're also hiding behind Novo [Nordisk]. We let them take the lead with NGO discussions." The firm has discussed having an open house for local residents to increase their understanding of Pipeline's business, but has not done so because of some employees' concerns regarding safety and security. The firm's founder, however, indicates he is willing to have visitors if they ask to come:

> If anyone wants to see what we're doing, we will give them a tour of the facility. Things have changed a lot in the last ten years. My old breeding firm was very traditional; the doors were absolutely closed to the outside, they were scared of journalists and the media. We're very open; we choose the exact opposite approach. That's not something we invented, we have just seen it as the way to do this kind of business these days... We could even bring in a kindergarten class. They would see cozy little mice and not see any harm being done to them.

Another ethical issue that Pipeline, like all animal research organizations, faces is the potential clash between the scientific goals of the experiment and the welfare of the animals. "In all my jobs, I have seen that tension," said Bryndum, speaking about Pipeline and her previous companies. "Scientists want to push to the limit, while the vet says, 'We need to stop here.' I would get in touch with a client to explain the situation and why we needed to terminate the animal. They always agreed." Kold, Pipeline's current vet, feels that this is not an issue for the firm: "We don't have old-school problems here regarding narrow-minded scientists who only see an experiment and what they can gain from it. There is always a better way [than mistreating animals]." A different tension can emerge for CROs between the interests of good science and the desire to satisfy customers. Observed Bryndum:

> It can be a difficult issue because many clients may want results that differ from what you give them. You have to stand strong on your views, but it can be very difficult, more difficult than animal-welfare issues. It helps a lot when you have the GLP rules. Individuals can always tamper with the results, but it is less likely to occur if you have to follow those guidelines where everything is signed and counter-signed, checked and rechecked.

Changes in the type of work that Pipeline performs have raised some new ethical issues. Said Kristensen:

> We decided originally not to do wound healing. But now we're Scandinavia's biggest firm in this area. We hired an expert who knows how to handle this kind of study and showed us that this doesn't harm the animals. We can see that the mouse doesn't care whether it has a wound or not. Once the first test experiment was okay, then we started to grow it as part of our business.

The company has also begun conducting pain studies that were not part of its initial activities. Noted Laursen, "You think more about ethics when it is a pain study, but it hasn't brought up any large problems. People are not very comfortable when mice are in discomfort." When asked how Pipeline balances the needs of the experiment with animal welfare, Laursen paused and then continued, "We try to do it in the best way we can. Our dilemma is that we want to treat the animals the best possible way and at the same time, we need to induce pain for the study. We use our experience to balance the dilemma and we try to keep animals as pain-free as possible."

In addition, the company has also broadened its offerings to include experiments in larger mammals, such as dogs and pigs. Kristensen explained the reason for the change: "Many of my early decisions were because I didn't want to do that kind of work myself; it was not something I had experience with. When you're an amateur, everything is an ethical issue. It is important to have experts, professionals who have done the work before."

Not everyone was pleased with the decision to begin conducting dog experiments, which was made without consulting the laboratory technicians. "We did a dog study where ethics was an issue," said Bryndum. "Doing research on dogs is not well thought of in Denmark. If doing larger animal studies, we usually use pigs, since that is a huge industry here, with 22 million slaughtered each year… [But we] had no problems with outside stakeholders." Staff members, however, are not compelled to work on experiments with which they are uncomfortable. "Each person has his or her own ethical standards," said Laursen. "Some people don't want to work with dogs, while other people don't mind. We have a policy that if you don't want to, you shouldn't do it. But even those who don't want to do it themselves are not trying to stop it."

More controversial still is whether Pipeline should offer to conduct studies in primates. "I'm sure the CEO would say, 'Yes'" said Bryndum, who then continued:

> He wants to expand the firm. I'd say no. I worked with them at Huntingdon and it was hard emotionally because they look so much like humans. It's hard seeing them in small cages. Also they are very dangerous, because they have diseases that can affect people. We had one brief discussion about doing work on primates, when a client asked if we did, and I advised against it.

Currently, Pipeline has partnered with another laboratory that will perform primate studies for Pipeline's customers that request it, but Kristensen acknowledged that he would "consider doing primate studies in the future." He was unsure why his attitude has changed since Pipeline was founded, but cited the greater knowledge he's acquired about the kinds of research conducted on primates as one reason:

I see an enormous need for primate studies. When we go to clients, we find many produce humanized MABs (monoclonal antibodies) that they can only test in primate systems for toxicology. Dogs or pigs would have immune responses against the human proteins. What I discovered through doing business is that most primate studies are about vaccination, something we do to our kids, which is not harmful to anyone. They are not like studies that split up the monkey into many parts.

The relevant ethical distinctions among species are less to do with their native intelligence or how experiments are conducted than with creating the appropriate environment for them, according to Kold:

There is not really any difference in the use of different species. It is more an issue of where to accommodate them. The facilities required are much larger [for larger mammals]. This is basically true of primates as well. There is always a demand to try to give an animal something like their normal surroundings; this is a lot harder to do with primates... Mice, for example, are happy with litter-mates around. With primates, you need something to actuate them and enhance the environment; if you don't have it, there might be a problem. You can't give them the right physiological parameters as you can with a cat, dog or rat rela-tively inexpensively.

CONCLUSION

With its growing business in the United Kingdom, Pipeline has clearly benefited from the differences between England and Denmark in regulatory regimes and attitudes to animal testing. As the globalization of the biomedical industry continues, however, the firm may face growing competition from countries with lower costs and fewer restrictions on animal testing. "From our perspective, so far we have only gained business due to ethical concerns," said Nelund. We haven't lost to countries with lower ethics, but I'm not sure how it will develop." Already McLean has observed a growing trend among British biopharmaceutical firms to outsource manufacturing to India and China, where the costs may be half of those in England. To date in animal testing, she has...

...come across some [outsourcing to Asia], but not as much [as manufacturing], because people don't trust them to do a good job. I would have to go there to inspect facilities to feel confident... [I know one firm that used] an Indian firm because it was a requirement to find developing country partners in their grant from the Gates Foundation that was paying for the research, but the company wasn't very happy with the work. It wasn't terrible; they could use the results for their own guidance, but not for a regulatory filing. They would need to redo the experiment from scratch... But I think it is only a question of time. Give them another 5–10 years to get better educated on just what the standards are and then they will be very competitive.

8

TGN Biotech: A Start-Up with Ethical Roots[1]

"If you don't agree, we promise we won't come," was François Pothier's guarantee as he addressed an auditorium full of local citizens in rural Quebec on June 17, 2002. TGN Biotech (TGN), a cutting-edge start-up bioscience firm, was ready to move out of temporary facilities and to expand to a larger and more permanent location. Pothier, President and co-founder of TGN, ran a 3-hour public engagement session in each of two different Quebec communities to discuss concerns and gain community support for TGN's proposed pilot farm, to be its biopharmaceutical manufacturing site to produce recombinant proteins in the seminal fluid of transgenic pigs. A year later, in August 2003, TGN announced its choice of Saint-Tite-des-Caps as the location for its new farm, but not before giving the community all the information and receiving community support.

This is one example of how TGN is proactively trying to develop a transparent and stakeholder-focused company. President and cofounder Pothier says that ethics at TGN is so much a part of the culture that it is just "what you do." Coming from an academic background, he feels that "transparency, sharing of ideas and ethical research processes are a given, not a choice." From day one, the founders knew they would have an internal ethics committee for animal welfare. Other mechanisms that TGN has put in place include a mission-and-values statement and a 100-page disaster plan. TGN also has a full-time Regulatory Assurance Director to communicate with regulators, and a full-time Quality Assurance

[1]This case was written by Jocelyn E. Mackie under the supervision of Abdallah S. Daar and with the assistance of Cécile M. Bensimon. All information, unless otherwise stated, was gathered from personal interviews of employees at TGN Biotech in August 2003 conducted by Mackie, Daar, and Bensimon. This case study is one in a series produced under grants funded by The Seaver Institute and Genome Canada to study models for ensuring ethical decision making in bioscience firms. Grateful appreciation is extended to all who participated and assisted in this case study, without whose efforts this case study could not have been written.

Director to ensure that internal quality control is developed at the same time as the science. With these and other mechanisms, TGN Biotech is one example of how a start-up bioscience company has tried to address ethical issues from its inception.

WHAT NOW?

In August 2003, Pothier pondered how he and his management team would continue to maintain TGN's exciting, supportive, and ethically driven culture. These questions become more complicated as TGN continues to grow; it has now received 2 rounds of financing and has 20 employees working out of 3 separate locations. How will TGN balance freedom of creativity with a need to structure processes like ethics? How will TGN deal with public criticisms about its transgenic technology, if and when it happens? TGN's management team recognizes that these questions will need to be answered in the coming years. Pothier, however, is confident that the committee, documents, and policies that TGN has set up so far have created a solid foundation on which to build.

TGN BIOTECH

TGN, based in Ste Foy, Quebec, was founded by Drs. Francois Pothier and Marc-André Sirard in November 2000. Pothier received his PhD in cellular biology from the University of Montreal in 1986. In 1990, he became a professor at Laval University, where his research focused on the production of transgenic domestic animals. Sirard received his degree in veterinary medicine in 1981 and completed his PhD, focused partly on in vitro fertilization, from Laval University in 1986. Both Pothier and Sirard have dealt with bioethical issues throughout their careers. Pothier is involved with several provincial and federal committees that define regulations for genetically modified organisms (GMOs), and he is a member of several ethics committees, such as the Science and Technology Ethics Committee of Quebec. Sirard sits on the Board of Management of the Institute of Nutraceuticals and Functional Foods.

One of their early interests involved the production of recombinant proteins in mammary glands, but they discovered that they were blocked from pursuing this because of a patent held by an American company. Looking for alternatives, they recalled that pigs produce an unusually large quantity of semen, and that pigs' seminal vesicles might well be a more efficient site for the production of proteins. Their first step was to test their hypothesis by producing proteins in the semen of mice.

Along with several other scientists from the animal science department at Laval University, they published their method for producing foreign proteins in the seminal vesicles of transgenic mice in a 1999 issue of *Nature Biotechnology*.[2]

---

[2]Dyck M, Gagne D, Ouellet M, et al. Seminal Vesicle Production and Secretion of Growth Hormone Into Seminal Fluid. *Nat Biotechnol.* 1999;17:1087–1090.

These proteins, once purified, could be used as injectable therapeutics (medicines) for both animals and humans. This led to the development of this technology in pigs, and it became the platform for developing TGN Biotech in November 2000. Pothier and Sirard decided to develop the company for three reasons: to keep the technology in Quebec, to allow their graduate students to continue working on the technology they helped develop, and to be able, one day, to market their final product as medicines for various human diseases. TGN received its first round of financing of $2.75 million[3] in November 2000.

TGN has an important partnership with Laval University, as TGN's two founders are still full-time professors in the animal sciences department.[4] TGN has a contract with the University to sponsor research in their laboratories in return for rights to license the technology. The University is thus the holder of TGN's patents. Both Pothier and Sirard have signed transparent and explicit contracts with the University outlining both their role as professors and researchers and how it differs from their role as business owners. They spend roughly 4 days a week at the University; the rest of their time is spent managing TGN.

The University and TGN have one United States patent, granted March 13, 2001, for the production of recombinant proteins in semen in non-human transgenic animals. They have a patent pending, and others to follow, for their purification methods and for specific animals and proteins. A recent Supreme Court ruling in Canada has made it illegal to patent any living organism or part of an organism.[5] For this reason, TGN cannot file a patent for its transgenic mice or pigs in Canada but intend to do so in other countries.

TGN is growing rapidly. There were several personnel additions between June 2001 and October 2002, including a Director of the Pilot Farm, a Director of Regulatory Affairs, a Director of Process Development, a Director of Quality Assurance, and a Director of Finance. TGN announced its second round of financing in June 2003 for $5.35 million. The company is currently funded by a combination of public and private venture-capital funds (see Exhibit 8.1).

TGN'S MARKET

Recombinant proteins are a growing share of the biotech sector, with more than 50 approved products currently on the market. There are also more than 500 products in the development pipeline but, under current production methods, worldwide manufacturing is close to capacity. The mass-produced biotech drugs currently on the market are grown in bacteria, yeast, or mammalian cell cultures, expensive methods of protein production. The high-volume production facility needed to produce these proteins can cost more than US$250 million, and take about 3 years

---

[3] All currencies are in Canadian dollars unless otherwise stated.
[4] At the time of writing this case study.
[5] Patenting Pieces of People. *Nat Biotechnol.* 2003;21:341.

EXHIBIT 8.1   **Current TGN Biotech Investors**[6]

Foragen

Foragen Technologies S.E.C. is a Canadian investment fund dedicated to the creation and launch of technologically advanced companies in the biofood sector. Its mandate is to promote start-ups by offering entrepreneurs seed capital, business and technology management expertise, and networking opportunities that will allow them to concentrate on developing and commercializing their technologies. Foragen is owned equally by RBC Technology Ventures Inc., Crown Investments Corporation of Saskatchewan (CIC), and SGF Soquia Inc., a subsidiary of Société Générale de Financement du Québec. Foragen has offices in Québec, Ontario, and Saskatchewan.

Innovatech Quebec

Innovatech Quebec is a Quebec government venture capital corporation specializing in pre-startup and start-up financial assistance for technology companies. With a $150 million fund, Innovatech's investment portfolio includes companies and projects in information technology, telecommunications, biotechnology and life sciences, industrial technology, optics and photonics.

Fondaction CSN

Founded in 1996, Fondaction is a financial institution working in venture capital for the development of Quebec's economy. Investments made by the Fund to date have contributed to keeping and creating over 5200 jobs throughout Quebec. Fondaction is active in approximately 100 Quebec businesses in all economic sectors. It may intervene in any stage in the development of businesses and gives priority to businesses engaged in processes of participative management, businesses in the social economy sector (including self-controlled businesses, cooperatives, and others) and businesses that are environmentally aware.

---

to build.[7] As new products enter the market, TGN feels there is a strong need and opportunity for cheaper methods to produce high quantities of complex proteins.

TGN is currently working on three proteins that have an estimated market value of US$1 billion.[8] Although the company has not yet released what these proteins are, all three will become the raw material for biogeneric products. *Biogeneric* means that the products are currently approved and are being produced by other companies, but TGN explains that these companies are using other, more expensive, methods of production. The first protein is a biogeneric product for veterinary use; the second is a biogeneric product for human use. TGN has created transgenic pigs that are already producing these proteins in their semen; only 100 pigs would be needed to meet the total world demand of 800 grams (g). This is considerably cheaper than using current production methods. The third protein is another biogeneric product for human use that is expected to have a multibillion dollar market. The first semen sample of this protein is expected near the end of 2003.

---

[6]Press Release. Bertrand Bolduc. Foragen, Innovatech Quebec and Fondaction Invest $5.3 Million in TGN Biotech Inc.

[7]Watson T. (September 2, 2002). The Pig as Factory. *Canadian Business*. Toronto. Vol. 75 Issue 16 Pg. 52.

[8]Hamann J. TGN Biotech Démarre. *Le Fil des Événements*. November 23, 2000.

SEMENESIS™

TGN's licensed technology is called SEMENESIS™ and involves the genetic modification of pigs to produce biological therapeutics in their seminal fluid. The pig becomes a biological drug-manufacturing plant. The process[9] begins in the lab with the production of a DNA fragment, known as the "transgene." This fragment codes for the protein of interest, and when inserted into the genome of the pig, it will instruct the cells of the seminal vesicles to manufacture that protein in the seminal fluid. TGN has developed and patented specific promoters. Promoters are DNA sequences that regulate the expression of the genes of interest in a desired anatomical site, in this case, in the seminal vesicles. The business plan allows TGN either to develop these promoters itself or to license the sequences from a corporate partner.

The next step involves microinjecting this DNA code into a pig embryo. Microinjection uses a microscopic needle to introduce DNA molecules into the genetic material of a newly fertilized egg. This integrates the desired transgene into the DNA make-up of a proportion of these young embryos, which are then transferred into a surrogate mother pig.

The piglets are then born naturally and are tested to identify those who carry the transgene. Those who do carry it are then reproduced naturally. The semen from these transgenic pigs is collected regularly (about every 2 to 3 days). The semen is tested for the desired protein and is then highly purified, which results in the raw material for drug formulation.

WHY PIGS AND WHY SEMEN?

There are several cost advantages to using pigs instead of other animals for protein production: pigs have a shorter reproduction timeframe, produce 10 to 12 piglets per litter, reach sexual maturity at 6 to 7 months and, compared to all domestic livestock, produce the largest amount of semen—about 250 to 500 ml per day, all year round. Compared to other mammals, this results in a quicker and more efficient process from the beginning to the end of the production cycle. Pigs also are relatively cheap to buy, easy to take care of, and have long lives (up to 15 years).

The advantage of using semen is that proteins are naturally produced in the seminal gland. In milk, for example, there are natural enzymes that attack protein (the product), but semen is designed to protect sperm cells, thus protecting the product. Another important benefit of using semen is that it is completely excluded biologically from the blood circulation. This means that there is little chance of the biologic getting absorbed into the circulation and reaching other parts of the pig (as is more possible with milk) and adversely affecting the animal. Pothier feels that another benefit is the reduced risk of spreading diseases such as bovine spongiform encephalopathy (BSE – Mad Cow Disease) when using pigs compared with cows.[10]

[9]www.tgnbiotech.com.
[10]Watson T. (September 2, 2002). The Pig as Factory. *Canadian Business*. Toronto Vol. 75 Issue 16 P. 52.

By ensuring that its farms have only specific-pathogen-free (SPF)[11] pigs, TGN feels it is easy to maintain healthy animals and minimize the risk of virus transmission. Now that the pilot farm site has been chosen, and the new in-house laboratory is built to develop the purification process, TGN is well on the road to having a product ready for market. Pascale Duchesne, Director of Regulatory Affairs, thinks it may be as long as 10 years before the purification process and clinical trials are complete and regulatory approval is granted. Although the expected date for going to market is still a long way away, TGN has taken several steps to try and create an ethically driven culture to support and sustain the development of its unique technology.

THE ETHICAL ISSUES

Employees at TGN were asked what they considered to be the ethical issues present in their company. The most prominent answer was always the issue of animal welfare. Issues of language were also mentioned, as well as the issue of the science itself, environmental issues, and future drug-development concerns.

ANIMAL WELFARE

TGN works with animals on a daily basis. Although some testing is done with laboratory mice, the majority of the work is with pigs. Interaction with the animals first occurs when TGN scientists genetically manipulate DNA and perform in vitro fertilization to produce transgenic pigs.

Nicole Gravel is the Director of TGN's pilot farm and is a trained veterinarian. She obtained her MSc in animal in vitro production under Sirard's supervision. Gravel also has past regulatory experience, having practiced with the Canadian Food Inspection Agency before joining TGN. Gravel explains that during the in vitro process at TGN, the surrogate mother pigs are appropriately sedated and the pigs' health and comfort are top priorities.

Another process involving the animals is the collection of the semen. The pigs currently live at a small, SPF test farm where they are kept healthy in small chambers or, for the sows in gestation, in cages. With the new farm (also SPF), no pigs will be in cages, and all will have either a single (for the sows) or shared pen. Although the semen could be collected once a day, TGN has decided to only collect every 2 or 3 days to minimize disrupting the pigs.

Depending on the product that is being developed, there is the possibility of the animal becoming ill. TGN explains that this is the last thing it wants. Therefore, the animals are screened regularly to ensure they are not absorbing the protein back into their own blood-stream.

---

[11]Specific-pathogen free: A term applied to animals reared for use in laboratory experiments when the animals are known to be free of specified germs that can cause disease (pathogenic microorganisms). Obtained at www.medterms.com.

Pigs that do not carry the desired transgene are euthanized by injection, being put to sleep first so that no pain is felt. Pothier originally wanted to give the non-transgenic pigs as food to shelters for the homeless, but he was advised by TGN's board members that there could be a fear about transgenic pigs getting into the food chain accidentally. There is also the ethical issue of giving food to the poor that is not considered safe enough for the rich. It is for this reason that the non-transgenic pigs must be euthanized to ensure none of the pigs from TGN enter the food chain.

LANGUAGE

Because of the negative perception of pig farming in Quebec and the stigma associated with testing on animals and working with semen, TGN recognizes that it needs to be especially careful with the language and communication methods it uses. In Quebec today, the pig business is seen as smelly, a pollutant to the environment, and an annoyance to surrounding citizens.[12] As will be narrated below, local citizens' main concern with TGN's proposed pilot farm was not about the use of genetic animals, but about the smell of a pig farm. TGN said openly that it was aware of the issue and explained how it planned to deal with it.

Because TGN is genetically modifying pigs, it feels it must appropriately explain the science, stating that it is not producing GMOs merely for the sake of producing a new technology or even to improve pig production yields. The goal is to one day improve and save lives. Bertrand Bolduc is TGN's chief operating officer (COO) and the President of BioQuebec (Quebec's bioindustry network).[13] Bolduc explains, "The worst-case scenario is to be considered a GMO company. In the past, GMO companies did not explain risks to consumers. The benefits were merely for farmers and not for the public. It is our strategy to differentiate ourselves from other GMO companies."

That TGN is working with semen creates another problem, as Pascale Duchesne, Director of Regulatory Affairs explains: "What we are doing is strange for some people, and the sexual connotation adds to the difficulty." TGN recognizes that its science is different and the language used to explain it—whether it is to investors, business partners, or the general public—must be analyzed and thought through very carefully.

THE SCIENCE

So far, TGN has not experienced much objection or concern about the actual science of transgenics. TGN finds that the average citizen is happy that it is producing transgenic animals to help save lives, not just for reasons of efficiency or scientific discovery alone.

---

[12]Spears T. (March 19, 2002). Huge Pig Farms are Health Menace. *The Ottawa Citizen*. Ottawa, ON, P. A.1. FRO.

[13]At the time of writing this case study.

There are several activist groups who are keeping a close eye on TGN, but as of yet, there has been no public backlash. TGN chose not to clone its transgenic pigs, judging that the sensitivity of the subject, and its incomplete safety, would probably cause a backlash; besides, TGN's SEMENESIS™ technology is efficient enough not to require cloning to perpetuate the transgenic line. TGN therefore sees no reason to risk the animals' health.

TGN recognizes that there will be some opposition to all testing on animals, to any genetic manipulation of living beings, and perhaps to in vitro fertilization as well. Pothier explained that there will always be some who simply do not agree. As will be discussed below, TGN's approach is to be as forthcoming and transparent as possible and to engage in dialogue with those who express concerns.

THE ENVIRONMENT

Environmental issues for TGN concern the pollution associated with farming pigs. When originally selecting the site for the new farm, TGN analyzed in detail how much land it would need to spread the manure produced by its pigs effectively without affecting surrounding water sources and other natural environments. TGN's site was chosen partly because of its distance from other farms, from water, and from human dwellings. TGN also performed a technical analysis of surrounding wind patterns to minimize the spread of odors.

Because of recent changes to environmental regulations, TGN has decided to invest in a $1 million manure treatment system. This system separates the solids from the liquids; the solids are composted and the liquids purified. This system is a large investment for a start-up firm still years from profitability, but it will dramatically reduce smells because manure will no longer be spread around the farm. Bolduc explains, "It will minimize any agricultural issues. We are a pharmaceutical company at the end of the day. We are going to sell drugs and want to act as a good corporate citizen. So, yes it is worth a million dollars." TGN does hope, however, to receive some financial help from the Quebec government for this investment, since it would be the first of its kind.

FUTURE DRUG DEVELOPMENT

TGN is 10 years away from having a drug on the market, but it has considered some of the issues it may face. Duchesne feels that the drugs will have to go through strict clinical trials, even though these drugs (produced by different methods) have already been approved and exist on the market. She explains that there may be some exemptions, but because their development process is different, full testing and trials will likely be needed. TGN also recognizes that there will always be opposing forces between saving lives and making money, and employees speak of how this is something they will have to balance continually to be successful.

MECHANISMS FOR A TRANSPARENT PROCESS

Ethics for TGN employees is, Duchesne explains, "not just about following regulations. It's more global; something that everyone in the company must think about." Michael Dyck, Director of Transgenics and Cellular Biology, explains that "ethics has been present at TGN from the beginning. Being academics, it is just natural for us to want to talk about our science and to do it in an ethical manner." It is this transparent culture, instilled by TGN's founders and first employees, that has spurred the development of a variety of mechanisms that address ethical issues at TGN. These mechanisms are: the Animal Welfare Ethics Committee; engaging in stakeholder dialogue; the Disaster Plan; developing a regulatory framework; and internal culture and processes.

ANIMAL WELFARE ETHICS COMMITTEE

Besides running the test farm and performing many of the surgical operations on the pigs, another of Nicole Gravel's important jobs is to set up and chair the Animal Welfare Ethics Committee (AWEC). The purpose of TGN's AWEC is to discuss animal welfare issues and to make recommendations to the company on how to best take care of and ensure the safety and well-being of its animals. When Gravel joined the company, TGN was working under the University's ethical protocols, but there were no official animal welfare regulations in place within the firm. She explains that "being professors, it was natural for Marc-André and François to want to develop an ethics committee for TGN."

The process began with discussions with a veterinarian at Laval University who assisted the committee at the beginning. In selecting its members, Gravel wanted to include some internal and some external members, with at least one member representing the general public. Pothier attended the first meeting, but was advised by other members that his position on the committee could be a conflict of interest, so he is no longer a member.

The committee met for the first time early in 2003, and currently the members include three TGN employees and five external members (Exhibit 8.2). Its President is an external member, Dr. Renée Bergeron, PhD, who works on animal welfare issues as a professor at Laval University. The other external members are a physician, Dr. Bernier, who represents the public, two members of the Canadian National Research Council (NRC-IRAP),[14] and one other person who acts as a representative for technical personnel.

The decision to form a committee at TGN was hardly a decision at all. Bolduc explains:

> We knew we would have an [animal-welfare] ethics committee from the start. It wasn't a requirement but we wanted to be recognized by the CCAC.[15] We also

[14]National Research Council of Canada—Industrial Assistance Research Program. http://irap-pari.nrc-cnrc.gc.ca.
[15]Canadian Council on Animal Care.

EXHIBIT 8.2   **Members of TGN Biotech Ethics Committee**

External Members

Dr. Renée Bergeron, Ph.D., President of Committee
• Animal welfare researcher

Dr. Michel Bernier, M.D., LMCC
• Representing public

Mr. Alain Michard, CNRC-PARI
• External members where day-to-day activities do not involve or depend on the use of animals

Mr. Georges Lagace, CNRC-PARI
• External members where day-to-day activities do not involve or depend on the use of animals

Ms. Claudia Coulombe, TSA
• Representing technical personnel

Internal TGN Members

Dr. Marc-André Sirard, D.M.V., Ph.D., Founder and Vice President of Research at TGN
• Experience in using animals for experimentation

Ms. Pascale Duchesne, M.Sc., Director of Regulatory Affairs at TGN
• Representing views of regulatory bodies

Dr. Nicole Gravel, D.M.V., M.Sc., Director of Pilot Farm at TGN

• Manages installations and in vitro projects

---

knew we wanted representation from the public; someone who had a general understanding of biotechnology. A doctor is a good choice. Dr. Bernier asks good questions, like how what we are doing could affect human health as well.

Gravel is currently developing a mandate for, and giving structure to, the committee. The committee will meet twice a year, in addition to a yearly visit to the farm. They have already visited the pilot farm once, and made one suggestion: to put the sows that are in gestation in a pen with more room to roam around than the current cage. TGN plans to implement this suggestion at the new farm site. The committee will also receive an official evaluation from the CCAC and, if all is in order, the CCAC will act as auditors and grant the committee a certificate of good practice.

STAKEHOLDER DIALOGUE

When Pascale Duchesne was asked how TGN gets the public to understand about transgenics and its business, she responded, "By transparency!" She gave the example of the community meetings held in the two prospective communities for the new farm. TGN chose two sites based on several criteria: distance from other farms, proximity to Quebec City, and total area available. The next step was to approach the mayors of these two communities, who at first were very skeptical. Bolduc asked for 30 minutes to explain the project to them and after that time promised he would leave if they were not interested. He was asked to stay.

At TGN's request, the mayors organized community meetings to propose the project to local citizens. An announcement was placed in the local newspapers

EXHIBIT 8.3   **Examples of Questions from Saint-Tite-des-Caps Community Members at Meeting**

---

1. Will there be pollution caused by the transgenic products or by-products?
2. Will there be many trucks driving by?
3. How much smell will come from the farm? Compared to a chicken farm?
4. Will my seasonal property devalue?
5. Have you already bought the land? If the majority of the population of Saint-Tite says no to you coming, will you come regardless? Will your project create jobs for biochemists?
6. Where will the euthanized pigs be incinerated?
7. Can I get a guarantee that I will always be able to drink my water without turning transgenic?
8. Will residue from the technology get into our streams and affect the fish population?
9. If TGN were to sell the business, what would happen to the buildings?

---

and all who were interested were invited to come. Each meeting began with TGN's slideshow about the company, the science, and the proposed farm location. The presentation explained the science in an understandable yet thorough manner. TGN addressed several important issues: how it was dealing with the smells the farm might produce; that TGN would only allow its farm to reach 1000 pigs (at which point it would open a new facility elsewhere). It also pointed out that allowing TGN to move into their community would mean that no other pig farmers could move in nearby because of zoning bylaws.

The meetings lasted over 3 hours, and TGN members answered all the questions the citizens had. The majority of the concerns were about the smell and pollution created by a pig farm, and there were only a few concerns about the genetic modification of animals (See Exhibit 8.3 for examples of community questions). One woman said that she did not personally agree with the project, but that she was very happy that TGN came to answer their questions. Many citizens were actually pleased about TGN coming to their community because of the exclusion of other pig farmers (whose farms would likely be much larger), and the creation of jobs for local citizens. TGN chose its site in Saint-Tite-des-Capes, Quebec and will hold another town meeting to discuss TGN's plans for building on the site and to announce the recent decision to purchase the manure treatment system, which will further decrease odors.

Bolduc explains that getting community support for this project was vital. He says that "aggressive companies get recognized for being so because they do the cheapest thing versus the right thing. We have not done things this way. Marc-André and François are not like that, and it is written in our internal documents to not be like that."

DISASTER PLAN

One of TGN's internal documents is the company's disaster plan. Several of TGN's investors and board members asked questions about how TGN would handle untoward events. TGN discussed hiring someone to write a crisis-communication plan but, by chance, a former employee who was doing a part-time MBA offered to develop one for them as part of a course. In communication with key people at TGN,

EXHIBIT 8.4   **TGN Biotech's Disaster Plan (Condensed Table of Contents)**

the former employee organized a team that reviewed questions and issues that could affect TGN and came up with answers and processes to deal with them. It developed a list of people to contact if a certain issue arose and determined who should be the spokesperson in each case. This list included contingencies that any firm might face (like fires and natural disasters), as well as preparation for issues directly related to being in the transgenic pig business. See Exhibit 8.4 for a condensed Table of Contents for TGN's Disaster Plan.

REGULATORY FRAMEWORK

Duchesne remembers the feeling she had when she was being hired as the Director of Regulatory Affairs for such a young company. "The fact they were looking for a regulatory person right from the beginning showed me that they are conscientious," she says. Duchesne has a Masters in Microbiology and practical experience in United States, Canadian, and European regulations: before coming to TGN, she worked on regulatory approval for ozone sterilizers at one company, was responsible for quality control and environmental monitoring of biological insecticides at another, and was a manager of regulatory activities for vaccines and biological products.

Duchesne's full-time job at TGN is to communicate with the many regulatory bodies under which TGN's science, products, and processes fall. She explains that "the regulations are developing at the same time as the science." TGN's science is so new that there are currently no specific regulations under which TGN's processes can be classified. Therefore, Duchesne works to educate and collaborate with regulators both in Canada and abroad while TGN continues to develop its science and methods. Duchesne feels that Canada lags behind the United States in developing new regulations. She is in regular communication with several regulatory bodies, such as the CCAC for animal welfare, Health Canada for the eventual product, the International Conference on Harmonisation of Technical Requirements for Registration of Pharmaceuticals for Human Use (ICH),[16] Environment Canada and Environment Quebec for matters concerning the farm, and the Canadian Food Inspection Agency (CFIA). Duchesne explains that there are a lot of gray areas, and that coming up with the right regulatory approach is a learning process both for TGN and for the regulators themselves.

CULTURE AND INTERNAL PROCESSES

"There is some kind of feeling at TGN that people just want to do the right thing," explain Gravel and Duchesne when asked about the atmosphere at TGN. There is currently no formal process at TGN to guide ethical decision making but, with constant communication and close contact with the founders, a supportive and ethical culture seems to be well established at the company. Some of TGN's internal processes that help create this culture include its vision and mission statement, its hiring practices, its full-disclosure strategy, and its proactive approach to quality assurance.

Vision and Mission

The visions and values of TGN came from the first employees: Pothier and Sirard, Michael Dyck, Director of Transgenics and Cellular Biology, and Dan Lacroix, Director of Molecular Biology. Bolduc then initiated the process of writing them down and drafting a vision and mission statement (Exhibit 8.5). Pothier explains that "the morals of TGN come from our employees. We developed the vision of the company with our employees, and they continue to grow with us."

Hiring Practices

Michael Dyck explains that ethics is part of the selection process. "We consider attitude, personality and ethical issues. That last one is a key one," he says.

Duchesne explains the process of her own hiring:

> At first I was afraid to join because I didn't know the people involved and they were working with animals. But then after asking lots of questions and seeing the procedures and people they had in place already, I saw that these people are caring towards animals and I was comfortable with the way the products were being developed.

---

[16]ICH—International Conference on Harmonisation of Technical Requirements for Registration of Pharmaceuticals for Human Use. http://www.ich.org.

EXHIBIT 8.5 TGN's Mission and Vision Statements

Mission

TGN Biotech's mission is to create superior value for its customers, employees, and shareholders by applying its experience and expertise in the transgenic field to generate therapeutic products aiming to improve quality of life for humans and animals.

The SEMENESIS™ represents the first phase of TGN Biotech's transgenics expertise developed and focuses on the expression of biologics in the seminal fluid of pigs.

Vision

TGN Biotech will become a leader in biological therapeutic production. TGN Biotech's goal is to become a public company and in the process, create numerous highly skilled employment opportunities, both in research and in business development, expanding into newly defined fields and markets. These objectives will be achieved through the development of the SEMENESIS™ technology platform, as well as the development of new applications and processes for transgenic technologies.

Company Values

Founded by François Pothier and Marc-André Sirard, TGN Biotech has put into practice values based on respect for individuals, honesty, professionalism, harmony, and efficiency. Respecting individuals has the aim of developing a healthy and harmonious relationship with the members of the company, the scientific collaborators, and investors, as well as the citizens along their statutory values. Honesty at TGN has the aim of creating a truthful culture for scientific research purposes and plays an essential role in our interactions with our shareholders and our customers. Our professionalism aims to improve our scientific, financial, legal, and commercial methods and approaches. Harmony is an essential tool for a successful company and is practiced through respect, tolerance, and team spirit. Finally, efficiency acts as the engine that allows the objectives to be produced according to an established plan resulting in good progression of the company.

We all are significant:

TGN Biotech hires the best personnel and supports internal promotion, when possible. People are remunerated according to their performance without regard to other factors. TGN Biotech considers that its past, current, and future success rests primarily on its personnel, and that they are the company's most significant credit.

We are leaders:

We are all leaders in our fields of responsibility and we commit ourselves to providing high-quality performance. We have a clear vision of our role, our objectives, and the contribution expected from us. We are dedicated to attaining concrete results. We are active in the community and our specific sectors.

We are professionals:

We are honest, open and respectful with our customers, colleagues, and collaborators, and we treat them as we would like to be treated. We respect scientific and intellectual principles in the pursuit of our objectives, keeping in mind our evaluation of the risks.

We are winners:

We are determined to make a success of what we undertake. We strive to improve constantly. We deliver the goods and achieve the goals within the intended deadlines. We enjoy the work we do.

We have a concern for ethics:

We are convinced that the objectives that we pursue are important and that we must respect the scientific method and the rights of the animals and the environment. We adhere to ethical principles recognized by the federal councils of research and we take part in educating the public.

Transparency

Full disclosure is a policy that TGN follows as much as it can without compromising its patent applications or giving away its research methods and findings too early. Transparency and disclosure was TGN's strategy throughout the entire farm site selection process. Bolduc explains that "negative perception would be very hard to overcome. Reality is perception. Therefore, we are transparent." TGN has published its science and openly discusses it as much as it can for peer review. Pothier, Sirard, and Bolduc frequently engage with the media and are all very visible in the biotech community by sitting on Boards and being members of committees.

Pothier explains: "We are scientists and researchers. We are happy to discuss honestly. But, from time to time, we learned to be more cautious in order to protect our ideas." The founders of TGN have faced the harsh reality of the tension between academic publication (full disclosure and discussion) and the need for patent protection in the business world. However, TGN still feels that discussion is vital. It is more natural for these employees to be open with what they are doing than keep anything hidden.

Quality Assurance

Another key employee at TGN is the Director of Quality Assurance, Nancy Jolin. By making quality assurance a central process early in its existence, TGN is hoping to ensure that its processes will be ready to meet any regulation when the time comes. The question is whether these mechanisms—the animal welfare ethics committee, a process for communicating with the public, the detailed disaster plan, personnel solely in charge of regulatory and quality issues, a clear vision and mission statement, and establishing a strong ethical culture—will suffice as TGN continues on its path to product development and as new issues and collaboration opportunities arise.

FUTURE POSSIBILITIES

TGN is currently at the purification stage in the development of the process. Bolduc explains: "We do have money in the bank, which is very different from other colleagues with companies at similar stages. We've raised enough to maintain our development but not go much faster." He says that for now TGN will produce up to the raw material, but downstream would like to develop all the way to the end-product. In time, it would also like to go beyond generic drugs and develop therapeutics that currently do not exist. He explains that this may mean changing the structure of the company and forming a partnership with an established drug developer. He adds, "We are in love with the company, but not so attached that we won't pursue opportunities, such as an exit strategy."

WHAT NEXT?

TGN clearly has many years to go before its products will be ready to market and probably several hurdles, such as developing the purification process and successfully completing clinical trials, to overcome first. Pothier says frankly, "It will be a challenge to maintain ethics in the future. The ethical issues will likely shift and could move from what we have right now to different questions as we move toward therapeutics." TGN has not yet faced any opposition to its business model or scientific processes, but is it only a matter of time?

How should TGN move successfully from a small start-up firm based on two scientists' vision to a mid-sized pharmaceutical manufacturing company? How will the open communication continue as employees are hired who have not been at TGN from the beginning? What other ethical protocols should TGN implement as it grows? And, how will TGN balance structure and freedom to continue to be an exciting yet ethically driven biotech company? TGN realizes that it will face these issues in the coming years. Based on the mechanisms that TGN currently has in place, and depending on the success of these mechanisms to meet TGN's needs, the question will be whether and when will these approaches need to be revised, joined with, or replaced by, new strategies.[17]

---

[17]By the time this case was completed, Bertrand Bolduc had left TGN to join a company closer to where his family lives. Jean-François Huc has since joined as the company's Chief Executive Officer.

9

INTERLEUKIN GENETICS AND ALTICOR: AN UNLIKELY PARTNERSHIP[1]

INTRODUCTION

In April 2000, Dr. Phil Reilly left his post as Executive Director of Harvard Medical School's Shriver Center for Mental Retardation to become Chief Executive Officer (CEO) of Interleukin Genetics, with the dream of turning this small biotech firm into the world's "first personalized medicine company."[2] Reilly was nationally recognized as one of the leading ethical thinkers regarding the medical use of genetic information and had advised many of the top biotech companies[3] seeking to commercialize this technology, but had never worked full-time in the private sector. "I thought it was an opportune moment to make a change," Reilly recalled. "I have long been known as a proponent of genetic testing, and this was an opportunity to put into practice what I've been saying."

His timing could not have been worse. Just after he agreed to join the firm, the dot-com-driven stock market bubble burst, with a resulting sharp drop in investor confidence in the high-tech sector. In biotechnology, a harsh reality set in after the excitement over the sequencing of the human genome, as scientists and investors realized that turning the knowledge derived from mining the genetic code into profitable businesses was going to be a long and arduous process. The stock prices of the leading genomics companies—Celera, Incyte, Human Genome Sciences—collapsed,

[1]This case study was written by David Finegold based on interviews conducted on October 1, 2003 at Interleukin in Waltham, Massachusetts. Unless otherwise specified, all quotations that appear in the case were obtained during these interviews. Interviews were conducted by Finegold, Jocelyn Mackie, and Cécile M. Bensimon. This case study is one in a series produced under grants funded by The Seaver Institute and Genome Canada to study models for ensuring ethical decision making in bioscience firms. We would like especially to thank Rahul Dhanda without whose help this case study would not have been possible.
[2]Interleukin Genetics, *A Healthy Evolution,* 2002 Annual Report.
[3]For example, Millennium, Genzyme, Myriad.

EXHIBIT 9.1   **Key Milestones in Interleukin's History**

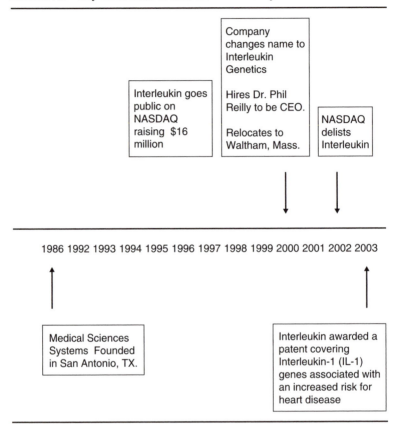

and they were forced to reinvent their business models. Overall, the Biotech Index of publicly traded companies lost more than 80% of its value from its peak in 2000 until it began to recover in 2003.[4]

Interleukin suffered along with its peers. Nearly out of money, it dropped to a low of two employees in the spring of 2000. With Reilly and a new Chief Financial Officer (CFO), Fenel Eloi, on board, however, the firm was able to raise $5 million to supplement a $5 million financing round that had been completed when he agreed to become CEO. By the end of 2002, however, this money ran out. Interleukin was delisted from the NASDAQ and needed some new source of funding to survive (see Exhibit 9.1).

The company had been trying unsuccessfully for several years to interest a large pharmaceutical company in setting up a pharmacogenomics partnership. Pharmacogenomics was an exciting new technological advance that offered the potential for companies to develop drugs based on new knowledge of how variants

[4]Burrill S. *Biotech 2003*. San Francisco; Burrill & Co.

of particular genes influenced drug efficacy and metabolism, enabling therapies to be tailored to individuals' genetic profiles and ushering in an era of more personalized medicine. Interleukin was seeking to use its expertise and intellectual property related to genes associated with inflammation, specifically those found in the interleukin-1 (IL-1) gene cluster, to help a pharmaceutical company develop safer, more effective drugs by linking them with individual patients' genetic profiles. However, a big hurdle that few companies had cleared was figuring out how to make money from a therapeutic that could be sold only to subpopulations of patients with particular genetic profiles.[5] Unable to divert the pharmaceutical firms from their focus on developing blockbuster drugs (with markets of $1 billion or more) to a more targeted approach, Interleukin's top management team brainstormed other potential ways to get its technology to market. Chief Medical Officer (CMO) Kip Martha, described the process:

> We made a list of stakeholders who might drive the move toward more pharmacogenomics technology. Pharma was a key stakeholder, but didn't want to change because of the negative impact on their bottom line. So we listed others: patient-advocacy groups, lawyers, regulators like the FDA [Federal Drug Administration], medical societies, small biotechs who wanted a niche market. We also put nutriceutical firms on our list…since they look for technology that captures people's attention and makes them believe they really should be using their product. [Nutriceutical firms] have much faster product life-cycles, regulatory hurdles are much lower, and they are all looking for some type of proprietary technology to differentiate themselves from competitors. Instead of just selling the same products, like St. John's Wort or green tea, they could also offer a test that allows consumers to customize a product-line tailored to their needs.

Nutriceuticals was the new term applied to drug-like products derived from food plants that have claimed health benefits. The recent interest in nutriceuticals had spawned an even newer field of study called nutrigenomics, which applies genetic knowledge to the development of nutriceuticals. Dr. Ken Kornman, Interleukin's co-founder and Chief Scientific Officer (CSO), explained Interleukin's interest in this field: "There were already some nutritional compounds that were known to modulate inflammatory genes. It not only made sense conceptually, but also had practical compounds that could modify the target."

Following this shift in its strategy, Interleukin began approaching companies that might be potential partners in the nutritional area. Kornman described what happened next:

> One large firm with a big pharma division became very excited about [becoming a partner with us]. We worked out a contract with them to develop the area of nutritional genomics. At the same time, unknown to us, a separate part of their upper management was discussing a deal to completely sell off the pharma division.

[5] One notable exception was Genentech, which had developed and marketed the anticancer drug *Herceptin*. This drug is prescribed for patients whose gene test shows that the cancer is susceptible to the drug. Only about 30% of breast cancers fit this description and, in women with this cancer variant, the drug is remarkably effective and can even cure the disease. For the other 70% of breast cancer patients, the drug has little effectiveness and is not prescribed.

Our deal would have been huge for us, but it was not large enough to catch the attention of their top management. The same week we were due to sign this agreement, they made the other announcement and our deal was off. We'd developed a whole business model for nutrigenomics with this firm, but weren't able to go forward.

Just when it looked as if Interleukin's options might be running out, the company was approached by a new potential partner: Alticor, the parent company of Amway. Representatives of Alticor had heard Kornman speak at a conference in the Netherlands about how personalized medical information would transform healthcare and nutrition. They approached Interleukin with an intriguing offer: to form a relationship that would combine Interleukin's proprietary pharmacogenomics technology with Alticor's capability to develop nutritional products to deliver customized nutritional solutions to consumers.

As the discussions progressed, Alticor offered to purchase a controlling interest in Interleukin and to make the substantial investment needed to complete the development of the technology. Alticor recognized that the new business of nutrigenomics was potentially large but also filled with ethical challenges that it had not previously faced. It was interested in Interleukin not just for its technology, but also the ethical approach that Dr. Reilly and his colleagues had adopted in this emerging industry.[6]

Interleukin's leadership team was intrigued by this offer but had reservations about joining a company known for its multi-level, direct-to-consumer marketing of a wide array of household goods and nutritional products. Explained Kornman:

> Our board has fiduciary responsibilities to shareholders. It has to look at the very practical meeting of those duties and, therefore, supported the concept as the best deal available. Our management team naturally had a somewhat different perspective: we have to be committed and excited to come to work each day. To do that, we have to feel we're doing something valuable. We had to get comfortable as a team that it was an ethical group to deal with. We looked at issues with multi-level marketing and tried to sort out what was real and what was not.

Alticor made the deal contingent on Reilly, Eloi, Kornman, and Martha remaining with Interleukin, leaving them with a difficult personal choice. Said Reilly: "They would not invest in our company unless each of us signed a three-year employment contract. Never before had I faced a situation where my decision about a job would directly affect so many other people." What should Reilly and his leadership team do?

COMPANY ORIGINS

Interleukin was originally founded in 1986 in San Antonio, Texas as Medical Sciences Systems, with a focus on developing computer simulations for chronic diseases. At the time, Kornman was Professor of Microbiology and Immunology and Professor and Chairman of Periodontics at the University of Texas Health

---

[6]Interview with Rahul K. Dhanda.

Science Center in San Antonio, and a consultant to several major oral care and pharmaceutical companies. The company was based on observations Kornman and a few colleagues made regarding the powerful role that inflammation plays in the pathology of disease and the variability in inflammatory responses among people with similar diseases.

They looked for an appropriate genetic technology, which they found in the molecular biology lab of Professor Gordon Duff at the University of Sheffield in England. The company entered into a sponsored research agreement with Sheffield and Duff, a world expert on cytokine genetics like the IL-1 gene cluster. Under its terms, Duff's lab became, in effect, Interleukin's outsourced research arm, licensing to the firm selected patents arising from its research in return for research funding, royalties, and a small equity-stake in the company. Duff's research and others' revealed that the IL-1 gene modulated the body's inflammatory response, and that inflammation is part of a critical pathway for common diseases. For instance, it is now widely accepted that inflammation plays a significant role in cardiovascular disease.[7] The standard indicator for this disease has often been considered to be cholesterol, but research shows that high cholesterol accounts for less than 50% of heart disease.

Using the intellectual property generated by Duff's group as a foundation, Interleukin raised money from angel investors and some contract research agreements, and began searching for the right commercial application. Said Kornman:

> By 1997, we had sufficient data to develop a genetic test in one of two areas: either periodontal disease or inflammation related to cardiovascular disease. Our business people kept saying, 'Focus on cardiovascular disease,' because the market was so large. But Myriad [Genetics] was just releasing a genetic test [for the BRCA1 and BRCA2 genes linked to breast cancer] and we were hearing such negative comments about it in the marketplace, we were terrified. So we chose to go with the dental market that was at other extreme in terms of risk to the consumer.[8]

With its first product—a test for genes associated with severe gum disease—now identified, and the promise of many further offerings based on its discoveries that IL-1 played a key role in the inflammation process, Interleukin was able to go public in November 1997, just before investor interest in biotech initial product offerings (IPOs) waned (see Exhibit 9.1). The company raised $16 million to fund its clinical trials and marketing efforts. This money was used to hire a sales force for the first product and to increase the workforce to around 40 people.

---

[7]Libby P. Atherosclerosis: The New View. *Scientific American.* May 2002:47–55.

[8]A number of ethical concerns were raised regarding Myriad's test. Among them were that not enough was known about the penetrance of the genes and, hence, the reliability of a gene test in predicting cancer. Neither was there enough assurance that physicians could adequately interpret test results. Sufficient privacy and antidiscrimination protections were not in place to keep the test results from being used against women to deny them health insurance or employment. Finally, there were no proven medical options available to women to reduce the risk of breast cancer detected by the test. This last problem was called the "so what" or the "therapeutic gap" issue—a woman, learning that she carried a breast cancer gene she could do nothing about, was likely to experience more suffering than medical benefit.

EXPLORING THE ETHICS OF
GENETIC TESTING

As Interleukin pushed ahead to develop its first genetic test, its leaders recognized that the characteristics of the IL-1 genes meant that its product development approach needed to differ from many of the other firms entering this new field. In contrast with some rare genetic diseases, such as Tay Sachs and cystic fibrosis, in which a single genetic mutation was associated with a distinct and severe illness, the diseases targeted by Interleukin were influenced by many genes acting together that produce less distinct differences in inflammation among individuals. Furthermore, IL-1 genes were associated with inflammation abnormalities in many common diseases, such as heart disease. Not only were Interleukin's leaders uncertain about how to develop genetic tests to predict these pathologies, but they also wanted to avoid the ethical criticism leveled at Myriad's marketing of the breast cancer gene tests

> The more we got into genetics in the mid-'90s, recalled Kornman, the more we realized we were looking not at rare genetic disorders, but rather looking at more subtle differences in common diseases. We weren't sure what to do with this, since there were no examples out there on how to develop commercial products in this area, or how to deal with regulation and other issues. Frankly, we were all excited about the science and not at all knowledgeable about the ethical and commercial area.

Cautioned by the controversy surrounding Myriad's breast cancer test, Kornman sought ethical advice.

> We benefited from the fact that we were naïve. We were ignorant of the practical issues associated with privacy, and concerns regarding discrimination, but we realized quickly we needed to understand them… I started contacting people quoted in the press who were cited as potential experts in this area. But every time we got into specifics on testing and the issues of what information was appropriate for consumers, all their thinking was predicated on rare genetic disorders, where a single gene can be a diagnostic as opposed to what we were working on, a prognostic test where identifying a gene might be useful for well people in maintaining their health. We knew enough to know that what we were hearing didn't make sense for us, but also knew we didn't know how to go forward.

Kornman spent close to a year "picking people's brains" and during that process was guided to Dr. Reilly at Harvard. Reilly was a JD/MD who had become an internationally recognized expert on the legal, social, and ethical issues of genetic testing. He was a founding fellow of the American College of Medical Genetics and had acted as advisor to many of the first pharmaceutical and biotech firms to enter the genomics field. Kornman went to visit Reilly:

> I quickly realized that he was working with Myriad to help them try to deal with their issues, and also, more importantly, detected that he understood the difference between genetic variations that modify common diseases and those that are sources of rare disorders. He also conveyed a strong belief that our area was where genetics was going to go. In the mid-1990s, he seemed to be a rare knowledgeable and rational voice in this field.

Reilly became a consultant to Interleukin in 1997 and, 2 years later, was asked to join the Board. Kornman explained the unusual decision to make an ethics consultant a director:

> We quickly realized how important his expertise was (to our business model). We decided we needed someone in a governance position. Board members are treated very differently from consultants. A firm may listen to what an expensive consultant says, but often doesn't do anything with this advice. They are much more likely to listen to the opinions of a board member.

Kornman went on to point out the difference between having an ethicist on the Board of Directors and creating a separate Ethics Advisory Board (EAB):

> I pay a lot of money for advisory boards. They can be very valuable, but it is a totally different dynamic. As soon as Phil moved onto the Board, the firm had a totally different approach to how we viewed these issues. You don't get that value from an advisory board. It doesn't permeate the thinking of everyone in the firm. Even with a small firm, usually an EAB only relates to a few people. It is rare to have the CEO or CFO attend. At the Board of Directors we have all key leaders there and their resources and authority come from the perspective of being on the Board. Seeing how Phil worked as a consultant, it seemed like such a rational thing to bring him on to the Board for a firm that wanted to do genetic tests and be in the genetics area. I saw no difference from having a world-class scientist on the Board. We wanted to have world-class advice in all key areas: science, business and ethics.

FAILED EFFORTS TO BUILD A PERSONALIZED MEDICINE BUSINESS

Interleukin launched its first product, the gum disease test, in 1998, but it failed to generate significant revenue. "The dental community wasn't quite prepared for it," said Rahul K. Dhanda, Manager of Business Operations. "It was the darling of conferences, but putting it into practice was much harder." Kornman agreed: "It didn't make sense from a business-model standpoint. We have a test that science has strongly validated, [but] we marketed it to specialist periodontists who were already seeing people with severe disease and struggling to know what to do with this additional information. The consumer saw the value, but the dentist couldn't figure it out." Interleukin continues to sell the genetic dental test in the United States and Europe and is working on improvements that will make it more marketable.

When Reilly became CEO, however, the firm turned its focus to pharmacogenomics, seeking to partner its genetic knowledge of inflammation with drug companies, to help them determine where their drugs were most effective. This also proved a difficult sell, explained Kornman:

> Like other early players in this area, we had the experience of sitting down with people from R&D [research and development] and having them get very excited about the technology, only to run into the reality that their firm's business model is dependent on blockbusters. They understand the need to move in a different

direction, but it is very hard to make the transition in their system, with the large pharma bureaucracy and current blockbuster mentality.

Pharmaceutical companies typically wanted drugs to reach the largest possible market and designed clinical trials to prove that a drug was safe and effective for wide populations of diseased patients. Population averages governed the drug-approval process, and differentiating groups of patients who responded differently to products could only reduce the size of markets. This raises a major ethical issue, according to Martha, the CMO:

> The pharmaceutical companies know they are making drugs that only work for a small percentage of people,[9] but their business model is to get as broad an indication as possible from the FDA and then to spend hundreds of millions of dollars marketing it to as many people as possible, even when they know it won't work for many of them. For a small fraction of what they spend to market these drugs, they could use this technology to analyze genetic variations to see for whom the drugs would really work best.

Martha expressed the frustration and personal ethical struggle that has resulted from the pharmaceutical companies' position:

> What is the right balance between focusing on things that will bring in revenue to help the firm survive versus fighting the battle against pharma to push our technology forward, because I think they should be forced to adopt it? I know I could beat my head against that wall for a long time and not make progress, but we need to take longer term view.

PUTTING ETHICS EXPERTS IN CHARGE

Recognizing how important ethical issues would be to its success, Interleukin's Board made the unusual decision to offer the CEO position to Reilly. Through his work as a consultant and director, the Board knew that Reilly brought an unusual mix of medical, legal, and medical ethics knowledge.[10] Running the multimillion dollar Shriver Center and consulting for many biotech and pharmaceutical companies also meant that Reilly had significantly more managerial and industry experience than most ethics experts. However, he still had never worked in, much less led, a for-profit company. The Board might have been expected to have some reservations about appointing an academic ethicist as CEO, but Kornman said this was not the case. He explained how the decision to turn over the reins of the company he founded to Reilly was made:

> I had more concerns than the rest of the Board. I know what a CEO has to go through on a daily basis and had great concerns that it wasn't part of his experience. I had seen him operate long enough to be impressed by his dedication,

---

[9]An academic partner of Glaxo SmithKline admitted in 2003 that most drugs sold by pharmaceutical firms are only effective for 30 to 60% of the population, depending on the indication.

[10]Reilly was a past president of the American Society of Law, Medicine, and Ethics, and ex-Board member for the American Society for Human Genetics.

brainpower and communication skills and to know he'd be excellent at key aspects of the business. But I had an explicit conversation with him before he was hired that addressed this issue. I asked him how he would cope. He said it was simple: 'As soon as I'm appointed I plan to go and recruit a highly experienced CFO who has the business experience I lack.' Everyone was very comfortable with this solution. He brought other key skills that more than made up for the experience he didn't have.

Reilly was true to his word. Soon after joining Interleukin he hired Eloi to be CFO, followed by Martha to run medical operations. Together with Kornman, they essentially relaunched the company, moving it to the Boston area to be closer to other biotech firms, renaming it Interleukin Genetics, Inc. to signal its focus on the genetics of inflammation, and raising needed capital. Reilly's presence also had a rapid impact on how the company was run and how it approached ethical issues. "It immediately changed the perspective of the firm," said Kornman. "All of our officers were hearing at every meeting from someone whose life's career had been focused on these issues. It imprinted the company forever. I personally feel very fortunate we stumbled in that direction."

Reilly gave his own perspective on the difference in his influence on the firm in his roles as consultant, director, and then CEO:

> Although I first joined Interleukin as a consultant in ethics and public policy, I would say my influence in that arena was significantly greater after I became CEO. I was probably least influential as a Board member because ethics issues did not tend to command the Board's attention, as it is focused on financial problems.

Another effect of hiring Reilly was to make the firm, which was still struggling to find a way to reach profitability in an emerging, high-risk segment of the biotech industry, more attractive to other scientific and business leaders. Martha explained:

> When Phil Reilly joined, it sent a positive signal to me. Then the fact he hired Fenel as well made it more attractive… Phil was interesting because he had no business experience but lots of legal and ethical background that would be vital in this area. I felt I knew about the scientific applications and would always try to do the right thing personally, but didn't know the specifics of the ethical area. I knew there were a lot of naysayers out there that we would have to deal with. I wanted to be able to understand that culture and its implications for us.

With his top leadership team in place, Reilly then made the decision to hire Rahul Dhanda in May 2001 to play a key business development role and deal with day-to-day ethical issues. Dhanda, the author of *Guiding Icarus: Merging Bioethics with Corporate Interests,* recalled how he made the transition from science to a focus on the ethical issues in bioscience business, combined with work on business development:

> Right out of college I began working in a lab at Harvard, and while I was there, I spent most of my time developing DNA-screening technologies for cancer. The technology we developed made it too simple to look at DNA, I thought. I began to ask myself a lot of questions about what I was doing, and the better I got at science, the worse I got at answering them, so I spent a lot more time reading about

the history and philosophy of science. At some point later on, when I was at a company doing research, I realized that the field of science studies was starting to look at corporate work in biotechnology. I became concerned with the social impact offered by corporations, and while bioethics was not on point, I felt it was closest—if the discipline could expand its interest to be more thoughtful and less cynical about industry. It seemed a natural place to develop my background to contribute to the social debate through a voice in industry.

Reilly noted that Dhanda's blend of responsibilities was inspired by how Steve Holtzman had very effectively combined ethics and business development roles during Millennium's start-up phase. Reilly recalled:

Rahul approached me with the idea of starting a bioethics consulting firm. I didn't think it would work, or if it would, I could do it on my own. But I thought, here's a guy I can help. Hiring Rahul gave us a disproportionately high level of attention to bioethics in this small company. But we are in a field that has relatively weak scientific credibility… We're hampered already by fly-by-night companies. We must distinguish ourselves from them. We can't just have an appearance of bioethics, but need to have it incorporated into product development.

With Dhanda's hiring, Interleukin has been in the unusual position of having two ethics experts making strategic and operational decisions. Perhaps because it has Reilly and Dhanda in business leadership roles, or perhaps because it does not yet have a major product on the market, Interleukin has thus far not attracted the same level of criticism as some of its competitors who are already marketing genetic tests directly to consumers. The additional ethical questions that these leaders raise during strategic discussions are not perceived by other members of the top-management team to have slowed Interleukin's progress. Just the opposite appears to have been the case, according to Kornman:

We structured our thinking based around the practical issues Phil taught us that we'll confront when we get to a product. None of the issues were ever presented as barriers, stumbling blocks or even challenges, but rather, because of Phil's perspective, we discussed them as opportunities to develop viable, valuable products that could be done in such a correct manner that the firm has long-term sustainability.

Dhanda contrasted this approach with the way ethics is seen by many firms in the bioscience industry: "Companies look at ethics as the mode by which you'll be restricted. The way we look at it is as the filter by which you solve a problem. At no time should it keep you from doing something. The approach is to look at how to do it right. Ethics can't prohibit you from doing something, but there are interests that you have to take into consideration." Martha agreed, explaining how Reilly and Dhanda have affected day-to-day decision making at Interleukin:

They are very open to talking about things. I will ask them if ethics is just about knowing the right thing to do, and they'll rise to the challenge. When we get to specific issues like cloning, it's a wonderful environment to be in. It's great to be able to engage in informed discussion about the issues. As we think about our trials and the studies we do, there are a lot of negative media stories out there about getting into people's privacy and genes. It's easy, if you're someone who wants to do the right thing, to get frozen and paranoid about not wanting to move forward.

It sounds so scary when all the publications are pointing to the dangers. Anyone I talk to at dinner parties or on the soccer sidelines jumps to these issues right away. They are worried that an insurance company will find out about their genes. They have only sound-bites from the media to go on. Having these guys here really helped me think through where I stand on these issues... and to be able to explain that to others. We have the ability to help people understand at a level we've never had before what personal vulnerability to disease is, and how best to deal with it. Without them here, it would be much harder. Do I really want to be a pioneer in this area? It's very valuable to have them here to raise our confidence that we're doing the right thing.

He went on to give a concrete example of how Reilly and Dhanda had influenced Interleukin's decision to approach one of the largest HMOs, Kaiser Permanente, regarding a possible partnership:

I got in my mind that medical-insurance companies and HMOs like Kaiser have a wealth of information about the outcomes for patients of different therapies based on their medical records. Otherwise you have to go to pharma to get information on how people responded to different drugs and treatments. But the idea of approaching them when everyone is saying insurance companies are the bad guys seemed crazy; I thought it might be like sleeping with the enemy. Without Phil and Rahul, we might not have done it at all.

Dhanda led the negotiations with Kaiser that resulted in an agreement that provides Interleukin with access to Kaiser's rich medical records for research purposes. In some cases, however, the impact of the ethical perspective that Reilly and Dhanda have brought has caused Interleukin to not pursue potential partnerships, as Kornman recalled:

We were presented with an unusual situation a few years ago. We had technology that we knew was valuable for a large biotech firm that had already submitted an application for FDA approval of a drug. We had data that was very important and the company understood how important the data were to the marketing of their drug and application. It was potentially a very important business opportunity for us. Somewhere along the way, it became clear to us that there were decisions that the big firm had made that in our minds weren't consistent with welfare of patients and ethical conduct. With Phil's ability to crystallize issues for us, we didn't walk away, but we raised issues very explicitly with the company, with the implication that unless they were resolved to our satisfaction we weren't comfortable working out a deal with them. And with that they blew us off and walked away. Our Board could have said, 'Boy, are you guys crazy? This could be a company-building deal and you blew it.' But they didn't. They accepted that this was the only way for us to do business.

Reilly and his leadership team's approaches to ethical issues were also clearly reflected in Interleukin's internal operations. Eloi recalled when the company faced a financial crisis in early 2003, and how the team handled it:

I came from an environment where you don't give employees too much information. Dr. Reilly said, 'Let's tell the employees everything there is to know.' It was the right thing to do. They knew when we were going to run out of money. The pressure was on me! We didn't lose anyone. Everyone felt we were in this together.

Martha has found the working environment is very similar to Genentech, one of the world's first biotech companies, where he started his career.

> It starts with things like the egalitarian nature of the organization and internal culture. At Genentech the CEO, Art Levin, and Curt Robb before him, set the tone. It was a place where the CEO competes for a parking place with a lab tech, and stands in line for lunch with you... There's a lot of ability for people to speak up and express opinions, to use their knowledge and expertise. Not a lot of artificial structures or walls. It's a very dynamic environment, where there is a lot of opportunity and visible rewarding and recognizing of success, with parties at the end of the day where top management shows up to pat people on the back... I found that for certain types of people, it was a very successful environment. A lot of that has spilled over to how we run this firm. Phil is the right kind of guy for it: we all know his wife and kids. We can all walk into his office to chat. That's how things are run here, very positive; it gets people through the day. I've worked in places where the prime motivation is fear; there is essentially none of that here. We're a team and all in this together, sink or swim.

APPROACH TO ETHICAL ISSUES

With ethics so much a part of how this small firm operates, Interleukin has not felt the need to establish some of the more formal mechanisms for ethical decision making used in other bioscience firms. "It has not been necessary to put in an official scheme," said Dhanda. The firm has allowed Reilly and Dhanda to spend a significant percentage of their work hours on external ethics activities, enabling them to serve on important ethics advisory and policy committees, to publish extensively on ethical issues, and to give dozens of talks at ethics conferences, courses, and workshops. Dhanda also runs an internal ethics-speaker series. According to Dhanda, "The most recent speaker was Chris McDonald (a Canadian professor and bioethics expert) who discussed ethics at the corporate level and how that relates to bioscience."

Even without many formal mechanisms in place, Interleukin has had to be prepared to deal with a number of ethical issues inherent in its business. "Ethics was a concern for a number of reasons," said Reilly.

> In the background were concerns about the risks of discrimination [that can result when people test positive for a disease-related gene]. I believe the risks are overplayed in medicine, but they are very real in the minds of many people. There are also concerns about informed consent. The biggest question concerns what to do when a test designed to test for one risk [identifies] results for another risk.

Many of the scientific and ethical questions that Interleukin and its competitors in the emerging nutrigenomics and nutriceuticals field needed to address were identified in a 1997 meeting of Canada's National Institute of Nutrition:[11]

Efficacy: Does the product as consumed retain the active form of the functional component? Does consumption of the food or food component produce the beneficial effect that is claimed?

[11]http://www.nin.ca/public_html/Publications/Rapport/rapp2_97.html.

Active Agent and Dose: Is the chemical nature of the active component known, as well as the amount required to produce the desired outcome?

Evidence: What evidence is required to establish the benefit and support the claim? Would the amount and type of evidence vary with the type of claim, such as claiming that a component aids digestion versus helps reduce risk of a disease?

Harmful Effects: Might consumption of the material produce harmful effects, including such undesirable short- or long-term side effects as increased risk of food intolerance or allergy?

Public Health: Do claims for these foods assist or hinder public health strategies aimed at promoting the health of Canadians? Is the claim compelling enough to lead to action in the right direction, without losing appropriate perspective relative to other risk factors?

Displacement: Does the functional food present new hazards by displacing traditional foods from the diet? Does it discourage a "total diet" approach that takes into account the balance of all foods in the diet?

Self-Treatment: Are the risks associated with self diagnosis and treatment increased? Will availability of the product encourage self-treatment that may delay an individual seeking professional medical help?

To prepare for these and other ethical challenges, Dhanda developed a list of the ethical issues Interleukin could face and the best ways to respond to them (see Exhibit 9.2). For example, the company could protect patient privacy through encryption of medical records and help patients cope with the decisions following the results of their tests by providing access to trained genetic counselors. These types of approaches are particularly important because there are no clear regulatory guidelines for nutrigenomics firms to follow. This meant, for example, that the Food and Drug Administration (FDA) did not require the company to conduct clinical trials, as it does with drugs that claim medical benefits. "We are self regulating," said Eloi. Dhanda explained further:

> This is a new field. We could have had a product out and marketed already, based on the information that already existed [but we chose not to]. The only regulations for home-brew tests[12] are that you have to have an accredited lab. Accreditation for genetic testing doesn't exist. The AMA has set standards for how they would like testing done that will guide us. We are trying to avoid the problem with other groups, like Genewatch [an industry watchdog], that our competitors have run into... If we are going to bring the right level of science to this field, it needs the same scientific validity as other fields. We need industry leadership and we want to differentiate ourselves from others in the industry. If there isn't industry leadership, you invite regulation, and you deserve it.

---

[12]The term "home brew" refers to diagnostic gene testing services by commercial firms and other laboratories that test tissue samples using their own probes and reagents manufactured in-house. These testing services are not regulated by the FDA, in contrast to gene-testing kits that are considered medical devices and regulated by the FDA as such.

EXHIBIT 9.2 **Bioethics Issues and Responses**

Bioethics at Interleukin

Privacy	Encryption
Confidentiality	Informed Consent
Discrimination	Personal access/release
Gene Patenting	Academic Collaboration
Population Targeting	Diverse trial inclusion
Corporate Rivalry	Develop relationships
Psychological Impact	Counseling
Distributing rewards among stakeholders	Rapid development of product for patients

Dhanda believes this approach could provide Interleukin with a competitive advantage: "There is no *Good Housekeeping* seal of approval on these products," he said. "We want to be the company that has that seal."

PROPOSED ALTICOR DEAL

It was Interleukin's ethical approach, as much as its technology, that attracted Alticor. "Reilly immediately stood out as a huge asset in the acquisition," said Alticor Senior Marketing Manager, Bill Ardis. "His attitude about nutrigenomics meshed so well with Nutrilite's [the name of Alticor's nutritional supplement product line] attitude about nutrition and products in general—simple things, like not recommending products just because they may lead to more sales, and that we can't sell anything with a dosage level that wouldn't produce benefits."

Alticor's merger and acquisitions unit had set up a team to look at firms in the nutrigenomics field that identified Interleukin. Ardis explained the rationale for Alticor's move into this field:

> We have traditionally been a very vertically integrated firm: we run farms in Brazil, Oregon, even a small one in California. That was our competitive advantage.

The problem is it's easy to duplicate the products we offer and we were losing the uniqueness of our story and our competitive edge. The area of nutrigenomics was seen as the wave of the future, the possible next big step in nutrition and a way to differentiate ourselves.

As with pharmacogenomics, the idea behind nutrigenomics was that by understanding differences in individuals' genetic profiles, health histories, and behaviors (exercise, drinking, smoking), it would be possible to identify disease risk factors and tailor a diet, including specialized supplements, to help reduce these risks. In a new industry that was scientifically controversial, Alticor was concerned that the opportunity might be compromised by competitors it perceived to be behaving unethically. Said Ardis:

> There are a lot of smaller supplement companies out there who are 'poisoning the well' for the whole industry, making outrageous claims. We have to compete with them for business without doing that. When we do customer surveys, the main thing that comes out is the trust the end-user has in our product. We would never do anything that would damage that trust.

Perceiving such a close fit, Alticor made an offer to Interleukin to purchase a controlling interest in the company in 2003. With Interleukin about to run out of cash, the offer—to purchase $7 million in stock, a 2-year, $5 million R&D agreement, a $1.5 million line of credit, and a possible $2 million milestone payment based on validating the efficacy of nutritional supplements—represented a badly needed lifeline. In return, Alticor would gain control of Interleukin's Board and run the company as a separate subsidiary.

Interleukin's leadership team met to debate the merits of joining Alticor. Reilly described the context: "Alticor had found us. There were not a lot of choices as a struggling biotech company. Many on Wall Street said, 'We like your story but we can't fund you.' The initial reaction to the deal was not positive. They (the analysts) didn't know Alticor." Some Interleukin team members also had reservations: "A nutritional company is not really what I was trained for," said Martha. "But we want to use our technology to prevent disease and we need to have a distribution channel to reach people and this provided one." However, he was left wondering: "Is this is the best way to spend my time?"

Should Interleukin's leadership team agree to the deal? If Alticor owned a majority of the company, would they be able to retain control over how their technology was developed and sold? As Eloi summed it up: "The thing that kept me up at night was worrying whether their independent distributors, individuals without our experience in the field, would have our commitment to ethics."

INTERLEUKIN GENETICS AND ALTICOR: AN UNUSUAL PARTNERSHIP

In Spring 2003, Alticor purchased a controlling interest in Interleukin Genetics and replaced four of the directors on Interleukin's Board, including the founder,

Ken Kornman, with its own executives. In effect, Interleukin became a separate R&D subsidiary of Alticor. Interleukin, however, set certain conditions before going ahead with the deal. "Our integrity and ethics are not for sale," said CFO Fenel Eloi. "The only way we would do this with them was if they signed up for complete scientific rigor, agreeing that we [would] put their products through all the tests. If they wouldn't sign up for that, we wouldn't do the deal."

Interleukin also retained its status as an independent, publicly traded company. "This was done for business reasons," explained Eloi. "For us to continue to attract good people, we have to be a public company... We told them we were not interested in being totally acquired, that we would have to stay public and keep some freedom."

Interleukin CEO, Phil Reilly, described his own reasons for proceeding with the deal: "I was the CEO of a publicly traded company; my job ethically was to find a sustaining company to join with. I didn't have a choice... I drew certain lines. I promised I would stay for 1 year. There also had to be enough money so we could do the studies we wanted to do and we had to have an important voice in product launch." He noted that the way Alticor had approached the negotiations, despite the dire state of Interleukin at the time, gave his leadership team confidence that their partner would keep its promises: "They sustained us with loans during the process. They could have let us go out of business, bought the stock, and offered us money to start a new company. But instead, they did negotiate."

He was also encouraged when he did some checking on Alticor and discovered that several of the members of his Scientific Advisory Board were already familiar with the company, thanks to the Van Andel Institute. This was established in 2000 by Amway co-founder, Jay Van Andel, and his wife, Betty, in Grand Rapids, Michigan, to conduct research and education using the latest genetic technologies to improve human health, with an initial focus on cancer.[13]

CREATING AN ETHICAL APPROACH TO NUTRIGENOMICS

Together, Interleukin and Alticor have set out to take as ethical an approach as possible to developing their first products in the new and highly sensitive area of nutrigenomics. This strategy begins with Reilly's leadership, according to Alticor's Bill Ardis: "When there are several ways to do it, he will always ask, 'Which way would be the high road?' and that's what we'll take. It's a perfect fit with how we do business."

The biggest ethical commitment from both sides was the decision to invest 2 years and millions of dollars in clinical trials for their first product, even though such trials are not required by the FDA and some of their competitors have launched products without trials. They have chosen to focus the trials on 6 potential

[13]www.vai.org.

ingredients that laboratory tests suggest may reduce inflammation for individuals with certain IL-1 genetic profiles. The trials will test whether the combination of Interleukin's genetic test and specific nutritional supplements produce the intended reduction in inflammation in humans. Said Martha:[14]

> We're pushing from our side to have the trials be as convincing as possible. They are sort of like what one would do for the FDA: they are placebo-controlled and randomized. But they may not be as big or cover all the ethnic and demographic groups that the FDA would require. But we are making the scientific point that they are well designed, and that we are spending more money and time to get it done than is required. A lot of our motive to push for it comes from our ethical commitment to do well, but the pitch we make to our partner is that ultimately it will be better from a business point of view, since we will both be under hard scrutiny, and need scientific results to fall back on. I personally feel much better from the ethical point of view. I need clear data to show it does work to feel comfortable with what we're doing.

Alticor shares a similar philosophy, according to Ardis: "We don't do a lot of clinical studies, but we do some. Most supplement companies do them rarely, if ever... We're very conservative: we don't want to make leaps in logic if we can help it. Also, this is the first product of this type we've launched, so we want to be extra careful."

While the first product was still over a year away from launch, Interleukin and Alticor took a number of other steps to try to prepare for any ethical issues. One key part of the process was establishing cross-functional teams to deal with privacy and marketing, including any ethical issues in these areas, as part of the product-development effort. In addition, Interleukin's Rahul Dhanda[15] ran a workshop at Alticor for members of the project teams, where he showed the film *X-Men* as a way of stimulating discussion of ethical issues associated with genetics. "It worked wonderfully," said Ardis. "In the film, there is legislation discussed in the House [of Representatives] regarding the registering of mutants. It was a great conversation starter. Within 10 minutes we had shifted to issues closer to home... Everyone came away feeling that it was a very serious issue, and that we couldn't take any risks with the ethical side or privacy issues."

Dhanda recalled some of the specific topics raised in the discussion: "We first made the transition that genetic testing isn't that different from other types of testing. Then some expressed concern that if you sell types of products linked to types of genotypes, people can link back [to someone's genetic profile based on the products they are using]. But they had dealt with that issue already." Some of the other conclusions from the discussion, according to Dhanda, were that "we are not going to sell a product to a consumer that doesn't trust us implicitly" and that "maybe [Alticor] should be getting involved in lobbying and understanding the health care industry."

---

[14]Martha decided to leave Interleukin at the end of 2003 for an attractive offer at another firm.
[15]Dhanda left Interleukin in the summer of 2004 to pursue an intensive 1-year management course.

Perhaps the greatest ethical challenge that Alticor and Interleukin will face is how to ensure that the large, global, self-employed sales force that will be introducing this new type of product to consumers presents the information in an informed and appropriate manner. As Eloi observed, "We don't want someone in Kansas to do something that we don't approve of. You can't have 100% proof... there are always going to be some risks. But we will have some risk mitigators." Among the mitigators that they are considering are: requiring that all people who want to sell the product first be trained and certified; having independent experts, such as the Mayo Clinic, review and approve all the promotional materials; and setting up an 800 number and website so that consumers who are interested in the product can obtain information and speak directly with a genetics expert.

Some of the other ethical questions that both Interleukin and Alticor have identified as needing answers are:

- What is the best way to obtain the genetic samples needed for the tests?
- How can they protect the genetic information of consumers who take the tests?
- How should they label and market the products to indicate their potential health benefits without making claims that would lead the FDA to classify the nutritional supplements as a drug?
- How should they engage with external stakeholders, including potential critics of nutrigenomics, such as Genewatch, as the firm prepares to go to market?
- Is it appropriate to test for and share information with consumers on genetic risk factors for which there is no direct proven treatment, along with the test information that is linked to supplements?

Ardis summed up where the partners stood at the end of 2003: "We're still 14 months away from the planned launch, and frankly a lot of the questions still remain to be fleshed out. In my opinion, we have more questions than answers at this point." But he seemed confident that the partnership they had formed would generate the right outcomes:

The Interleukin team [have] been wonderful people to work with... When I went back for an MBA, it was great to be back in school and be intellectually stimulated again. That's just how I've felt working with Interleukin. We're dealing with issues that are challenging and stimulating. They may not always have answers, but they're determined to discuss the issues and find the best solutions. I feel like we really made the right decision to work with them.

10

SCIONA LTD.: A PIONEER IN NUTRIGENOMICS: THE PATH TO CONSUMER ACCEPTANCE[1]

INTRODUCTION

Sciona Ltd., a 10-person British biotech start-up on the cutting edge of technology in personalized medicine and nutritional genomics, had already remodeled its business strategy merely a year after launching its first product in May 2001. The new company stirred a lot of controversy almost as soon as it brought its product to market and quickly became the focus of much attention in the United Kingdom, both reflecting and reinforcing the public debate that was taking root on genetic testing. The British public was not yet familiar with issues of genetic testing and was, therefore, deeply influenced by activist opponents[2] who published material warning the public of the potential risks of genetic testing. The British public was familiar, however, with many of these advocacy groups who had spearheaded the opposition to genetically modified (GM) foods that eventually led to an unprecedented anti-GM-foods movement in Europe in the late 1990s. Although Sciona's leadership team had suspected that they would encounter some resistance—they were pioneering a new concept that had yet to be regulated and posed a number of ethical dilemmas—they had not anticipated the storm of controversy that engulfed them. From the early formation of its business strategy, Sciona management had recognized the need to address ethical issues associated

[1]This case study was written by Cécile M. Bensimon under the supervision of Peter Singer, following interviews they conducted with Sciona executives and ethics consultants in November 2002. We would like to acknowledge Martin Tierney's work on the study, and Abdallah Daar, Margaret Eaton and David Finegold provided constructive comments and edits on earlier drafts. This case study is one in a series produced under a grant by The Seaver Institute and Genome Canada to study models for ensuring ethical decision making in bioscience firms. Unless otherwise specified, all attributed quotes were obtained during the case study interviews. We would like to thank all who participated in the case study, especially Rosalynn Gill-Garrison, whose invaluable assistance enabled us to complete this study.

[2]*Activist opponents, interest groups,* and *advocacy groups* are terms used interchangeably in this case.

with the new field of nutrigenomics and, more specifically, their direct-to-the-public business strategy, which had been their original business model. They had attempted to address these and other issues early on in planning discussions, and with the guidance of an ethics expert. But they had not invested in public engagement. The issue of exactly how much the public actually knew about genetic testing and, more importantly, what its attitudes were at the time Sciona started marketing its product and services remains a matter of conjecture. The fact is that Sciona was faced with fierce opposition almost immediately.

Convinced that their products and services embraced the future of personalized medicine, the leadership team took another look at what the company was doing, how it was doing it, and why it was accused of being unethical by advocacy groups, in particular Gene Watch and the Consumer's Association. Company executives continued to seek the advice of an ethics expert to help them navigate through the ethical dilemmas surrounding them. Since then, Sciona has restructured its business model and accompanying approach to ethics, introducing the following changes:

- It now distributes its product via health practitioners, rather than direct to the public
- It created a dialogue with the quasi-regulatory body, the Human Genetics Commission, which played an important role in leading to the public consultation process on genetic testing
- It is working with the British In Vitro Diagnostics Association (BIVDA) to draw up a regulatory framework

This case study will review how Sciona managed ethical dilemmas and concerns, and the powerful campaign against them. It will also show how the company moved from responding to criticism and opposition to building a proactive ethical decision making framework, both internally and externally.

BACKGROUND

The founder and Chief Executive Officer (CEO) of Sciona, Chris Martin, wanted to capitalize on new advances in genetics research by offering customized dietary advice based on an individual's genetic make-up. As a result of the Human Genome Project and the advances in microarray technology, he saw a business opportunity in delivering personalized healthcare services by way of a direct-to-the-public business model. This business model was based on the principle that people should have the right, if willing, to access their genetic information without the involvement of an intermediary, such as a health practitioner. To "put the power of gene technology directly in the hands of individuals,"[3] Sciona had to implement an unprecedented strategy that spurred internal debate around three challenges:

1. Adopting a direct-to-the-public business model that was a radical departure from the industry's traditional strategy

[3]Available at URL: www.sciona.com. Accessed January 2003.

2. Defining and marketing a product that presented several ethical dilemmas
3. The absence of a regulatory framework in the field of nutrigenomics to structure the issues.

PRODUCT LAUNCH

Sciona faced the challenge of commercializing the concept of personalized medicine and translating the new science of nutritional genomics into a viable business strategy. Sciona's strategy was to first commercialize the concept of nutritional genomics, with plans to expand to other areas in genetic diagnostics such as heart disease, skin care, drug metabolism, weight management, and sports. Although advances in genomic and genetic research have led to an early understanding of some of the links between genetics and nutrition, the commercial application of genetic testing to determine the relationship between nutrition and genetic processes was unprecedented. Until this time, the role of predictive genetic testing, carried out by clinicians, had been limited to the detection and management of disease genes—for the purpose of identifying gene abnormalities that may make a person susceptible to certain diseases or disorders. Sciona's approach was thus a distinct shift. Its proprietary process in nutritional genomics was to look at the way diet and genetics influence, and lead to, optimum health. The company offered a genetic nutrition analysis, combining an understanding of an individual's genes with information from a lifestyle questionnaire to provide personalized dietary recommendations. By marketing to health-conscious consumers, Sciona's executives sought to offer a service that could, explained Senior Technical Officer Dr. Rosalynn Gill-Garrison, "empower the average person to take responsibility for his/her own health." With the direct-to-the-public business strategy, Sciona challenged the notion that the average person would be unable to understand such information and, therefore, should not access it without the assistance of, and interpretation by, a trained health professional. The management believed that there was an unmet need to use genetic information to improve nutrition and health that it could fill.

The new company was originally funded by friends and private investors. By September 2000, the company had raised £700,000. Its business plan was well received by commercial investors seeking new opportunities. By February 2002, the company had raised almost £3m in venture-capital funding from a group of commercial investors led by The Prelude Trust. As part of the transaction, Sciona acquired Genostic Pharma Ltd.,[4] The Prelude Trust's first funded company in March 1998. Genostic Pharma Ltd. had identified and patented a core group of genes (totaling 2500 genes) and their sequence variants that provided critical clinical

[4]World Intellectual Property Organization. *International Application Published Under the Patent Cooperation Treaty (PCT).* December 1999, WO 99/64626.

prognostic information.[5] With this large gene database, Sciona intended gradually to establish new frontiers for the research and application of genetic diagnostics. The company launched its product in May 2001, limiting its application to the screening of nutritional and alcohol metabolism with the prospect of later developing and introducing a wider range of products. (Currently, Sciona tests 7 [initially, 9] genes that are vital to enzyme activity related to nutrients and alcohol; see example below in "Ethical Considerations.") Designing and delivering this product, however, had been fraught with questions, concerns, and dilemmas. "The strength of the science and ethics drove discussions from the beginning," recalled CEO Martin. Indeed, Sciona wrestled with ethical issues and concerns from day one. What it had not done, however, was to engage the public in its ethical discussions. It was in response to the challenges the company faced as it attempted to penetrate health markets that company executives began to publicly address ethical concerns

ETHICAL CONSIDERATIONS

To better understand the service Sciona was providing, and to catch a glimpse of the interplay of ethical issues it faced, consider the following real customer case.[6]

Peter, a research biochemist at a large chemical company in the eastern United States, is in his early 50s and has a strong understanding of the science of health and diet. After a Sciona presentation, he requested a package containing more information about the service, a lifestyle questionnaire, and a cheek swab used for collecting DNA. A few weeks after having mailed the completed questionnaire and the swab containing cells taken from his cheek, Peter received Sciona's advisory report providing personalized dietary recommendations. If Peter had expressed some serious concern to the company about something in his advisory report, Sciona would have advised him to seek counsel from a qualified professional.

Sciona's technology allowed it to test for 19 genetic mutations on 9 genes. Of particular interest to Peter was the identification of a polymorphism in the gene coding for an enzyme called methylene tetrahydrofolate reductase, or MTHFR.[7] As a biochemist, Peter was aware that the frequency of the mutation was relatively high, but he was still surprised to learn he had it. He said, "As a scientist, I know the frequency of alleles. Emotionally, I didn't make the connection." As a precaution, which Sciona recommends, Peter asked his physician to look at the results of his test.

---

[5] Genostic Pharma's logic is that not all genes are of equal biological importance with regard to the physiologic functioning of humans. The company focused its research on the identification of genes central to the induction, development, progression, and outcome of disease or physiologic states of interest. Moreover, the identification of core group of genes would enable the invention of a design for genetic profiling technologies. Ibid., pp. 6–7.

[6] Anonymous customer, interviewed by the authors, July 19, 2002.

[7] MTHFR is an enzyme involved in regulating folate and plays a part in lowering the levels of homocysteine (Hcy) in the body. Too little folate, or too much homocysteine, has been linked with many health problems. There is a common form of the MTHFR gene, C677T, which reduces the activity of the MTHFR enzyme.

His physician tested for the amino acid that is regulated by the enzyme coded by the MTHFR gene. This test determined that Peter's levels were slightly higher than average, for which he now takes a prescription-strength folate supplement. This diagnosis was consistent with Sciona's recommendation that, because Peter had "variations on the MTHFR gene, meaning that the related enzyme may need some help in maintaining the growth and repair of tissues in your body," he should "increase consumption of foods rich in folate, vitamins B_6 and B_{12}, and take a B-complex vitamin that contains the recommended amount of folic acid." Peter understood the results of both tests. He said, "I understand that I am homozygous with the C677T mutation with borderline high homocysteine levels; I am now taking supplements." Peter's reaction to the test was favorable, but he expressed concern that there was no discussion in Sciona's report of "what you, as a consumer should do with this knowledge; for example, contact family." He adds, "Being tested positive makes you responsible for family." What Peter meant by his observation was that he was left uncertain whether he should notify his family members in case they also carried the same genetic predisposition. The question of responsibility for informing family members is the same question that applies to the diagnosis of a wide range of diseases, such as heart disease, cancers, and diabetes.

This case illustrates how the information provided by Sciona can be useful and yet create uncertainty, even with a sophisticated customer like Peter. If he asked, "What do I do with this information?" and "What does this means to me?" it was certain that other customers would ask these questions and more. Sciona realized that its service would raise these and other uncertainties among its customers and, for this reason, engaged in internal discussions about the role that ethics should play in the marketing of its product. The management recognized that doing so would be core to the success of the company's strategy and had been committed early on to addressing ethical issues. The company's history reflects management's early commitment to ethics. The following relates how Sciona had an ethics platform from the beginning, long before the company launched its first advisory product to the public.

THE CASE FOR ETHICS

Anthony Leung, Commercial Director, described three reasons why the executive team wanted to explicitly incorporate ethics into Sciona's business strategy. First, nutrigenomics is a "new science that is ahead of our thinking of what can be done with it, yet it is essential to talk about what to do with it; otherwise, science is developed in a vacuum. So by its very nature, this new science forces the incorporation of ethics into strategic decision making." Second, ethics directly relates to the *raison d'être* of Sciona: "Ethics is the core fibre of our firm. We are good people who want to do good work, so we want to think things through to reflect our good intentions." Thirdly, incorporating ethics "is a selling point for commercial marketing purposes. We want people to feel safe and trust that we are safe." He adds, "We want to differentiate ourselves as standard-bearers." Rosalynn Gill-Garrison, Senior Technical Officer, reinforced this last point: "We felt that it is our responsibility as pioneers to shape the regulatory environment because there are currently no clear regulations."

Although Sciona had been able to avoid lengthy and expensive regulatory compliance protocols, the lack of a regulatory framework proved to be, said Martin, "a big barrier" from a public standpoint and fueled many difficult questions and much internal debate. To reassure the public about its product, there was a consensus within the company that its product should always "empower people and not alarm or scare them," explained Gill-Garrison. It was for this reason, further explained Martin, that Sciona decided it "wouldn't look at anything that someone couldn't do anything positive about." This goal was reflected in the strategic decision to promote its product as a preventive approach to health, and not as a treatment for unhealthy people, by only identifying polymorphisms that lack a strong correlation to specific disease risk, including polymorphisms where there is a strong link to disease with known treatment. Having made this decision, the executive team felt that they had effectively addressed, and mitigated, potential ethical risks.

WORKING WITH AN ETHICS EXPERT

Despite these decisions, Sciona foresaw other ethical issues that continued to spur intense internal debate around the issues of privacy, discrimination, language, and scientific validity. Conflicting views within the executive level made it increasingly harder to resolve any of these questions. As part of the process of managing ethics, Sciona sought the advice of an ethics expert, who agreed to work with them on a confidential basis. Sciona entered into a consulting arrangement with the expert, who remains anonymous and independent. It was felt that the anonymity of the ethics advisor benefits both the firm (the advisor is not publicly critical and can, therefore, be more candid) and the advisor (who doesn't risk losing credibility because of a perceived conflict of interest).

Sciona expressed the view that the ethics expert played a key role in enabling the firm to address, and in some cases resolve, ethical dilemmas. The advice has helped it to define terms and conditions, develop reporting structures and, in general, provide a framework to identify and address ethics. There is consensus among Sciona employees that managing ethics by way of an external ethics consultancy model has been very effective from a standpoint of timeliness, accessibility, and financial viability (fees are paid on an ad hoc basis). Both the company and the ethics expert agree that the impact was direct in two ways. Sciona was:

- Encouraged to engage in the public debate and with regulatory bodies, which has allowed the company to learn from and educate the public at the same time
- Persuaded to sell its products and services through health professionals

Perhaps out of humility, the ethics expert feels "My input was minimal. Any changes in practices have been a combination of public debate and what I told them." The role was one of consultation, not a built-in role: "The company sends me documentation which I review to provide independent ethical critique… I managed a process of dealing with one issue at a time, each of which had many contingent factors to consider." Sciona executives were sensitized to the principles of harm and benefit and given perspectives on the ways in which these might conflict. The consultant described using a framework that is "deliberative and interactive, principled and process-based at the same time," to guide the firm on the following issues.

PRIVACY

Sciona identified the issue of privacy as the first ethical concern it needed to resolve. According to Senior Technical Officer Gill-Garrison, "Privacy is the biggest issue; it involves questions of database, insurance, police access… it is integral to capital raising; it is raised by every investor." She recalled that privacy issues posed several challenges for the company in connection with the development of technologies that deliver information safely and securely, and decisions about the storage of samples and of information, so as to protect patient privacy while at the same time safeguarding company information. To alleviate ethical concerns regarding privacy, Sciona recently decided no longer to store individual reports—more specifically, not to store customers' DNA samples. The company described how DNA samples are used and discarded:

> Sciona collects two fibre-swabs from each customer, extracts the DNA, and runs the samples through the laboratory to measure the polymorphisms. Any DNA that is not required for the first pass of screens is held until the complete set of polymorphisms for the individual has passed Quality Control. After the data set for the individual has been approved, any remaining DNA is destroyed by incineration. DNA samples are then identified by bar code.[8]

Sciona continuously upgrades software to address ethical issues that arise. For example, Sciona reviewed its procedure for storing reports:

> In the event of a lost report, at the customer's request, we were able to regenerate a report by re-entering the personal information from the database, which contains the personal identifying information and the code linking to the anonymised genetic information held on a separate database. The personal information was only kept for a period of 1 year, at which time it was purged from the database. Therefore, anyone who had lost a report after 1 year would have to resubmit samples. Upon review of this procedure, we decided to upgrade our software to allow the storage of reports with an identifying hash mark to ensure that any report re-sent to an individual would be exactly the same as the original. The software encrypts the report, stores it in a temporary database, which is regularly downloaded on to DVD and stored off site to ensure security of the private information.[9]

DISCRIMINATION

Sciona's leaders were also aware of the argument that individuals' genetic information could be used by insurance companies or employers to discriminate against them. Sciona, therefore, met with the Association of British Insurance (ABI), the insurance industry association, to discuss issues of confidentiality and

[8]Sciona document, given to the authors May 2003.
[9]Ibid.

discrimination. When the ABI told Sciona's CEO that what Sciona was doing was of no interest to insurers, the company leadership felt confident that the question of discrimination would not pose any ethical dilemma and determined that this was not a substantive ethical issue in the United Kingdom.

LANGUAGE

Another important ethical consideration was the way language was used in the company report to customers of the test results. Because this advisory report is the main content of Sciona's product, company executives worked to make it understandable and, in each new version, more intelligible. At first, there was intense internal debate about how much information to provide in this report and how to convey it. Martin recalled that the leadership team grappled with the questions of how to translate the information in a way that could be easily assimilated; how to communicate the possibilities or the implications of this information; how to use the information on probability; and what information to give to their customers and in what context.

Company executives knew that they did not want to oversimplify the information nor misrepresent the test results, but they were not sure how or what to communicate. What they were clear about was that information was to be communicated in a responsible manner. The challenge, recalled Leung, was "to find the right language to provide dietary and not medical information. The language we use is an ethical dilemma because you're damned if you do, damned if you don't. If we medicalize the report, we're using scaremongering tactics. If we don't, people wonder if we are hiding something."

The Sciona executive team turned to the ethics consultant to help them resolve this dilemma. The advice was that the company should disclose all the information available because, the consultant explained, the issue is not *what* you tell people, but *how* you tell them. Based on this advice, Sciona also sought the ethics expert's feedback and guidance on the appropriateness of the language used in all stages of product development, marketing literature, and policy statements.

THE SCIENCE

Another ethical dilemma that Sciona continues to face is the still-infant state of the science of nutrigenomics. This problem is reinforced by the lack of regulatory framework. Some of the questions are:

- What is the level of evidence or statistical significance of the test results?
- If results are inconclusive, how do you make use of the science?
- Is it responsible to convey information derived from a science that remains uncertain?

The ethics advisor cautioned the company to state explicitly that the science is at a preliminary stage.

APPLICATION TO THE HUMAN
GENETICS COMMISSION

At the suggestion of the ethics advisor, in October 2001 Sciona submitted an application under the Voluntary Code of Practice to get a seal of approval from the Human Genetics Commission (HGC). The Code of Practice was drafted in 1997 by the HGC, a quasi-regulatory body that acts as advisor to the U.K. government on developments in genetic research. In December 2001, the Genetic Services Sub-Group (GSSG) of the Human Genetics Commission and Sciona met to discuss this application. The HGC praised Sciona for consulting with an ethics expert, as well as for the procedures the company had put into place to safeguard privacy and ensure accuracy of the testing results.

On the other hand, the HGC expressed some concerns with Sciona's science. In response to these concerns, Sciona prepared a review document about the science of nutrigenomics and referred the HGC to academics who had research programs investigating the links between nutrition and genetics. Sciona hoped that these professors would be able to advise the HGC members on the state of the science. This failed to allay all the HGC's concerns, and it made a number of requests for additional scientific information. Sciona continued to defend the strength of the science underlying the company's product. Company executives expressed the view, and continue to contend, that "we employ a number of highly qualified and specialized scientists who base the science of the service on peer-reviewed, scientific studies. We believe that the advice provided to customers is scientifically robust and this is upheld by a number of respected academic research scientists that are specialized in this area."

By then, Sciona had been on the market with relative success for 5 months; it had been awarded a sum of £130,373 by the Department of Trade and Industry's Small Business Innovation Program; the company was working toward expanding its product range; and finally, in November 2001, it had announced a partnership with The Body Shop as part of its marketing strategy to distribute its product through retail outlets. All this, in the executives' opinion, corroborated the viability of their business plan.

PUBLIC OPPOSITION

In addition to internal and regulatory concerns, Sciona was faced with other scientific and social barriers, including the uncertainty of the science and consumer fear over gene technology. Among activists, genetic testing engendered strong opposition, and Sciona became the focus of groups such as Gene Watch and the Consumer's Association. Protests against Sciona created controversy about the risks and benefits of the new company's product and of genetic testing in general.

The company first became aware that opposition against them was gaining momentum during a radio interview that the CEO gave in November 2001. Chris Martin was taken aback when the announcer introduced a former Greenpeace scientist, the current director of Gene Watch, during their live interview and gave her a platform to condemn Sciona. She warned her audience that if they provided the company with their genetic information, it could be used to discriminate against them and result in loss of employment opportunities and insurance coverage, to which Sciona responded that the company had considered the issue, reviewed it with the ABI, and essentially laid it to rest.

Around the same time, Gene Watch, a public interest group in the United Kingdom, launched a passionate campaign against Sciona, identifying five main concerns.[10]

1. Sciona's products and services were misleading:

   For most people, eating a healthy, balanced diet, getting enough exercise, and not smoking are much more important in determining their health than their genes are. For example, Sciona tests for a gene linked to a vitamin called folate, but folate levels depend much more on age, diet, and whether people smoke than what genes they have.

2. The consequences of the tests results were unknown and could be used to discriminate against insurance or employment opportunities:

   Many scientists are exploring possible links between the genes Sciona tests and serious diseases like heart disease, cancer, and mental illness, including workplace-related illnesses and birth defects. Although genes are usually poor predictors of future health (which depends on many factors), people may learn something they don't want to know. This could also be information an insurer or employer might ask for in the future, and one day use to exclude people from insurance or employment or compensation for a work-related illness.

3. Participants' genetic information could be used for research purposes without their consent:

   Sciona plans to keep the samples and to link personal genetic information and lifestyle information in a database (known as "Biobank") as long as customers remain subscribers. Although Sciona has a "privacy policy" regarding names and addresses, an individual's personal genetic profile is unique to them and could be used to identify them—and who is related to them—in the future. Many companies will pay to do genetic research on these kinds of samples and want to claim exclusive rights to genes by patenting them. Once a gene is patented this means no other company can sell new genetic tests or treatments using the same gene. The law does not require Sciona to inform people if it decides to patent or sell their genetic information. There is growing evidence that privatizing genes in this way can stifle medical research and increase costs. Furthermore, many people believe it is immoral.

4. Nutritional genetic testing was unnecessary and designed to reap corporate profit:

   A number of the big pharmacy and biotech companies are promoting the idea of "predictive medicine." This means using genetic tests to predict the chances that someone will get serious illnesses like heart disease, cancer, or mental illness,

[10]Available at http://www.genewatch.org/humangen/tests/bodyshop.htm, November 2001. Accessed January 2003.

and then offering either lifestyle advice or medication. The benefits for the companies are that they can sell genetic tests and expand the drug market to "pills for the healthy ill."

5. Such new, complex, and unproven technologies should not be used for commercial interests; they require safeguards, such as:
   - the creation of a statutory body to assess the validity and utility of all genetic tests before they are used (other than for medical research)
   - medical oversight (through general practitioners or genetic counselors) of genetic testing to ensure that those taking tests have given fully informed consent and have access to counseling if needed
   - legislation to control the collection, storage, and use of genetic information and to prevent genetic discrimination

Sciona publicly countered the arguments made by Gene Watch.

- Refuting the activist group's first allegation, Sciona pointed to the results from its questionnaire analysis that revealed that greater than 70% of the individuals had low levels of folate in their diet. The company expressed the view that this is of particular concern in the United Kingdom, where the Food Standards Agency, against the recommendation of the majority of experts in nutrition, has decided not to fortify grain products with folate.
- In response to the second allegation, Sciona pointed out that the company had considered the issue of discrimination and had determined that it was not a substantive ethical issue for the company.
- What was a substantive ethical issue for Sciona was the question of confidentiality. The company stated that it would always maintain patient confidentiality and made it clear that it would never divulge information to a third party without a subscriber's consent. In other words, any discrimination in employment would be the result of the subscriber informing an employer. Sciona considers that it is far more important for an individual to be able to make an informed decision regarding employment and believes that individuals have the right to know whether they are at increased risk of illness caused by exposure to agents to which they might be particularly sensitive.
- Refuting the argument that predictive medicine is designed to reap corporate profits, Sciona argued that it is absurd to suggest that predictive medicine is a trend that is mainly in the interests of companies, because most people would prefer to avoid illness rather than cure it after onset. Sciona further argued that predictive medicine could have a major public health policy benefit, reducing costs of critical care and developing healthier populations.
- Finally, in response to Gene Watch's fifth allegation, Sciona publicly defended the strength of the science underlying the company's service. "We circulated the company's review literature to recognized experts in the field of nutrition and genetics and received confidential reviews from these individuals in support of the choice of genes that we were using for our nutrition product, and the research which has been published on the interaction of diet and polymorphism for these particular genes."

Gene Watch did not respond. Sciona's responses did not diminish the controversy over its product. In January 2002, Gene Watch and the Consumer's Association asked pharmaceutical companies to pledge not to carry Sciona's testing kits and urged The Body Shop to remove them from the 11 stores already carrying the kits. (They eventually did so 4 months later, citing lack of demand.) In February 2002, Gene Watch agreed to publicly debate with Sciona on BBC Radio's *Woman's Hour*, only to vehemently condemn the company. In the ensuing months, the group vigorously pursued its campaign against Sciona and genetic testing and continued to draft Parliamentary briefings and push for regulations.[11]

Adding fuel to the claims made by Gene Watch, the British newspaper, *The Guardian*, published articles targeting Sciona. The first article, titled "Public Misled by Gene Test Hype: Scientists Cast Doubt on 'Irresponsible' Claims for Checks Offered by Body Shop,"[12] quoted scientists, including the Chair of the HGC, who questioned the value and scientific basis of Sciona's products. The second article, "Genetic Testing Rules 'Unenforceable': Watchdog Calls for Suspension of Unusable Code of Practice"[13] warned the public against the "crisis that had been triggered by the development of unregulated gene sequencing technology." Both articles reinforced the controversy over the strength of the science and generated more critical publicity about genetic testing and Sciona.

RESPONSE TO OPPOSITION

By then, both company sales and morale were decreasing. Martin and his executive team felt impotent in the face of these difficulties. The CEO conceded, "We were very unsophisticated at how to communicate, how to influence people." Yet, according to Gill-Garrison, "Any problem we've encountered has been one of PR management and not ethical issues." Company executives felt that controversy over their product stemmed largely from misinformation disseminated by activist opponents and from consumer fear that, in their view, had been manufactured by activist groups. As the debate on genetic testing has developed, explained Sciona, "We have come to understand that public fears and perceptions and the views of anti-genetics activists are not the same thing."

Because the company believes that most people are capable of weighing the risks and benefits on their own, Sciona made two strategic decisions:

1. In dealing with a consumer-activist group such as Gene Watch, we declined to enter into situations or publicity which would allow them to continue to air their alarmist messages, and instead, when engaged in public debate, we chose to point

---

[11]Available at www.genewatch.org, March to November 2002. Accessed January 2003.

[12]Meek J. (March 12, 2002). Public Misled by Gene Test Hype: Scientists Cast Doubt on Irresponsible Claims for Checks Offered by Body Shop. *The Guardian* http://www.guardian.co.uk/uk_news/story/0,,665777,00.html. (Accessed February 2005.)

[13]Meek J. (June 4, 2002). Genetic Testing Rules "Unenforceable": Watchdog Calls for Suspension of Unusable Code of Practice. *The Guardian* http://www.guardian.co.uk/uk_news/story/0,,727208,00.html. (Accessed February 2005.)

out the areas in which we shared concerns with Gene Watch, such as issues concerning privacy and regulation. We believed that any head-to-head debate would simply provide further opportunities for this organization to generate publicity

2. We also made the conscious decision not to engage our academic experts in this debate, out of concern that if they spoke out publicly in defence of our private company, they might jeopardize their own positions.

In view of the level of critical publicity that activists had achieved, Sciona chose to focus its efforts around gaining consumer confidence and acceptance. To achieve this, the company defined terms and conditions, drafted a mission and ethics statement, and established voluntary compliance measures, all of which are posted on the company's website. Sciona executives felt that acceptance would be best achieved through education to raise public awareness and understanding of new technologies and suitable regulatory regimes to protect consumers against unscrupulous activities.

To implement the first strategy, the company sought to develop research collaborations. To this end, Sciona has since been awarded two research grants: the first is with the University of Reading, a CASE studentship that is a 3-year study of fish-oil intervention in women with ApoE4 genotypes; and the second grant is a LINK award with the University of Southampton, which is a 2-year study of vitamin E intervention in men with early-stage rheumatoid arthritis. Additionally, the company has itself funded a 3-year project at the University of London's School of Pharmacy that will be exploring the implications of regulatory and ethical strategies in genetic testing for scientific and industrial development and health improvement in the United Kingdom.

To implement the strategy of creating a regulatory framework, Sciona worked to rekindle its dialogue with the HGC. "Without regulation," stated Gill-Garrison, "it becomes very difficult to do anything. We find ourselves in a situation where we are taking every critic by the hand. Putting in place regulatory guidelines and industry best-practices would enable players to have a robust defence." Martin added, "We want to be regulated because appropriate regulation is essential to ensure consumer confidence."

Dealing With the Regulatory Vacuum

In April 2002, Sciona met again with the HGC, with whom it had not talked since their initial meeting in December 2001. The purpose of the meeting was to review Sciona's application under the Code. The HGC concluded that the "current service offered [by Sciona] does NOT comply with the Code of Practice,"[14] primarily because the tests offered by the company do not fall within the Code. Sciona agreed with the HGC's logic because the Code was in fact ill-designed to deal with its product, lacking any regulations specific to nutrigenomics. Because Sciona felt that it was in everyone's interest to regulate this new science, it continued to engage the HGC in a dialogue to work toward a system of appropriate regulation.

Martin felt that it was "essential that the Commission is made aware of the emerging potential benefits of "well-being" screening and does not inadvertently

---

[14]The Human Genetics Commission. Draft statement, April 2002.

frame recommendations that could remove or limit these benefits."[15] However, discussing the benefits of Sciona's product with the HGC proved a challenge, perhaps reflecting the deeply entrenched debate on public/private healthcare in the United Kingdom. Although the Chair of the HGC was open to Sciona's service and remains constructive in her dealings with the company, the HGC was unconvinced at that time of the proven value of genetic tests as initiators of dietary change to reduce health risks. The HGC concluded, in a confidential document, that the scientific evidence of the interaction between genetic variations and diet as a determinant of future health was inconclusive, and further research would be required to show a clear link between the science and the benefit.[16]

The public debate on genetic testing intensified when Gene Watch posted an unpublished copy of this HGC document. Sciona later pointed out that this document had been drafted a month prior to the HGC's plenary meeting in May, where it was agreed that the real issue was that the HGC did not have the tools to determine the validity of the service offered by Sciona. Martin echoed this perspective in the article published by *The Guardian* in June 2002: "The science has progressed to the extent that the service we offer, aimed at individual health and well-being, is quite different than what the HGC had in mind when they drafted the Voluntary Code of Practice in 1997."[17] He continued to draw attention to the fact that Sciona had willingly addressed ethical dilemmas related to the service it offered, and had drafted ethical guidelines specific to its product with the guidance of a "leading, independent U.K. medical ethicist." This did not, however, dispel the controversy surrounding Sciona. It was apparent to the company that it was losing, or perhaps had lost, the public debate.

Restructuring the Direct-to-the-Public Business Model

At the suggestion of the ethics consultant, Sciona yielded to intense pressure by restructuring its business model in July 2002. Company executives were well aware, recalled Sciona's ethics consultant, that the success of "the business necessarily relies on long-term relationships, and thus the company is vulnerable to perception." Initially, Sciona had not viewed its decision to sell directly to the public as an ethical consideration, but rather as a commercial strategy. It had become apparent, however, that it was not a viable business strategy.

Given the level of critical publicity that activists achieved, Sciona decided that it would be prudent to sell its product only through medical and other health practitioners until the regulatory debate developed and more educated, informed views emerged in the media. The leadership team continues to believe that "a product based around a nutrigenomic test and lifestyle advice can be ethically supplied to

---

[15]Press release, December 2001. Available at www.sciona.com. Accessed December 2002.

[16]Human Genetics Commission Genetic Services Subgroup. Draft statement (not for publication): Code of Practice on Human Genetic Testing Services, April 2002.

[17]Meek J. (June 4, 2002). Genetic Testing Rules "Unenforceable": Watchdog Calls for Suspension of Unusable Code of Practice. *The Guardian* http://www.guardian.co.uk/uk_news/story/0,,727208,00.html. (Accessed February 2005.)

the public, directly or through a range of health practitioners, pharmacists, and health stores, as long as the science is robust, the advice is carefully worded, and appropriate support mechanisms are offered to those seeking further advice."[18]

RESULTS

This new marketing channel prompted Sciona to review its internal processes. In particular, Sciona reinforced its commitment to "working to the highest ethical standards" by taking steps to "commission independent ethical reviews of its activities by external bodies."[19] Sciona sought ISO 9001 certification from the international certification body, Lloyd's Register Quality Assurance Ltd (LRQA), which acts as an external auditor for quality assurance by reviewing a company's procedures and regulations. In October 2002, Sciona received such certification for its provision of genotyping services and products based on genetic and lifestyle data, thus confirming that the company's quality management system meets the requirements of the international standard for the process.[20]

PUBLIC CONSULTATION ON GENETIC TESTING

Around the same time that Sciona changed its business strategy, the HGC undertook a 3-month public consultation into the supply of genetic tests direct to the public, to examine the issues of genetic tests offered directly to consumers. Sciona welcomed this initiative. In a press release, Martin stated, "From day one, Sciona has argued that this emerging field of science must be regulated because it is the only way to guarantee consumer confidence. As the leading pioneer in this area, we are looking forward to working with the HGC and playing our full part in the consultation and its outcomes."[21] Leung said in the same press release, "This consultation will help broaden the debate on the regulation of genetic testing and assist public understanding of genetic screening and the issues it raises. It is imperative that we have this debate." Indeed, Sciona strongly believes that the HGC's public consultative process will serve Sciona's dual goals of gaining consumer confidence and acceptance.

In July 2002, the HGC published a consultation document posing "a number of questions to the public about the existing controls, possible future developments and the position in other countries or in comparable areas of health and consumer protection."[22] With the continued guidance of its ethics consultant (who did not submit a proposal to avoid a conflict of interest), Sciona submitted a draft proposal to the

---

[18]Sciona document, given to the authors, May 2003.
[19]Available at http://www.sciona.com/article.asp?id=33. Accessed December 2002.
[20]Available at http://www.sciona.com/coresite/about/quality_systems.htm. Accessed January 2003.
[21]Available at http://www.sciona.com/article.asp?id=15. Press release, July 16, 2002. Accessed January 2003.
[22]Human Genetics Commission. The Supply of Genetic Tests Direct to the Public: A Consultation Document. July 2002, p. 2.

HGC's consultation panel on regulation. At the same time, the company began working with BIVDA, the industry trade association, to draft an industry self-regulatory framework. Leung explained that Sciona decided to work with BIVDA because:

> (1) a code of good practice will carry a lot more weight if it's done within BIVDA because they represent our industry; and (2) it is not feasible to wait 2 years for government to legislate regulations; we cannot wait. It's absolutely necessary to have a standard to measure against now.

Sciona's views on regulation rest on the basic principle of the public's right to know. It thus argues that future regulation should balance this right with appropriate consumer protection. It supports voluntary regulation with restrictions on the type of tests that can be offered and agrees that there should be a system that allows for the classification of tests or services into two categories: tests that are approved for direct supply to the public and tests that are not approved for this purpose. The first category would include tests that can be interpreted directly by lay individuals, whereas the second category would include tests and services that require the intervention of a professional intermediary, such as a genetics counselor or a health practitioner. Sciona further supports the view that regulation needs to be tailored to what the product is testing for, the likely outcome of those tests, and how the tests are made available to the public.[23]

THE HGC RECOMMENDATIONS

In February 2003, the HGC completed the consultation process on genetic testing. It had involved the public, consumer groups, professional bodies, and commercial organizations. In April, the HGC published an interim report. It stated that it did not support direct-to-the-public supply of genetic tests. This was extremely disappointing to Sciona, whose CEO stated:

> We believe the HGC has missed an opportunity to find the right balance between protecting the consumer with a sensible degree of regulation while making sure that the United Kingdom's biotechnology industry is not held back by over-regulation in this emerging field. Instead of providing clear guidance, this has left the public in no-man's land.[24]

Most disappointing to the company, which continued to support the view that its products empowered people to make decisions about their own health, was that these new recommendations meant that the public would not be given the choice to use this new technology to help make informed decisions about the management of their own health and that "nearly 1 year from the beginning of this consultation process, the industry is still waiting for definitive guidance on how regulation will actually work."[25] The HGC, meanwhile, was preparing its final report that would be published a month later.

---

[23]Available at http://www.sciona.com/coresite/news/views.htm. Accessed February 2003.
[24]Available at http://www.sciona.com/article.asp?id=81. Accessed April 2003.
[25]Available at http://www.sciona.com/article.asp?id=81. Accessed April 2003.

In May, the HGC published its report, *Inside Information: Balancing Interests in the Use of Personal Genetic Data,* which outlined its main conclusions in the debate on genetic testing and made formal recommendations to Ministers.[26] As the title reflects, the HGC sought to produce a report that adequately incorporates different, often competing, interests and viewpoints on the broader implications of advances in human genetics. The HGC Chair, Baroness Kennedy, stated that balancing interests "…is an important aspect of our work. I am very much aware of the fact that people approach this issue of personal genetic information from varying perspectives. We have tried here to take account of a wide spectrum of views and have attempted to reach conclusions which are morally defensible and sensitive to the different interests involved."

In its report, the HGC made 37 recommendations to Ministers about policy or areas that its members feel should be further examined.[27] It is organized into six main chapters: what is genetic information and general principles; clinical practice and special cases in genetics; research and genetic databases; insurance and employment; forensic uses of genetic information; and DNA parentage testing. Recommendations that are of interest to Sciona pertain to the safeguards needed to ensure patient confidentiality and the requirements to share such information in appropriate ways. Of particular interest is the recommendation to prevent unfair discrimination on the grounds of a person's genetic profile. In this vein, the HGC urges the Government to "consider in detail the possible need for separate UK legislation to prevent genetic discrimination and that this evaluation form part of a long-term policy review on the use of personal genetic information in insurance and employment (6.41)."[28]

Although Sciona has not yet publicly responded to the HGC report, Gill-Garrison said, "We are pleased that the HGC has recognized that there are a range of genetic tests available. While the HGC profiled Sciona in its report, there are some inaccuracies which we were disappointed to see. We are currently engaged in dialogue with the HGC on how best to address these inaccuracies."[29] What especially matters to Sciona is that a regulatory framework is starting to take shape. And the U.K. government, in its response to the HGC's report, commented on the usefulness of the report in "providing a sound basis for developing future Government policy on human genetics."[30]

After the HGC report, and largely based on its conclusions and recommendations, the UK Department of Health published the government's White Paper on genetics: *Our Inheritance, Our Future: Realising the Potential of Genetics in the NHS.*[31] In this White Paper, the UK government makes the commitment to invest £50m (US$83m) in the development of genetic services and safeguards. Moreover, the government announced that the HGC's recommendation to render

[26] Available at http://www.hgc.gov.uk/insideinformation/index.htm#report. For a summary, see http://www.hgc.gov.uk/insideinformation/insideinformation_summary.pdf. Accessed June 2003.

[27] Available at http://www.hgc.gov.uk/insideinformation/iirecommendations.pdf. Accessed June 2003.

[28] Available at http://www.hgc.gov.uk/insideinformation/iirecommendations.pdf. Accessed June 2003.

[29] Company document, given to the authors May 2003.

[30] Available at http://www.doh.gov.uk/genetics/govresp.htm. Accessed June 2003.

[31] Available at http://www.doh.gov.uk/genetics/whitepaper.htm.

non-consensual genetic testing a criminal offense has been legislated. Significantly, the government's White Paper outlines the potential for gene therapy, thereby validating Sciona's *raison d'être*: "Genetics promises a more personalised approach to health care, with interventions tailored to each person's own genetic profile."

But although the HGC and the Government reports validate genetic research and further bolster the regulatory process concerning commercial genetic testing, they also reinforce—simply through the wide range of issues they address and the wider range of questions they wish to explore further—the complexities of managing and resolving the ethical issues inherent to genetic research. Both the HGC and the UK's Department of Health agree that there is still much to do to develop appropriate safeguards to allow the public to benefit from advances in genetic research and to set up a regulatory framework that adequately incorporates the implications of such advances.

TOWARD THE FUTURE

Sciona, however, is not waiting for legislation or regulatory guidelines, although it continues to participate in the discussions on regulations. The company plans to launch a series of genetic tests covering heart disease, skin care, drug metabolism, weight management, and sports. Company executives feel that they now have ethical processes that are, according to Leung, "as strong as a company this size can have." Some acknowledged that "their actions with respect to ethical issues seemed like an apology or rationalization" because we were not "proactive enough with respect to ethical issues" and "our ethics and policy statement was drafted in response to the battery in press," said Leung. Indeed, the transparency of the controversy about the risks and benefits of Sciona's products forced the company to think carefully about what it was doing and how it was doing it. Most importantly, it motivated company executives to address ethical issues outside the confines of the company.

In addition to working with an ethics consultant, Sciona feels that it "got other things right," such as engaging in a dialogue with the HGC. The company also talks about "what they got wrong." They say that they had underestimated the extent to which the views of anti-science activists are accepted as altruistic, honest, and representative of the public; they hadn't garnered any external support for their product, such as through a leading university group, and they were ill-equipped to argue against activist opponents' claims. "Today," explained Leung, "we've sufficiently reconciled our need to address ethical issues with our ability to do so." He added, "Ethical processes need to be constantly reviewed and looked at and renegotiated. It's a process."

Now that Sciona is currently preparing to launch its products and services in the United States, what issues is the firm likely to face? How are they likely to differ from the United Kingdom, given the difference in healthcare systems? What lessons did Sciona learn from its experience in the United Kingdom that it will apply to its United States launch?

It also remains to be seen what direction the public debate on genetic testing will take in the United Kingdom and beyond. Clearly, the debate is only getting started.

11

AFFYMETRIX, INC.: USING CORPORATE ETHICS ADVICE[1]

INTRODUCTION

Thane Kreiner, Senior Vice President of Corporate Affairs at Affymetrix, says that of all the cases he recalls reading in his MBA studies 11 year ago, he remembers best the cases about business ethics. They appealed to him as a student, and he has brought an interest in business ethics to his current position. Affymetrix is a pioneer in the commercialization of DNA microarray technology, which is used to produce what are commonly called DNA chips, genechips (GeneChip® is a registered trademark of Affymetrix), or biochips.[2] And Kreiner's interest in business ethics has served Affymetrix well: he was responsible for forming the company ethics committee, which has influenced some highly important corporate decisions.

DNA CHIP TECHNOLOGY

The relevance of ethics to Affymetrix's business has much to do with the technology the company is commercializing. DNA chips can perform massively parallel, miniature bio-analyses that allow scientists to carry out huge numbers of biology

[1]This case study was written by Margaret L. Eaton based on interviews with three Affymetrix employees and five ethics advisors in July and August 2003. Two of the interviews were conducted with David Finegold, who also gave valuable assistance in editing this case study, as did Abdallah Daar. This case study is one in a series produced under grants funded by The Seaver Institute and Genome Canada to study models for ensuring informed ethical decision making in bioscience firms. Grateful appreciation is extended to all those who participated and assisted, especially to Affymetrix's Katie Tillman Buck, without whose efforts this case study could not have been written.
[2]Affymetrix named its products GeneChip® microarrays.

experiments all at once.[3] With these chips, experiments that once took days or years to conduct can now be done thousands of times in a few hours. This revolutionary technology was invented in the late 1980s by a team of scientists led by Affymetrix's founder, Stephen Fodor, who is the company chief executive officer (CEO) and Chair. Using semiconductor manufacturing techniques coupled with advances in combinatorial chemistry, Fodor and his team were successful in creating microarrays: small glass chips containing vast amounts of genetic data.[4] Affymetrix developed a series of these DNA chips for sale to researchers who use them to acquire, analyze, and manage complex genetic information. The data they generate are used to improve the drug-discovery process and the diagnosis, monitoring, and treatment of disease. The company's GeneChip® microarrays are part of a system that also includes reagents and instruments to process the chips and software to analyze and manage the genetic information obtained from the use of the chips.[5]

DNA microarray chips are used primarily for two research purposes: to analyze DNA (to identify genetic sequences, genes, and gene mutations) and to analyze gene expression (identifying which genes in a cell are active in particular circumstances). Both of these capabilities will eventually have clinical applications in the diagnosis and treatment of diseases.

APPLICATIONS OF DNA CHIPS

Affymetrix's DNA chips have been used in these ways:

- To identify small differences in genetic codes. Thus, for example, scientists can identify genetic mutations in human immunovirus (HIV) linked to drug resistance, which will enable the development of new anti-HIV drugs and the monitoring of infections to determine when HIV mutation occurs.
- To study liver enzyme genes whose variations affect drug metabolism. These chips will allow physicians to tailor drugs and doses for maximum benefit to patients while avoiding serious and life-threatening toxicities.[6]
- To analyze the entire human genome for minor but important variations called single nucleotide polymorphisms (SNPs), which are associated with disease predisposition, drug effects, and other medically relevant information.[7]

---

[3]Potential for such experiments include basic research, human disease diagnostics, individualized medicine, human identity testing, food testing, livestock diagnosis or grading, agricultural biotechnology, and environmental testing.

[4]Fodor SP, Read JL, Pirrung MC, et al. Light-Directed, Spacially Addressable Parallel Chemical Synthesis. *Science.* 1991;251:767–773.

[5]Information about DNA microarray technology and DNA chips comes from Affymetrix (http://www.affymetrix.com), Shi L. DNA microarray (genome chip); Monitoring the genome on a chip (http://www.gene-chips.com), and Robbins-Roth C. Tools for Genome Studies; DNA chips and microarrays. In: *From Alchemy To IPO.* Cambridge, Mass: Perseus Publishing; 2000:73–81.

[6]Roche Diagnostics has partnered with Affymetrix to develop this AmpliChip™ CYP450 array chip.

[7]Affymetrix's Mapping 100k array can simultaneously examine 100,000 SNPs.

- To study the complete set of genes in the human genome. This chip (the Human Genome U133 Plus 2.0 Array) will greatly enhance genetic research because it can measure the activity of all known human genes in a biological sample.
- For rapid identification of bacterial or viral agents that might be used in a biological attack.
- For gene expression analysis, which makes it possible to monitor the activity of thousands of genes simultaneously and over time, thus providing important insights into which genes play a role in normal and disease-cell functions.[8]

COMPANY HISTORY

Affymetrix was founded in 1991 as a division of Affymax N.V., a combinatorial chemistry company. In 1992, Affymetrix was spun out to focus on DNA chip technology and considered itself a science company that supported genetic diagnosis research. Headquartered in Santa Clara, California, the company started by licensing the technology developed by Steve Fodor and his team while they were at Affymax, which retained a 65% ownership. In addition to traditional sources of capital,[9] Affymetrix was somewhat unique in obtaining a significant portion of its research and development (R&D) funding from a series of government grants totaling $30 million.[10] The government funding was especially welcome because the company did not have to relinquish equity in return.

Once the technology showed promise, Fodor formulated a business plan. The first step in the plan was to develop a standard microarray product that could give the firm a first-mover advantage in a fast-moving and immature market. Income then came from licensing fees and royalties on Affymetrix patents. The company's first chips were sold to public and private research institutions, after which chips were sold commercially—to large pharmaceutical companies, for instance. Some of the commercial chips were developed for general use, and some were custom-made for the requirements of a specific user, including those used for the screening of large populations to detect genetically influenced diseases.

Affymetrix has faced and overcome several business challenges along the way. Patent litigation needed significant attention during the company's early years and

---

[8]Using this kind of an Affymetrix chip, for instance, researchers identified a new class of leukemia with a chromosomal anomaly characterized by a highly distinctive pattern of gene expression that correlates with very poor prognosis. Detecting this kind of virulent leukemia can allow physicians to give patients more reliable expectations and tailor therapy appropriately. Eventually, knowledge of this chromosomal defect will also give researchers clues about how to develop new drugs that can combat the chromosomal defect.

[9]The company raised $21 million in a Series A round and $39 million in Series B, and its initial public offering in June 1996 raised $92 million gross.

[10]Most of this was from one 1994 grant from the National Institute of Standards and Technology's Advanced Technology Program given to an Affymetrix-led consortium, from which Affymetrix obtained $20.8 million. The grant was for the development of an industry standard for the reading and analysis of many types of microarray chips.

forced the company not only to become involved in defending its technology patents but also in the pressing question of the acceptability of gene patenting.[11] There were also complaints about the price of the early microarrays; many researchers felt that microarray companies catered only to the rich pharmaceutical and biotechnology customers, leaving the average academic researcher without affordable access to the technology. As Affymetrix grew, competition in the array markets became intense.

Eventually, many patent suits were resolved either in Affymetrix's favor or without significant negative impact, and those remaining were actively defended. Over time, Affymetrix was able to improve its processes, thus lowering the cost of the arrays.[12] The company also accommodated academic researchers with special pricing arrangements.[13] The company's competitive status improved when some important competitors dropped out of the market (notably Hyseq, Incyte, and the large advanced materials companies, Corning and Motorola), and when Affymetrix was able to produce products ahead of its remaining competitors.[14]

To complement and consolidate its technology platforms, Affymetrix acquired small companies,[15] and spun off a company[16] whose related technology did not fit within the core business. An early original equipment manufacturer (OEM) relationship with Hewlett-Packard in 1994 allowed Affymetrix to manufacture second-generation commercial scanners that would read DNA chip analysis results. Other industry partnerships followed with both small genomics companies and large pharmaceutical companies.[17]

The company launched its first commercial GeneChip® microarray product in 1994. The DNA chip was intended to aid in the understanding of how HIV mutates over time to develop resistance to anti-retroviral drugs. This common problem of HIV drug resistance was leaving physicians with little more than a guess as to which drugs might remain effective and for how long. In early 1997, Glaxo Wellcome started to use this chip to build a database of information correlating

---

[11]Affymetrix owns more than 230 patents and has 420 applications pending, and has been involved in IP litigation with competitors (and sometimes collaborators) Incyte, Synteni, Hyseq, Oxford Gene Technology, Perkin Elmer, and with Stanford University. Will Patent Fights Hold DNA Chips Hostage? *Science.* October 16, 1998;282:397.

[12]Fox JL. Complaints Raised Over Restricted Microarray Access. *Nat Biotechnol.* 1999;17:325–326.

[13]Affymetrix instituted its Academic Access program for publicly or charitably funded researchers in which up-front payments were generally waived and volume commitments resulted in substantially discounted pricing.

[14]Thiel KA. The Matrix: A Revolution in Array Technologies. Biospace.com, June 7, 1999. Available from http://www.biospace.com/articles/060799_matrix.cfm.

[15]Genetic Microsystems (DNA array instrumentation technology) and Neomorphic (computational genomics).

[16]Perlegen was founded to use Affymetrix's technology to survey 50 human genomes a year for variations associated with multigenic disease and for the genetic contributions to drug response. Affymetrix maintains a substantial equity position in Perlegen and has access to its discoveries.

[17]Partnerships included those with Genetics Institute and Incyte Pharmaceuticals. In 1995, Glaxo Wellcome purchased Affymax, although they have since sold out their position in Affymetrix.

EXHIBIT 11.1   **Company Financials**

Five Year Summary

Year	Sales	Net Income	Eps
2002	289,874,000	−1,630,000	−0.03
2001	224,874,000	−33,121,000	−0.58
2000	200,830,000	−53,990,000	−0.98
1999	109,074,000	−25,504,000	−0.54
1998	52,413,000	−26,800,000	−0.62
5-Year Growth Rate	53.3	NA	NA

HIV genetics with drug sensitivity and patient outcomes that will help to understand HIV drug resistance and what constitutes optimal treatment regimens. By 2003, with chips and partnerships such as this, Affymetrix reported its first profitable year: it had generated product and product-related revenue of approximately $280 million and a total revenue of approximately $300 million (see Exhibit 11.1); it had acquired 350 customers including pharmaceutical, biotechnology, agrichemical, diagnostics, and consumer products companies, as well as academic, government, and other non-profit research institutes.

AFFYMETRIX ETHICS ADVISORY COMMITTEE FORMATION

GENESIS OF THE CORPORATE ETHICS ADVISORY COMMITTEE

A company with these types of products was bound to encounter business questions that raised ethical issues. By all accounts, the person who first realized this was Paul Berg, a member of the company's Board of Directors. Dr. Berg, an eminent Stanford University scientist, won the 1980 Nobel Prize in chemistry for his fundamental studies in the biochemistry of nucleic acids. In the early 1970s, he was the first scientist to construct a recombinant DNA molecule (a molecule containing parts of DNA from different species), which was the genesis of the new field of genetic engineering. This background made Berg highly qualified to advise the young DNA chip company on scientific and technical matters. Dr. Berg did not limit his advice to this arena, however. He has always been interested in the larger consequences of genetic manipulation and was famous for having convened the Asilomar Conference in 1975, at which genetic-engineering scientists, policymakers, and journalists addressed the private and public concerns that had been voiced about the safety of the new recombinant DNA technology. Asilomar ended with the conference agreeing on a voluntary international moratorium on recombinant DNA research, and self regulation until the technology could proceed without undue risk. Asilomar is widely regarded as a rare instance of scientists independently questioning and successfully regulating their own cutting-edge work.[18] It was also the dawning of a modern awareness that ethical and social issues (such as accommodating public perceptions about appropriateness and safety) can intrude on genetic science. And in the process, Paul Berg had established a reputation that these related issues mattered to him.

Berg's specific concern on behalf of Affymetrix was based on ethical questions that had been raised about genetic testing for cystic fibrosis (CF). There had been a great deal of uncertainty about the ability of the gene test to predict disease severity. Consequently, doctors could not accurately tell prospective or expecting parents whether their CF-positive baby would experience mild and manageable disease, or die early after much suffering. Berg had followed this issue with interest, wondering whether such an imperfect test would help or harm decisions about procreation or pregnancy termination. "Since I knew about the CF-gene test controversy, this is the reason that I suggested to Affymetrix that they form an ethics committee," Berg said.

Thane Kreiner remembers that, in December 1996, "…we were at a company Board meeting and Berg leaned over to me and said, 'We should convene an ethics advisory committee to make sure that beneficial uses of the technology occur.' I talked about it with Steve Fodor and he said, 'Great, work on it with Paul.'"

That Fodor immediately assented to the idea was not surprising to those who knew him. Fodor wore several hats as a scientist and corporate CEO, and he was also known as a technology visionary. According to one company employee, "Steve always had a sense of context into which technology would slot. He has a grand vision for the good that technology can do, and ethical issues like privacy, equality, and access to technology have always been part of that vision." He also recognized that business people tended to view the social and ethical responsibilities of companies differently from scientists and lay people. At a 1999 Princeton University symposium on bioethics, Fodor observed that "having a commercial background brings a different bent to the ethics around the subject…." To illustrate this, he offered this wry anecdote:

> I was talking to a friend of mine whose father used to run a dry-cleaning company near New York City. Every day when the clothes came in, he would go through the pockets of the clothes and see what he found in there. One day he found a $100 bill. He said this raised a serious ethical question—whether he was going to tell his partner.[19]

Fodor's realization that duty and responsibility can be defined differently within a company from in the general community is a likely reason that he was immediately receptive to the suggestion that Affymetrix obtain outside advice about its business conduct.

[18]Russo E. Reconsidering Asilomar: Scientists See a Much More Complex Modern-Day Environment. *The Scientist*. April 3, 2000;14:15. Available from http://www.the-scientist.com/yr2000/apr/russo_p15_000403.html.

[19]Goertzel B. Steve Fodor: Pioneer of Massively Parallel Genomics. Available from http://www.goertzel.org/benzine/FodorProfile.htm.

Thane Kreiner's background was also mixed. He held a PhD in neuroscience and an MBA from Stanford University. He also readily appreciated the potential benefits of corporate ethics advice and, trusting Berg's motives, was pleased to have the opportunity to initiate the program. By the time that Berg had proposed the ethics committee idea to him, Kreiner had been at Affymetrix for more than 2 years, first as Vice President of Sales and Marketing and then as Vice President of Business Operations and Public Affairs. His experience at the company had also made him aware of the ethical issues surrounding testing for genes associated with diseases. Running an ethics committee seemed a good extension to his current responsibilities, and he was personally interested in seeing that the company technology was used for only beneficial and responsible purposes.

At the outset, Berg and Kreiner wanted an ethics committee consisting of an intellectually diverse group of people so as to obtain maximal insight and broad-based advice. Together, they decided that they needed expertise in ethics, law, and medicine. In the medical field they wanted both a physician and a genetics counselor who could provide a perspective on the impact of genetic technologies on patients. They also wanted a journalist who could educate the company about public perceptions of genetic issues, preferably a science journalist who also knew what stories would be interesting to the public and who was aware of public opinions. A religious advisor would also be beneficial. Finally, they wanted a scientist and, according to Kreiner, "You can't get any better than Paul Berg." Together, this mix of people was expected to produce rich and broad debates and informative discussions. A collection of these disciplines was impossible, and too costly to replicate internally. Outsiders also were more likely to be objective and lack the "blinders" that company insiders would have about corporate activities.

Knowing what they wanted was the first step in setting up an ethics committee. Kreiner and Berg had also thought about what they did not want. Their list included these items:

1. Avoid perceived or real conflicts of interest, which could undermine the value of the advice and create the impression that the company was purchasing its ethics advice. Obviously, this concern would be factored into the amount of money the advisors were paid. Berg and Kreiner also decided that the committee members (except for Berg) could neither be affiliated with the company nor financially or personally interested in it.

2. Avoid the perception that the ethics committee is a public relations tool. Berg and Kreiner decided that the company would not "market" the fact that they were using an ethics advisory committee and that the names of the members would not be revealed by the company.

3. Avoid putting committee members in a position where they made business decisions. Kreiner did not want to abdicate the responsibility of making ethical decisions for the company. As he said, "This is an ethics advisory *committee*, not a Board; the committee offers advice but has no decision making authority or veto power." Part of Kreiner's job would be to manage any objections of committee members when the company did not follow the committee's advice.

4. Beware "group-think." According to Kreiner, group-think occurs "when members of a group making a decision all come from the same framework. They may think they have produced a considered opinion, but really they have not because they are all looking at the question from the same perspective. We want ethics advice untainted by common views among the members." To avoid group-think, committee members needed to be strong and independent thinkers, not too easily swayed by the expertise or opinions of others.

COMMITTEE MEMBERS

Fortunately for Affymetrix, the local San Francisco Bay Area was a rich source for the disciplines and the kinds of people the company wanted to attract to the ethics committee. Berg approached two academic colleagues, a medical-technology law professor and a medical anthropologist, both of whom were affiliated with an academic ethics center. These two individuals then assisted in recruiting a genetics counselor from a large health maintenance organization (HMO), an academic physician with national ethics credentials, a retired bishop, and a science journalist. Despite some early skepticism about the sincerity of the company's intent and how ethics advisors would be used, all the people who were invited to join the committee did so. Kreiner and Berg were pleased with the assembled group. They were independent, intellectually diverse, and at least two of them had direct patient care responsibilities. Together, the group had the potential for collegiality and could address scientific, medical, genetic, social, ethical, legal, religious, and public affairs aspects of the topics likely to be addressed. By March 1997, the ethics committee was acknowledged and started to meet as a group.

The company insiders who are members of the ethics committee include Paul Berg, Thane Kreiner, Katie Tillman Buck, and Janet Warrington. Tillman Buck is a recent Princeton chemistry graduate who was attracted to Affymetrix because of her experience with a yearly Princeton bioethics forum. She had been especially impressed with Steve Fodor's interest in ethics, which he expressed at the university forum in Tillman Buck's senior year. She assists Kreiner on the ethics committee by setting the agenda, performing background research, preparing committee presentations, and ensuring that the appropriate corporate executives attend. She has also assisted in writing the ethics statements that appear in the Securities and Exchange Commission (SEC) filings and on the company website[20] (see Exhibit 11.2).

Janet Warrington is a medical geneticist who joined Affymetrix in 1996. Warrington calls herself the closest thing to a customer within the firm. She proposes projects that will move the company technology forward and attempts to find external funding and research partners for these projects. As a result, she collaborates with clinical oncologists and other physicians who can provide specialty expertise. Warrington was inspired to pursue a science career after a frustrating childhood experience witnessing a physician's inability to save a neighbor's child

[20]http://www.affymetrix.com/corporate/outreach/index.affx.

EXHIBIT 11.2   **Affymetrix SEC Filing Statement Regarding Ethical Ramifications of Its Business**

Concerns surrounding the use of genetic information could reduce demand for our products. Genetic testing has raised ethical issues regarding privacy and the appropriate uses of the resulting information. For these reasons, governmental authorities may call for limits on or regulation of the use of genetic testing or prohibit testing for genetic predisposition to certain conditions, particularly for those that have no known cure. Similarly, such concerns may lead individuals to refuse to use genetics tests even if permissible. Any of these scenarios could reduce the potential markets for our clinical applications products, which could have a material adverse effect on our business, financial condition, and results of operations.

Affymetrix SEC 10k Report, 2002.

who suffered and died from cancer. She had also had graduate work experience at the University of California at Irvine, dealing with families who were genetically tested for Huntington's disease; she knew how application of these tests could both harm and benefit, when family members learned that they did or did not carry the gene for this deadly disease. Affymetrix was attractive to her because the innovative technology would allow her to work on medical problems that had been resistant to solutions, especially for complex problems involving cancer and women's health.

Warrington's experience motivates her to consider her work in a broader context and she "sees ethical issues every day in the laboratory—any time I think about a new study, how I report data, how I submit a manuscript." She gives a lecture, "The Ethical Day in the Life of a Research Scientist," that includes asking how informed consent should be handled for studies and whether some studies should be done at all. She is also prone to ask the "so-what?" question in situations, such as when a gene test is developed that can detect a disease or disease risk that cannot be ameliorated. If nothing can be done to reduce the risk or prevent the disease, she imagines that patients will say, "So what if I know I have a gene defect? What good does it do me?" The anxiety, fear, and social harms that can result from testing can overwhelm the benefit of knowing. She questions the sale of such a test for patient use. Asking the question will often direct the developer of gene tests to limit their use until the good that can come from testing outweighs the harm. Because her ethics focus was evident to others, Warrington was asked by a company lawyer to join the ethics committee. Warrington reports that she was excited about the assignment and thought that her presence on the committee made sense. Since her appointment, she often brings discussion items to the ethics committee.

THE ROLE OF THE ETHICS COMMITTEE

Individual committee members are in accord about their roles on the committee. All who were interviewed describe the essence of their responsibility as thinking about the wider consequences of Affymetrix's corporate activity and advising about the ways those consequences can be addressed proactively. One member likens her role to that of the "good philosopher" described by William James, one who has attained "the habit of always seeing an alternative, of not taking the usual for granted, of making conventionalities fluid again, or imagining foreign states of mind." Another member said:

> What I have seen is that many people who start companies are so enthusiastic about their product that they cannot imagine that there will be objections to it. So, what we on the ethics committee do is imagine for them, for instance, if something went wrong or if the product did not work as anticipated. We think about the consequences and the implications of implementation.

There is also congruity between the company and committee members in the ways they describe their intended personal contributions on the committee. For instance, the two members with patient-care experiences (the physician and the genetics counselor) know they were chosen to discuss the likely reaction of patients to genetic testing and the uses of genetic information. The genetics counselor said:

> From me, they wanted to understand the public's perceptions about genetic testing and how to translate scientific information in lay terms. I bring to the committee insight into how information is understood and used in the community, how patients think about genetics, how it impacts their life, what's the state of what's out there with respect to applied science.

GROUND RULES

The first task of the new committee was to set up ground rules and operating guidelines. The topics included meeting frequency, operations, and attendance; the manner of decision making (e.g., consensus, majority rule, etc.); responsibility for decision making; confidentiality; publicity; compensation; and conflicts of interest. The guidelines set up by the early committee were sufficiently satisfactory not to have been changed since then.

The members decided that they would not have set terms, and that the nomination of new members would be an informal process, mutually agreed upon. Meetings would be quarterly or, occasionally, more often if an urgent question needed to be addressed. Occasional lack of attendance was expected and tolerated. The issues for discussion would be determined by Kreiner and Tillman Buck with input from committee members. No formalities for discussion were to be used and no formal votes taken. Members were more comfortable with consensus decision making. All members easily agreed that their input would be advisory only and that the company was free to accept or reject this advice. Members discussed how they would react if the company rejected advice, and all felt free to resign if they felt strongly about any particular matter. The committee also decided that committee work would take place primarily at the convened meetings, although some outside communication might be needed.[21] No formal reports would be written; advice would be given to the executives directly (when they attended meetings)

---

[21]At times, some committee members have agreed to do other services for the company. One member, for instance, reviews papers for them, has participated on ethics panels at professional meetings with Thane Kreiner, has given ethics talks at the company, and has arranged for other biomedical ethicists to do the same.

or through Thane Kreiner, who would relay the advice to other upper management. Regarding conflicts of interest, no one had family members who worked at the company and all members agreed that they would not hold any Affymetrix shares. Once these routine matters were set, the committee tackled the more weighty issues of confidentiality, publicity, and compensation.

CONFIDENTIALITY

All members agreed that committee deliberations were confidential unless disclosure was authorized by the company, and a non-disclosure agreement was signed to this effect. It was understood that full access to company information was crucial to a well-considered ethics opinion, and that the company needed the assurance of a written promise of confidentiality. In addition, free, open, and full discussions would not take place unless everyone trusted that the process was confidential. Total secrecy, however, was unworkable, because some members expected that they would want to use company information in their other work. The cutting-edge work of the company, and the thought-provoking questions it raised, could be very useful for the academics who taught about ethical issues related to medical technology, teaching, or in academic writings. Furthermore, academics were used to operating in an environment that valued the free exchange of ideas and unfettered publications, an environment not compatible with the requirements of corporate confidentiality. In the special case of the journalist, she would be privy to information that would make intriguing news, and her work could not but be informed by Affymetrix information. Kreiner did not want to foreclose all possibility of members using company information, as he wanted to make the committee work mutually enriching for all. He agreed, therefore, that confidential company information could be used so long as the committee members informed him first about their intended use and obtained his consent.

For the academics, this arrangement has posed no problems. All have used information and insights they obtained though the committee to augment their teaching. By agreement, company information used for teaching usually does not name Affymetrix. The anthropologist has used information gained from the race and genetics issue discussed by the committee (see below); the medical technology lawyer has incorporated company insights into his teaching about the legalities of intellectual property ownership of genetic information; the physician has also used company information in classroom settings:

> For instance, the Affymetrix experience has helped me when I discuss with medical students how responsible are you for the technology you develop, what are your responsibilities when you perform research for a company, or consult? I can easily write my teaching examples into a non-GeneChip example. I also use a website in teaching that came to me from the Affymetrix experience.

Members have also asked for, and used, Affymetrix slides presented at committee meetings. Unlike the agreeable arrangements with the academics, however, the journalist found that she could not manage the confidentiality requirements and perform effectively as a reporter. For this reason, she resigned her position and no journalist has replaced her.

Having decided on the confidentiality terms, it was easier for the company to agree to give committee members full access to information on matters under committee consideration. Access to information was important to the members, all of whom felt that providing advice based on limited information would be irresponsible, unhelpful, or even harmful. Kreiner reports that there have been no instances in which he felt confidentiality had been breached, and committee members report that they have never been denied access to relevant information. According to one member:

> They don't give us all the details on each matter. They will say something like 'This is a big pharma customer' without telling us the name when the deal has not been completed. They give us general information—where this deal fits strategically with the company. I don't get the feeling that they are holding anything back when they brief us. They always seem willing to give us information. But if they don't know the answer to our questions (they don't always have all of the information at the meetings), we can always get it. At the meeting, they will ask each other in front of us what they know. It would have to be carefully choreographed to seem like they were being forthcoming when they weren't. They never say, 'We can't tell you that.' That would make me suspicious. We ask about a lot of different topics, so it's unlikely that they would be able to anticipate them all in advance. Sometimes they may not recognize that there is an issue or information that is relevant until we ask about it. And they don't seem to keep pertinent information out if we don't ask about it.

PUBLICITY

Early on, the committee members decided that they did not want the company to make any public announcements about their formation or their work. For the public to know that an ethics committee had endorsed or approved company activity smacked of public relations, and the members did not want to be used that way. In addition, anonymity would free them from feeling constrained about giving candid advice. Affymetrix agreed and has not published the names of the committee members, but does disclose in the ethics section of its website that it uses an ethics advisory committee to address corporate issues. Some committee members list on their resume that they serve as an ethics advisor to Affymetrix and, at times, have made the same disclosure when speaking at academic meetings. For media inquiries, Kreiner will first approve of the nature of the inquiry and then inform the member that an interview is requested, after which the committee member can decide whether to contact the press. Committee members who want to discuss their committee work openly ask for Kreiner's permission first. According to all involved, this set of practices has worked well, despite the potential for backlash. Committee members recognize that different inferences can be drawn about a company that seeks ethics advice. According to one:

> Affymetrix has not been secret about having an ethics committee but the company doesn't "market" us either. Not only would we resent it but there is no good way for a company to do this. They can't say they have an ethics committee but the advice is secret. It would just raise too many questions and invite too much inquiry.

COMPENSATION

The question of payment for ethics advisory services was a tricky one. Too much money would make it seem as if the company was buying loyalty and acquiescence. As Kreiner put it, "If a major chunk of a consultant's income comes from the company work, we have to wonder about their objectivity." His worry was reflected in the biomedical-ethical literature, which contained many opinion pieces about the growing phenomenon of biotechnology companies seeking and obtaining ethics advice. It is commonly believed that the bias that money produces is real but underappreciated or not acknowledged in the medical community. Physicians, for instance, say that their prescribing practices are not influenced by gifts, honorariums, dinners, and fancy trips from pharmaceutical companies, despite the companies' belief that such perquisites are effective for this purpose. Even if the companies are mistaken in this belief, many commentators advise caution to avoid the appearance of conflict of interest when dealing with corporations. This view is especially prevalent in the biomedical-ethics community. The director of the premier US biomedical ethics center, Daniel Callahan at the Hastings Center, has warned, "Those of us in bioethics have a special obligation not only to avoid acting corruptly...but also to avoid being part of a culture of money that can force bad choices on us." Others believe that any relationship between an ethicist and a corporation is suspect. One academic from the Center for Bioethics at the University of Minnesota wrote, "I worry that each corporate check cashed takes us one step closer to the notion of ethics as a commodity, a series of canned lectures, white papers, and consultation services to be purchased by the highest bidder and itemized on an annual budget report."[22]

On the other side of the debate are those who believe that it is unreasonable to think that consultants to corporations should refuse payment simply because they provided ethics advice. Ethicists have training and expertise and can rigorously apply ethical and philosophical principles to help companies analyze ethical dilemmas. They also can educate about the prevalent ethical standards for endeavors such as human research, medical treatment, drug marketing, and organizational behavior. If ethics advice is considered valuable to a company, the argument goes, then advisors should be paid. Some ethicists have stronger views on the subject; Art Caplan who directs the Center for Bioethics at the University of Pennsylvania, said, "If I get a call from a Merck or a Pfizer asking me to personally look over some of their patient consent forms for medical research, I'm not going to do it pro bono. The last place I'm going to make a charitable donation is to a major pharmaceutical company."[23]

[22]Elliott C. Throwing a Bone to the Watchdog. *Hastings Center Report*, March/April 2001.

[23]Atkin JL. Educator and Experts: Bioethicists Traverse the Uncertain Ground Between Paid Consultant and Moral Teacher. Science & Spirit, November/December 2001. Available at http://www.science-spirit.org/articles/articledetail.cfm?article_id=263. Caplan reports in this article that he earns $20,000 per year in corporate consulting fees and converts the money into gifts to his Center.

Caplan's view is that so long as his center is not dependent on one corporate source of funds and he feels free to walk away from a corporate funder, he can maintain the objectivity of his advice.

Kreiner was aware of these debates; however, he also needed to consider that it was unlikely that many qualified advisors would agree to serve without some compensation. Most were busy professionals with other commitments, and many academics could not afford to work for free. Despite the need for compensation, he encountered the scruples of some advisors, who objected to taking any money for providing ethics advice. As one member put it, "I'd like to keep my conflict-of-interest virginity. That is worth something to me." Several other members, acutely aware of the persuasive power of money and concerned with their reputations struggled mightily with this question. After discussing the question at length at one of the first meetings, Kreiner and the committee members eventually decided that "fair market value" for the services of the members was $125 per hour, plus travel and other expenses. After 7 years, the amount was increased to $175 per hour, not because anyone requested it, but because Kreiner decided members were overdue for a raise. According to one member, the amount was considered appropriate because it was "not superlucrative, not so much to be an inducement, but not so small as to be meaningless." Members could choose to accept the fee, decline to accept any money, or have Affymetrix donate the money to a charity.

Members decided to accept the payment from a mix of motivations that included need (some members had salaries and financial needs greater than others—"I have kids in college!" said one) and the desire to avoid any suggestion that the money would influence their advice. One member said:

> My bottom-line personal test for how much money I should accept for ethics consult work such as this is that the money has to be so little that it is inconsequential for me. My moral test is if I took the job because the money was good, then I have a conflict. If the amount of money is low enough that I can take it, or leave it, then I'm comfortable.

In the end, some members refused any compensation. Others accepted the money, some asking for payment for committee time but not for time spent doing background work or traveling to and from the meetings. One member uses the money in part for personal use and in part to support professional work. Another member decided that it was not ethical for her to be paid by her employer while she was doing Affymetrix work, so she ensures that Affymetrix work is separate from her job.[24] No one chose the charity option—one member who refused compensation said he would consider donating to charity if the payment were higher. One last member had no choice since his institution keeps all consulting fees and controls his level of compensation.

---

[24]For the employers of others, consulting was considered an acceptable activity.

MOTIVATION

If money was not a motivating factor to join and stay on the Affymetrix ethics committee, what was? Committee members had different responses to this question. Some replied that the motivation actually went the other way, against working for a biotech company. No one wanted to be seen as a paid apologist for the corporate status quo. Despite any misgivings, however, many members declared that they were initially inclined to join because Paul Berg had asked them to. They admired Berg's reputation for scientific excellence, his sincerity and high principles, and they relished the opportunity to work with him. Thane Kreiner was also well respected by those who knew him. Together, members said, Berg and Kreiner convinced them that the company was populated by people who wanted to take the high road in business decisions and were genuinely interested in obtaining legitimate ethics advice. Members were also intrigued by the challenge involved in defining the high road in a biotech business. The educational opportunity was another consistent motivating force. Members anticipated that they would have access to cutting-edge scientific information and their expectations have been met. One member said: "The main benefit for me is the fact that I learn an enormous amount. I think I've learned more than I have contributed. For instance, when Steve Fodor went to the white board and drew a description of a haplotype—it was so good a 4th grader could have understood it."

Indeed, most ethics committee members do not view their Affymetrix work as totally selfless. Most feel that the information gained at the meetings about genetics, emerging technology, and how a biotech organization works helps them in their careers as teachers, physicians, and researchers. The committee physician feels he has a better appreciation of how cancer gene testing affects patients. Berg feels he is a better informed company director for having participated in the ethics committee discussions. The educators on the committee use the insights gained on the committee in their teaching and to advise students.

Finally, all members report that the opportunity to work with each other, the collegiality of the group, and the stimulating nature of the discussions is "fun." As one member put it:

> It's a long and often awful commute to get to these meetings. But the meetings are lots of fun and I learn a lot. I would not do it otherwise. It keeps me up to date in genomics. It helps me understand the policy arena, patenting DNA issues, bioterrorism. Of the things I do, this is more fun and interesting and also easier because we are not making decisions for the company; we just give recommendations.

Another academic, who agreed that the Affymetrix experience was "fun," also enjoys the opportunity to have a real impact: "I like having an influence. Some articles that I write go into a black hole and maybe no one reads them. But, here, what we do has an impact on company decisions." Another company member agrees:

> The benefit of having an ethics committee is that it becomes a place to have a quasimoral discourse where there is no other space for this at the company. There may be hallway discussions, but these will not change the culture. More than

what we say or recommend, the ethics committee becomes a forum to have the larger discussions. When the top executives have access to the ethics committee, which is so critical, that is where the impact occurs. If we only worked with PR, for instance, it can't affect corporate culture.

OPERATION OF THE ETHICS COMMITTEE

With these fundamental matters decided, the committee proceeded to meet between three and four times per year. Meetings last 2 to 3 hours and, because of Bay Area traffic, many members spend 2 hours commuting. Despite the travel difficulties, the meetings are usually well attended by most of the outsiders. Kreiner and Tillman Buck always attend, and Janet Warrington usually does. Depending on the agenda and availability, several of the company senior executives also attend, including Steve Fodor, President Susan Siegel, General Counsel Barbara Caulfield, Senior Vice President of Clinical Genomics Research and Development Rob Lipshutz, and Vice President of Governmental Affairs Robert Wells. With Fodor being the exception, these executives come to listen and, except to ask clarifying questions, do not usually participate in the committee discussion. Fodor, on the other hand, actively engages in the discussion when he is there. Company scientists also attend when the discussion issues impact them or when the committee needs a technical explanation about a product under discussion. All committee members report that they highly value the attendance of both company groups, as it shows that the committee work is taken seriously. Without the attendance by the high-level management, some members would feel their work unappreciated and not worth their effort.

Initially, Berg and Kreiner set the committee agenda. Later, Tillman Buck took over. Issues come to the agenda three ways:

1. Top-down from the Executive Management Committee (on which Kreiner sits), the corporate attorney, and business development staff who often ask the committee to address an existing or proposed business deal in which the company is making a chip for a use that raises ethical concerns;
2. Bottom-up from scientists (such as research collaborations, sample-sharing practices), sales and marketing (Is this deal appropriate? Should we turn this customer away?), business development, public relations/communications, and government affairs staff. Sometimes customers ask what position the company takes on particular issues;
3. From the committee—Kreiner sometimes asks members what they want to discuss and they have suggested spin-offs from the agenda. Janet Warrington said that she has asked for items to be placed on the committee agenda when she hears loud hallway debates about an issue.

NATURE OF ETHICS ADVISORY
COMMITTEE DISCUSSION

At first, committee discussion consisted of theoretical or "what if?" discussions dealing with products, partners, collaborators, and customers: for example, what if a business partner wanted Affymetrix to develop a chip that could enable bioterrorism? Related early discussions dealt with the extent to which the company would justifiably be "tainted" by the questionable activities of their business partners. Sometimes, members responded that the better option was to become engaged in the business and set high standards to prevent low-road competitors from taking the business. At other times, the committee adopted the position that, even if other competitors would do so, Affymetrix should not develop chips used for objectionable purposes.

From these initial discussions, the committee adopted a decision matrix that could be applied to future similar decisions (see Exhibit 11.3). The matrix, developed primarily by the attorney member, is invoked, for instance, when addressing questions

EXHIBIT 11.3   **Affymetrix Ethics Committee Decisions Matrix**

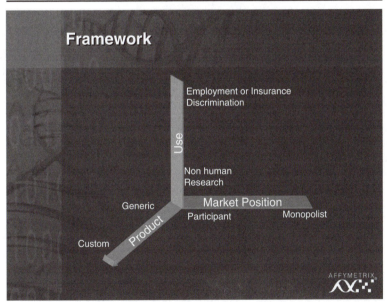

This decision matrix indicates that the company should take more responsibility for the use to which its chips are put when the PRODUCT is custom-made for a particular purpose (as opposed to a generic use), when the chip USE has a greater negative impact on humans (as opposed to using the chips for animal research or restricting the chips to human research where patients' interests are protected), or when Affymetrix has monopolist MARKET POSITION (as opposed to participating in a market with many other competitors). Starting from the central axis of the matrix, the possible actions that Affymetrix can take range from: do nothing (i.e., business as usual), make recommendations to protect against harmful or inappropriate chip use but do not seek to enforce them, seek to encourage federal or state legislation on some uses, put contractual limitations on the use of the chips (as permitted by law), or avoid the market entirely.

Decision matrix conceived by Hank Greely, Professor of Law, Stanford University. Available from: http://www.affymetrix.com/corporate/outreach/ethics_policy/ethics_policy.affx

about how Affymetrix should pick business partners and how much the company should be responsible for what business partners do with Affymetrix products. The "product" arm of the matrix is based on the fact that, when Affymetrix makes a custom chip, the company knows exactly what the chip will be used for. In such cases, applying the matrix would prompt Affymetrix to take more responsibility for what the customer does with the chip. (See the diagram in Exhibit 11.3 for a more detailed description of how the decision matrix is applied.) Examples of what Affymetrix might do to ensure responsible chip use include imposing contractual restrictions or lobbying for regulations that will prevent the chips from being used in harmful or medically inappropriate ways. On the other hand, less responsibility is required for what customers do with the chips if they are for a general use and sold to many different customers. According to Tillman Buck, this decision matrix is now fairly ingrained in how the company insiders think about their responsibilities in any new business association.

Soon after the basic arrangements had been agreed upon and the decision matrix had been developed, the committee began to address more substantive matters and was asked for advice on questions such as: should Affymetrix do business with a particular customer? How much investigation should the company perform when the potential customer is unknown to them? Is a proposed chip use legitimate or responsible? Is a particular technology or chip ready to be applied to patients? What many members immediately liked about these discussions is that they were not theoretical—they were about real decisions that the company must make. Members report that their being asked to deliberate on "real" issues was important in the development of their interest in the work and trust in the company.

By all accounts, the ethics committee's meeting discussions are open, free-wheeling, lengthy, sometimes contentious, but never rancorous. The issues raised are often so interesting that some members cannot help but expand the discussion into nonbusiness realms and need to be dragged back to "ground level" and reminded that the matter at hand is related to a business decision. Other times, the discussion will get side-tracked into the science if members feel they need a full understanding to come to a conclusion about ethics. For questions about doing business with a new customer, the committee members often want to know about the people involved: in addition to the science, are the people credible? Is this a reputable business group? Despite the tendency for discussions to range widely, meetings are uniformly described by company insiders as informative and productive.

When asked to describe the nature of their ethics discussion, the committee-member responses varied:

> We don't do a straight ethical analysis. Any sensible outsider could give the same advice. No one on the Board has a PhD in philosophy. Instead, what happens is that we use our expertise to spot issues.
> It is never pure ethics. Social issues come into play. Or the law. There is always an interest pulling to other factors that are not really ethics.

In meetings, someone will say, 'I am not sure how comfortable I am with the project,' and then they make the switch to discussing whether doing something will make the company look bad. So the discussion veers away from ethics to PR and we discuss that for a while. The discussion is not so applied that the advice is completely situational—we do talk about holding to certain values. But a mix of things gets discussed.

In some discussions, the ethical concerns have been over the top. They are simply not feasible to address in a corporate environment.

No single person tends to dominate discussions, unless they are expert on a particular issue. Likewise, there is less participation from members on topics outside of their area of expertise. According to one member, "No one BS's or pretends to know something when they don't." Every one of the committee members interviewed expressed appreciation for the contributions of the other members. They say that the group has formed a close collegiality, and members feel comfortable in the give-and-take environment of the discussions and debates. Some of the members have developed reputations among the other members—there is a "pot stirrer," a "maverick," one that is "cautious but always thoughtful and wise," another who is "always impatient about dwelling on issues you can't do anything about." One member always wants to "get somewhere" with the discussion. Some do not see the use of including a religious viewpoint and others do. One member is noted for his ability to sum up and make sense of the discussions. None tends to argue from any particular ideology, although, over time, the members have come to anticipate the general positions others will take. Some are more pro-technology, or favor unfettered scientific freedom. One member is concerned about the tendency to believe that technology can solve all medical problems and argues that many surrounding problems need to be solved to make Affymetrix technology work for patients. Others always present the patients' point of view. Some usually try to dampen corporate enthusiasm or encourage the company to reach further to obtain an ethical result (for instance, ensuring proper wording in a research consent form does not ensure true informed consent). Despite these tendencies, most members say that they cannot uniformly predict how members will decide any particular issue. Also, despite these various approaches and viewpoints, the committee members have usually been able to reach consensus recommendations when asked to do so.

The attendance of some people at meetings is always influential. Paul Berg's presence is especially important. He is described as outspoken and forthright but also very good-natured when the discussion goes against him. The fact that Berg has important scientific and policy connections and is often asked to testify in Washington on science policy matters makes him an important source of information and an important voice for the company views. Members also enjoy the meetings with Steve Fodor because his bullish view of technology and vigorous input makes for a lively discussion that the members find "fun," which often means fun to challenge. According to one member, the discussions that take place when Fodor is present force everyone to step back from their preconceptions and think in a focused way about the benefits and risks of commercializing genetic technology.

POTENTIAL CONFLICTS WITH THE EXECUTIVES AND THE BOARD

Everyone involved with the ethics committee recognizes the possibility that the company may take an action with which the committee strongly disagrees. Some ethics committee members worry less about this possibility than others. Those who worry less are more interested in knowing that the company listened to their advice and respected their opinion. According to one of the members, "It is [the company managers'] engagement at the meetings, and the fact that they are concerned, that matters to me. It is less important for me to know what the company did specifically on a matter that they brought to us." These members are less apt to inquire about what the company did following an ethics opinion.[25] Others do want to know what company actions are taken and would consider quitting the committee if they felt strongly that the company was disregarding sound ethical advice. Committee members gave examples that could trigger this reaction: for one member, it would be if the company moved away from the "technology vision mode" and instead took the position it had a right to sell any legal product without regard to any ethics input; another member said that she would become "terribly disappointed' if the company were to abrogate the committee's decision making authority and cite an ethics committee opinion to justify controversial action; a third hypothetical example was if the company were to sell microarrays on the Internet claiming that it was the best way to implement patient autonomy, using an ethical justification for a business decision. Unless there were data to support it, this kind of action would be considered a misuse of ethics that would lead to strong objections by at least some of the committee members. If objections were ineffective in such a scenario, said one member, "I can always quit. I wouldn't go to the newspapers but the company would know what I thought."

Paul Berg, who sits both on the ethics committee and the company Board, does not know what would happen if the ethics committee advised on an issue and the company "went the other way" in a situation where there was a major business interest at stake.

> Although I guess that the Board would more likely argue in favor of profit for a legitimate use of the chip, it would be hard for the Board to deal with this and I don't know what they would do. And I don't know what the ethics committee would do. It would have to be something very offensive to create such a split. I would want the ethics committee to provide the company with a broader perspective and be willing to concede that the company has to make the decision.

---

[25]Committee members are allowed to receive follow-up information regarding matters discussed at meetings.

Fortunately, the company has not had to face such a situation.

MAINTAINING OBJECTIVITY AND DISTANCE

Ethics-committee members worry about whether they can maintain the outsider's perspective that they initially brought to the company. Members understand that they function most usefully when they bring a detached perspective and maintain a willingness to critique and analyze company action. They have also come to know that a company ethics committee cannot function well without mutual respect between the company managers and the committee members. Each must value the other's work, and mutual trust needs to exist to engage in the open and free debates that ethical issues require. This necessary trust and respect, however, can lead to further entanglements, where members start taking such a personal interest in the company and its managers that they begin to feel reluctant to challenge or question company activity in an objective fashion. Some committee members worry that getting too close to the company undermines the basic utility of the committee. As one member said, "There is a level at which we are co-opted since we care about the company and the company people. We end up not sitting outside. That might detract from our effectiveness."

This issue has been discussed at committee meetings. "It is never far from our minds that we are advising a business that needs to make profits to survive," said one member. All committee members interviewed feel an allegiance to the company and want the company to succeed. Some members are fully aware of the company financials and debate among themselves whether they should worry about whether they should raise ethical objections to company activity that is likely to undermine company viability. They struggle (some more than others) with wanting to recommend the right course of action versus recommending actions compatible with financial success. In a slightly different vein, most members also said that they recognize that the reputation of the company and the company brand is a critical part of the future success of the company. The brand would become tainted with unethical activity, and members disagree about how much the risk of "looking bad" is factored into their advice to the company. Some believe that they have a duty to consider and preserve company reputation. These members are much more apt to supplement their recommendations with further advice about how the company can mitigate criticism.

Thane Kreiner takes a pragmatic view of this issue. He wants and encourages the outsider's view since, in his mind:

> ... there is no upside for us if they pull their punches and no downside risk if they are critical of the company. They won't get fired for it. Over the years, they have seen that we want them to tell us what they think, so it's more valuable if they don't accommodate us. We don't want them to accommodate us. I think they like that. But also, they seem to like us as people and probably want us to succeed, so we all get along that way. But I can't think of a meeting where there hasn't been an argument about the issues, so I think they are still keeping an open mind about things.

Other relationships have developed since the committee was formed between ethics committee members and the company that have also raised questions of both objectivity and conflict of interest. In one case, one member was an expert on a topic relating to a potential use for one of Affymetrix's chips. Affymetrix had written a proof-of-concept paper about this new chip and asked this committee member to provide an ethics perspective on the proposed use. The member agreed to provide the input but would not accept money for the work, in order to eliminate questions about objectivity. In addition, the committee member was worried about the time commitment required for the work. Since the company wanted a comprehensive review of the issues, it decided instead to give a monetary gift to the committee member's institution to fund a conference on the topic, to which the committee member would contribute. The company wanted to fund the work and the committee member was eager to do it, but they stewed over the conflict-of-interest issue, especially because they were both mindful of the criticism in ethics literature of ethicists who advised corporations. "So, I am conflicted," said the member:

> I want to do the work. The company is seriously interested in supporting the work since it deals with the implications of the use of their technology. However, I have to deal with the conflict-of-interest aspect. I don't want to give back the money. I feel that, so long as I disclose the gift, it should be OK and satisfy the critics. But I also recognize that others may not think so.

IMPACT OF THE ETHICS ADVISORY COMMITTEE

Members of the ethics committee have modest and measured opinions about their impact on the company. They all recognize their role as advisors and know that they do not make decisions for the company. Neither is anyone under the impression that the committee is causing the company to make fundamental corporate decisions. "We are responsible for minor variations on their major business activity," believed one member. Most members believe that their worth comes primarily from an awareness they provide about the larger consequences of corporate action. One committee member put it this way: "We hold space and a focus and a safe place for them to have non-core business discussions. To ask questions like 'Is there anything wrong with this deal?' or 'How far should we go to be ethical?' We help them clarify why certain activity is acceptable and why other choices are not." Another member said, "We ask the questions no one at the company is asking." Committee members are convinced that this aspect of their work is valued by the company because the high-level executives continue to attend meetings and often specifically acknowledge the usefulness of the candid and different points of view. One committee member also thinks that the reason the senior executives attend is that "they welcome the opportunity to take a more intellectual perspective on their work."

Thane Kreiner agrees that the ethics committee "keeps that awareness present for company people who have business interests in the forefront of their minds."

In addition, he values the various perspectives brought out during committee discussion:

> Our ethics committee members debate, and this gives us a rich context [of the various points of view] and that is what is valuable for us. We don't just get their conclusions. When you hear their various points of view, you understand how others think about the question and about what the company does. If we understand the play of opinions, we make better decisions and we can never get [that diversity of opinions] with internal advice. The nuances are important. When there are shifts in public policy or public opinion, they tend to be subtle and you don't recognize them at the time until they become big shifts and then it is too late.

Janet Warrington verified that attending meetings is always an education and allows her to see issues through new eyes. The value for her is to "get more information and insight rather then persuasion to do one thing or another." She also reports that, on several occasions, the ethics advice has influenced the direction of her research.

ADVICE PROVIDED BY THE ETHICS ADVISORY COMMITTEE

Despite the general agreement among the ethics committee members about the intangible nature of their impact, there are many examples of situations in which corporate action has been directly influenced by the committee's recommendations and insights. According to Berg, it has had an impact on company financial decisions. For instance, ethical advice has persuaded the company to refuse business or to modify the terms of a business deal. Questions that lead to such decisions include whether the use of the chips by a customer would be offensive, whether a particular deal would reflect negatively on the company reputation, or whether the company can make money on the chips, given any ethically driven selling restrictions. Berg described several such matters that were discussed by the ethics committee:

> One scenario we dealt with had to do with a request from a foreign company to use a chip for something that was illegal here but not in the foreign country. How do you come down on this? Is it malevolent or is it just the flavor of the times with an energetic group condemning some use? Other scenarios that raise this kind of concern are when implementation of a diagnostic was too early to understand the full consequences of its use. Or when chips are used for non-medical purposes— for instance, using a GeneChip microarray to trace the genetic profile of a seed so that a bio-ag company can sue a farmer who has pirated the seed, or identifying people genetically predisposed to like a particular scent so that a soap company can sell more product.

It is amazing to some committee members that, before Affymetrix was profitable, the company decided, in part on ethical grounds, not to accept some proposed money-making deals. The fact that Affymetrix has placed limits on the use of their chips has won compliments from ethics committee members who have seen other companies ignore misuses of their products so as not to undermine sales.

There are other examples of instances in which ethics committee advice has impacted company action.

EXHIBIT 11.4   **Affymetrix Contract Language Regarding Ethics Conflicts**

Ethics. The Parties acknowledge and agree that the field is new and changing and that it is difficult to foresee potential ethical or legal issues, or other matters of public concern, arising from various potential applications or misapplications of the activities contemplated by this Agreement. Therefore, the Parties agree that, in the event a material ethical or legal issue, or other matter of public concern, arises or becomes foreseeable relating to the activities contemplated herein, the Parties agree to work together to take mutually agreeable steps to address such issues in a lawful, ethical manner. In the event the Parties disagree as to what course of action is appropriate in a given situation, the Parties agree to engage a neutral third party (e.g., a group of internationally recognized scientists and/or ethicists) to help the Parties decide on a proper course of action. For the avoidance of doubt, in no event will either Party, their respective Affiliates or licensees use, or permit to be used, the Chips or Collaboration Products in any illegal, immoral, unethical or other patently offensive purpose.

LOBBYING FOR GENE-TEST PRIVACY LEGISLATION

The development of the decision matrix led Affymetrix to lobby for legislation to ensure the privacy of genetic information gained from the diagnostic use of microarrays and prevent various types of discrimination that can flow from disclosure of genetic information. Kreiner took responsibility for this effort and is proud of the fact that Affymetrix was the first company to sponsor this kind of legislation. He saw it as a win-win situation, because eliminating any barriers to genetic diagnostic products would be good for the company and the privacy protections would benefit and protect patients. He recognizes that the company can be criticized for sponsoring social legislation when it was actually motivated by self interest. "My view is," he said, "what's wrong with that? It is also good for society, so we support it for that reason too. In the long run, if there is legislation in place, people will trust that genetics tests will not lead to discrimination and so they will be a lot more likely to ask for the test than if they feared discrimination."

CONTRACT LANGUAGE REGARDING
ETHICAL RESPONSIBILITIES

Ethical issues were expected to arise frequently in Affymetrix's technology and research affairs, so the ethics committee suggested that Affymetrix adopt contract language with collaborators and partners to address ethical concerns. This recommendation was the result of an ethics-committee discussion of Affymetrix's SNP discovery work.[26] SNP discovery work requires large-scale screening of human tissue samples sent to the company by physicians. Because this work is customized (meaning that the company is screening for a particular, medically relevant, genetic difference), the ethics committee's decision matrix analysis indicated that Affymetrix had a responsibility to do more than provide the technology and the testing service. Provisions preventing harm to the tissue donors were in order. A lengthy debate ensued at the ethics committee about what those responsibilities were.[27] Paramount in the minds of the members was that tissue donors provide informed consent before they donate tissue (as opposed, for instance, to using existing stored tissue from hospital laboratories without patient consent). The discussion about how informed consent should be obtained consumed a significant portion of the committee deliberations. Some members felt that it was sufficient to advise the donor about what would be done with the tissue and allow the donor to decide whether to donate tissue. Others objected to this "market model," which meant that, so long as there was full disclosure and patient consent, physicians could do anything. The alternate view was based on the belief that patients did not have the wherewithal to make optimal decisions for themselves in complicated medical research transactions, and that physicians had a duty not only to inform patients but also to protect them from the harm of what was being proposed. Eventually, the Affymetrix research contract contained language that required those who obtained tissue samples to adopt prevailing ethical standards to ensure the informed consent of the donor. All ethics committee members endorsed the language, but some did not think it went far enough to protect the interests of patients. Other contract provisions, however, were unanimously endorsed: that donor privacy be maintained; that no tissue sample would be sent to Affymetrix containing any information that could identify the tissue donor; that tissue samples would not be kept at Affymetrix but would be returned or destroyed; and that Affymetrix had the right to inspect the processes of those who collected tissue samples.

A contract clause that was more general was adopted after the ethics committee, at the request of a company development manager, reviewed a proposed collaboration with a new partner whose core business had raised ethical concerns in the past. The ethics committee suggested that the contract and others like it contain a clause saying that the parties recognized that ethical issues could arise, and that there would be some mechanism to resolve any ethical disputes. At first, the corporate partner balked at the language, which created fears at Affymetrix that it could lose the business because of their insistence on the contract terms. However, Affymetrix was able to convince its partner that ethical issues had existed in the

---

[26]SNP stands for "single nucleotide polymorphism." An SNP is a place in the genetic code where DNA differs from one person to the next by a single nucleotide subunit. These slight genetic variations between human beings may predispose some people to disease and can explain other medically relevant differences between people.

[27]The medical literature at the time was addressing the conflicts between the societal need for this scientifically valuable tissue and the potential harm to tissue donors. Loss of medical privacy was of concern, as researchers often needed to know the medical history of the donor in order to make sense of DNA information, and this sensitive information could be used by insurers or employers to deny benefits if inadequate privacy protections allowed the information to be disclosed. In addition, there were both positive and negative consequences to donors if they were told that researchers discovered a disease-linked mutation in the donor's DNA. Whether donors should be paid for their commercially valuable tissue is another subject addressed at the time in the literature.

past with corporate work and that Affymetrix legitimately needed to protect its interests by having an agreed mechanism to prevent and resolve disputes based on ethical concerns. Eventually, the new partner agreed to the language, and Affymetrix has included the clause in subsequent research-partnership contracts. (See Exhibit 11.4 for this contract language.) According to Katie Tillman Buck, "This provision throws people. They don't know what to think about an ethics clause. We tell them that we want to make money, but not at any cost. As far as I know, even though potential partners question the language, they have all agreed eventually to sign it."

PARTNERS AND COLLABORATORS

Initially, the ethics committee addressed this matter theoretically and supplied an overall guideline that stated that the company technology should be used in support of scientifically sound and ethically justifiable purposes. Company collaborators and customers should also adhere to this standard. In one case in which this guideline was invoked, the company received a request to develop a chip for use in pre-implantation genetic diagnosis (PGD) that could test for multiple inherited disorders on a single chip.[28] The committee felt that, in general, PGD was not illegal and that its use could advance science and alleviate patients' suffering; therefore, there was nothing ethically wrong with using PGD. However, there was a high potential for misuse, as the tests could be used for non-medical purposes, such as selecting desirable and undesirable traits, including sex, and discarding embryos that were less than genetically perfect. In addition, there was no uniform agreement about what medical conditions (for example, deafness) were appropriate to include in PGD testing. A particular concern in this case was that the party that had approached Affymetrix with the research proposal was located in a country with no legal or ethical protections in place to protect against the misuse of PGD. This was an important emerging area of medicine that needed careful development and agreement in the medical community about permissible uses. Janet Warrington, who has a special interest in this potential use of company chips, was surprised at the initial negative reaction of the committee. She pointed out that if Affymetrix did not work with this party, some other company would, and maybe less responsibly. The ethics committee, however, advised Affymetrix that the development of a multi-disease PGD chip should be done in countries like the United States, where standards are in place, primarily through institutional review boards (IRBs), to prevent misuses. According to Thane Kreiner, this was good advice, and the company rejected the foreign proposal. Since then, Affymetrix has continued to consider other requests to supply chips for PGD research and started hosting and sponsoring multiple in-house and external lectures on the medical, social, and ethical implications of multi-disease PGD testing. According to Warrington,

---

[28]PGD is genetic testing of embryos prior to in vitro fertilization. The purpose of PGD is to diagnose genetic birth defects such as cystic fibrosis or Down syndrome. Embryos that test positive for disease genes are not implanted into the mother's uterus.

"We want to be proactive, not reactive. We see this as a process, during which the technology evolves and our insights mature. 'No' means 'No for now.'"

CLINICAL USE OF GENECHIP ARRAYS

Most Affymetrix customers use the chips for research, and each contract with these researchers contains a clause stating that the chip cannot be used for clinical purposes because of Food and Drug Administration (FDA) regulations. In addition, Affymetrix has not rushed into the clinical space, in part because of the concerns of the ethics committee that any diagnostic use of the Affymetrix technology currently needs more research to demonstrate utility and reliability for use in healthy or sick patients. Because information on the impact of genetic influences on disease was still evolving, Affymetrix turned down a proposed business deal to develop a chip to detect the presence of genes (called BRCA1 and BRCA2) that predisposed women to breast cancer. Concerns existed at the time that not enough was known about the penetrance of the genes and, hence, the reliability of a gene test in predicting cancer. Neither was there sufficient assurance that physicians could adequately interpret test results. Sufficient privacy and antidiscrimination protections were not in place to protect the test results from being used against women to deny them health insurance or employment. Finally, there were no proven medical options available to women to reduce the risk of breast cancer detected by the test. This last problem was called the "so what?" or the "therapeutic gap" issue—a woman learning that she carried a breast cancer gene she could do nothing about was likely to experience more suffering than medical benefit. Such an array would be useful only for a limited number of women from families at extremely high risk for breast cancer. These circumstances led the ethics committee to advise that further research was needed before this diagnostic test could be used in patients. According to Thane Kreiner, although there were other reasons, the ethics committee's advice was the tipping point that led the company to refuse the business.

At the time of the case study, Affymetrix was beginning to work with another customer to develop a chip used in disease diagnosis, and the company was working closely with this partner to ensure that clinical use was not premature.

RACE AND GENETICS

A company approached Affymetrix to make a GeneChip microarray using 70 SNP markers to identify the race of people; for instance, whether a person or his or her ancestors came from Africa or Asia. The company claimed that its success rate in race prediction was good. However, uncertainty about the implications of identifying race led Kreiner to request ethics committee input. Some of the intended uses had been developed from a respected research program and seemed legitimate. For instance, criminal investigations and perpetrator identification could be aided by using such a race chip: a witness who incorrectly identifies the race of the suspect can be discounted. The ethics committee did not think that such a use was ethically problematic, especially because Affymetrix could include more markers on the chip and make the race prediction more reliable. But what did concern the ethics committee was that the company also wanted to use the chip to offer a genealogy service and market the service to the public. Since the advent of open access to the human genome, researchers had been speculating that gene tests could be used to decide claims about people's ethnic, political, familial and religious identity. Already, Y-chromosome tests had been used by some African Americans who believed themselves to be descendants of Thomas Jefferson and his slave mistress Sally Hemings. Other potential uses included determining who had sufficient native ancestry to qualify for government restitution programs; who could be considered a citizen; who had rights to affirmative action programs; what a person could claim about his caste or religious sect.[29] This interest and activity was just one step away from commercial gene-testing services. Some committee members felt that using the chip to trace ancestry could cause "mischief," depending on what the user made of the information. Two members on the committee regarded race identification as a gross deviation from science; race was, in their opinion, often meaningless, offensive, and misused. Others worried about the product's impact on the concepts of race and that the company would exaggerate what the test could do. The committee addressed various options, including whether the good that could be accomplished with the appropriate use of the chip could outweigh the possible harms of the scientifically suspect genealogy service. After two meetings devoted to this issue, the ethics committee and the company decided that not enough scientific and ethical research had been done to validate the integrity of genetic racial profiling, and the company declined to participate in the project. In addition, the company would participate in any future work in this area only if the scientists associated with the project were well respected and credible.

MILITARY IDENTIFICATION CHIP

Affymetrix asked the ethics committee to discuss a proposed collaboration with the military to develop a chip that could be used to identify military personnel. Mitochondrial DNA from the service member's mother (alive or dead) could be incorporated into the chip and used to identify a soldier who died in combat. Some committee members expressed general concerns that the military has a known history of subverting technology for questionable uses. This comment triggered several questions: Would the company reputation suffer if it were known that it collaborated with the military? Should everything that has a potential for malevolent use be condemned? Can we afford to pass up this business opportunity because of a misuse that is only potential? Could Affymetrix control the military uses for the chip, and if not, would it bear responsibility for all uses of the chip? Do you always have to trust your business partners? Can we try to explain our objections and hope they will accommodate us?

[29]Elliott C, Brodwin P. Identity and Genetic Ancestry Tracing. *BMJ*. 2002;325:1469–1471.

MISCELLANEOUS AND FUTURE ITEMS FOR DISCUSSION

Other issues addressed by the committee, but in less specific circumstances, include such topics as gene patenting, genetic studies of native populations, general impressions about a new genomics business that was an Affymetrix spin-out, and the appropriateness of the company's ethics website. Committee members also advised on the educational conferences sponsored by the company. In the future, the committee expects to discuss and advise on the consequences of marketing individualized medicine diagnostics,[30] legitimate uses of a whole-genome chip, direct-to-consumer advertising of genetic testing services,[31] and consumer uses of chips. As for the ethics committee function, both Kreiner and Tillman Buck want to preserve and improve on its contribution. Tillman Buck would like to see membership grow and wants to replace the religious member (an emeritus bishop on the committee had retired). Tillman Buck would also like to include a patient advocate and a member who can represent the "voice of the people," recognizing that there are no such ideal people and that, as with most religious advisors, the sophistication of the science discussions would deter many candidates.

THE FUTURE OF ETHICS AT AFFYMETRIX

Thane Kreiner recalls that in 1996, when the ethics committee was formed, Affymetrix employed about 90 people. Laboratory meetings were attended by all employees and company-wide communication was easy. Affymetrix was imbued with Steve Fodor's passion for science and technology and the need to "always do the right thing." According to Kreiner, the company was small enough that Fodor's leadership set the tone for the entire company. By 2003, there were 900 employees

---

[30]Individualized medicine involves the use of genetic tests that can identify which patients are likely to respond to certain drugs or which drugs will produce unacceptable toxicities. For instance, a new breast cancer drug, Herceptin, is prescribed only for patients whose gene test shows that the cancer is susceptible to the drug. Only about 30% of breast cancers fit this description and, in women with this cancer variant, the drug is remarkably effective and can even cure the disease. For the other 70% of breast cancer patients, the drug has little effectiveness. Some questions that this kind of treatment raises are: Because of the added expense of the test and the smaller markets for drugs that are used in subpopulations of patients, will the cost be prohibitive and access denied to all but wealthy or insured patients? How certain does the test have to be to deny a sick patient a potentially beneficial drug? Will desperate patients demand access to a drug that the gene test reveals has only a low likelihood of helping them? Affymetrix had recently developed a gene analysis chip for Roche, called the AmpliChip CYP450, which will be used to detect variations in DNA that are known to affect genes that control how the body metabolizes certain drugs. According to Paul Berg, Affymetrix's ethics committee will be discussing the consequences of marketing gene tests that enable doctors to make prescription decisions based on the genetic make-up of patients.

[31]Genetic testing services were already being marketed to consumers on the Internet, a growing number of which related to testing for health reasons. These services ranged from standard tests, such as hemochromatosis and cystic fibrosis, to more unconventional tests related to nutrition, behavior, and aging. Gollust SE, Wilfond BS, Hull SC. Direct-to-Consumer Sales of Genetic Services on the Internet. *Genet Med.* 2003;5:332–337.

in 7 locations around the world, making it much more difficult to maintain a uni-fied, company-values–driven culture. It was obvious that more effort was needed to ensure that a commitment to ethics and corporate values remained widespread in the company. Both the inevitable separation of many employees from top man-agement, and of modern, outside influences, threatened this goal.

For instance, Kreiner saw a worrying business trend that he thought could undercut Affymetrix's ethics culture. The federal sentencing-guidelines[32] and the Sarbanes-Oxley Act[33] had motivated companies to adopt compliance programs that, because they were geared to encouraging companies to obey the law, were also often called "corporate ethics programs." Kreiner believes that there is a dif-ference between the two—ethics extends beyond, for example, the principle that companies should not break the law or lie to shareholders. According to Kreiner's former business professor, "Deciding whether you should break the law is not an ethics' question, it's just wrong." According to Kreiner, ethics answers, unlike compliance, are not "yes" or "no":

> An ethics approach to solving business problems is context-specific and based on values. Business decisions have to be values-based because things will always come up that are not covered by the regulations. That is why, at Affymetrix, compliance personnel report to the legal department and ethics reports to me. We don't conflate them.

However, knowing that many companies do conflate them makes Kreiner alert to the possibility and vigilant in keeping the two separate at Affymetrix.

Kreiner and Tillman Buck realize that efforts to preserve the corporate culture need to be ongoing. High on their list of objectives is to continue supporting the work of the ethics committee. One way that Tillman Buck is attempting to preserve ethics awareness in the company is to support the development of new-employee awareness training. This effort includes informing new hires about the existence and availability of those company insiders who can address ethics issues and about the ethics committee resource. She is also working to improve communication about ethics and values so that employees know that the company takes ethical issues seriously and sincerely wants all of the employees to be responsible and honest. As a part of this effort, she and Kreiner started an in-house ethics speaker series, which has become very popular inside the company. Routinely, 60 to 75 people attend and plans are in place to webcast the events to the company's other sites.

Outside the company, Affymetrix has built education, especially ethics education, into its corporate outreach and philanthropy programs. On his own, Kreiner has

---

[32]The federal sentencing guidelines (see 18 United States Code sec 8A1.2) contains a list of requirements that, if implemented, are considered mitigating factors that reduce the risk of criminal prosecution and lower penalties for federal corporate crimes. Having an effective compliance program in place to detect and prevent violations of the law is crucial among these requirements.

[33]After a series of high-profile corporate scandals, the Sarbanes-Oxley Act of 2002 was enacted to protect investors by improving the accuracy and reliability of corporate disclosures made pursuant to the securities laws. The Act contains broad-based provisions affecting corporate governance, disclosure, and financial accounting for public companies.

spent time advising other biotechnology colleagues and companies on how to use ethics advice in their businesses. He also gives lectures on the ethical implications of the company's work.[34] Affymetrix frequently funds and sponsors technology ethics conferences at high schools and colleges and is involved with the ethics presentations at the annual BIO meetings. Kreiner gets especially excited when he talks about the ethics education programs. He described a collaboration with the Woodrow Wilson Center in Washington, DC, where a 150 people (scientists, academics, patent lawyers, Patent and Trademark Office executives, ethicists, judges, and science policy experts, along with 8 senior executives from Affymetrix) gathered to discuss the business, legal, social, and ethical aspects of gene patenting.[35] The event was moderated by a science reporter from *The Washington Post* and broadcast on a website at the Woodrow Wilson Center. The event, according to Kreiner, was "terrific" and so popular it crashed the Center's server.

CONCLUSION

At the end of 2003, Affymetrix was working toward operational profits, faced with strong competition, and operating in an environment of evolving intellectual property, technology, regulatory enforcement, and ethical concerns. Despite the challenges, Affymetrix considered itself to have many advantages: first-mover status, an industry-standard product, manufacturing leverage, scientific leadership, a strong intellectual property portfolio, strong technological know-how, and a broad customer base. According to Thane Kreiner, it also had the advantage of having a strong, independent, intelligent, and committed ethics committee to help the company maintain its excellent position.

[34]For instance, Kreiner gave a talk at the 2001 BIO Annual meeting entitled "Chips and Ethics," which presented the following questions: Who should have access to genetic information? How will genetic information be used? What forms of genetic testing are acceptable or unacceptable? Who owns the human genome? How will the privacy of genetic medical information be protected?

[35]See www.geneticage.org for a webcast of the event.

12

PHARMASNPS INC.: CREATING AN ETHICS ADVISORY BOARD[1]

INTRODUCTION

PharmaSNPs was one of the first generation of companies to use the decoding of the human genome to develop better targeted therapies for cancer and other diseases. Like many of its peers, it raised venture capital, developed its technology, and went public based on the promise of this new science. And like many of them, it completed all these steps for growing its business without any formal mechanisms for addressing the new ethical issues raised by its cutting-edge technology. As the technology progressed from the research laboratory into clinical trials, however, PharmaSNPs' leaders realized this needed to change. The Chief Medical Officer and Vice President of Clinical Development explained what prompted PharmaSNPs' clinical group to address ethical issues: "Much of what we do is unregulated or at least ambiguous as to what is the right thing to do. The more work we did, the more we realized our exposure and uncertainty. We could sense that we had ethical issues in our labs, but they weren't apparent to us." The company felt it had to address these issues to avoid "a disaster in the clinical trial area."

The solution the firm chose was to establish an expert Ethics Advisory Board (EAB). This case will discuss the formation, mission, and operation of the EAB and some of the issues it dealt with. It will conclude with a focus on the lessons

[1]This case study is of a real company that has been anonymized because the firm was acquired after the case was completed. The case study was written by Cécile M. Bensimon and David Finegold based on interviews they conducted on September 26, 2002 at Company headquarters. Four members of the company's EAB were also interviewed by phone. Unless otherwise specified, all quotes attributed to these people were obtained during interviews. This case study is one in a series produced under grants funded by The Seaver Institute and Genome Canada to study models for ensuring ethical decision making in bioscience firms. The participation, suggestions, and assistance of all involved are greatly appreciated.

for firms considering creating an EAB that can be learned from PharmaSNPs' experience.

THE COMPANY

PharmaSNPs was founded in the early 1990s, under a different name, to exploit genomic research's shift in focus: from the sequencing of the human genome toward the application of the raw genetic data gained from it, and thus to determining gene functions and gene sequences. The firm's focus was to develop a pharmacogenomic approach to cancer therapy.

Pharmacogenetics is the subset of genomics research that seeks to understand the genetic differences that make individuals unique and that affect a person's propensity for disease and response to therapies. Genetic differences between individuals are often caused by tiny variations in human DNA. These occur in one or more nucleotide bases of DNA sequence within the genome; they are called single nucleotide polymorphisms (SNP), known more commonly as *snips*. Because there was growing evidence that differences in SNPs explained patterns of inherited vulnerability to disease and response to drug therapies, researchers believed that genomic technologies could produce a new class of diagnostics and a major expansion in targets for drug therapies.

Prior to the decoding of the genome, all drugs that had been developed worked on about 500 known targets. The new technology offered the potential to increase the number of medically relevant or "druggable" targets to an estimated 10% of human genes (3000–4000 targets). It was also believed that pharmacogenomics[2] would transform the drug development process. But these changes represented a potential double-edged sword for pharmaceutical and biotech companies. On the one hand, a better understanding of the effects of genetic variation on drug response could, by identifying those patients most likely to benefit from the drug and excluding those likely to suffer side effects, increase the likelihood of clinical trial success. The new technology, therefore, could lower the average cost of drug development. On the other hand, subdividing patients into smaller groups based on their genetic profiles could turn what used to be blockbuster markets, such as all breast cancer patients, into much smaller, more targeted markets, such as those with a particular gene associated with a specific form of breast cancer.

In 1996, the company reorganized, renamed itself, and broadened its business model to serve a range of therapeutic areas to include oncology, cardiovascular disease, central nervous system disorders, and inflammatory disorders. Concentrating its efforts on genes of known function, PharmaSNPs collected and analyzed tissue samples to identify SNPs and haplotypes (groups of SNPs) for these disease areas and developed pharmacogenomic technologies (the application of SNPs to drug discovery) and diagnostics. A key strategy for PharmaSNPs was leveraging its knowledge base and technologies through strategic alliances to try

---

[2]The terms *pharmacogenetics* and *pharmacogenomics* are used interchangeably in this case.

to commercialize products, that is, applying pharmacogenomics to the discovery and development of new pharmaceuticals and diagnostic products. By 1999, PharmaSNPs had developed relationships with companies in the pharmaceutical sector, diagnostics/laboratory services, life sciences, and contract research organizations (CROs). According to its then CEO, its business model was to generate "short-term revenue through a combination of partnering and service contracts in the pharmacogenomic field," which could then be used to fund its own product development efforts.

Between 1992 and 2000, PharmaSNPs attracted more than $50 million in venture capital from some of the leading investors in the field. In 2000, it redefined its business model to refocus on one therapeutic area. The company's initial public offering (IPO) raised another substantial amount of capital, giving the firm cash reserves of more than $100 million that would sustain several years of operations. Although the field was crowded, PharmaSNPs was seen as a potential leader in pharmacogenomics because of its broad platform of offerings.

HOW A PHARMACOGENOMICS STUDY WORKS

For PharmaSNPs to provide this set of tools and services, it had to perform some preliminary SNP collection and diagnostic work. This process involved collecting DNA samples for SNP detection from an ethnically and geographically diverse panel of individuals. After a SNP is discovered, a genetic test (or assay) that allows rapid and repeated identification of the occurrence of that SNP in a targeted population is developed. If a SNP is shared by patients with a particular disease, it can serve as a diagnostic marker. Similarly, if the patients who respond to therapy share a set of SNPs, the SNP set can be used to guide therapy.

The process of identifying a patient by his or her genetic profile is called *genotyping*. Within a patient's genotype, a mapping of key SNPs can be identified. For PharmaSNPs' purposes, the goal of a genotyping analysis is to identify the presence of a small subset of pharmaceutically relevant SNPs in a patient. It is necessary to screen thousands of patients to determine whether particular SNP sets do, in fact, pertain to those who do or do not respond to a therapy. If the study yields positive results, the SNP set can then be used as a diagnostic *tool* to determine which patients should receive a particular drug treatment.

SEEKING ETHICS ADVICE

The process of gathering, analyzing, and turning the SNP data into commercial products raised a variety of potential ethical issues that PharmaSNPs was unsure how to resolve. The firm decided to seek the advice of an ethics expert and was referred to the Senior Vice President (SVP) of a software company that had developed a statistically verifiable method of anonymizing large databases, a way to address one of the primary ethical issues–patient privacy–for firms creating genetic data banks. The SVP instead suggested: "Rather than hire us as an ethical backstop, a more robust approach from a public perspective was to put together a high profile Ethics Advisory Board (EAB)." PharmaSNPs' CEO was excited about this prospect, but knew little about working with an ethics expert, let alone an EAB, so decided to work with the consultant first. After about 1½ years, the need for a more comprehensive mechanism to address the many ethical dilemmas surfacing in PharmaSNPs' clinical (patient-related) activities was clear. It was then that the CEO agreed to "go for" an advisory board; the software company SVP agreed to create and chair the EAB.

CREATING AND IMPLEMENTING AN EAB

Constructing an effective EAB is not easy. It is a challenge to create a group "that can work together *and* that can give practical advice on how to execute. This is in fact what is often missing from EABs," said the Chair, determined not to repeat at PharmaSNPs the problems of other EABs: bringing together academics too theoretical to produce well-defined and workable solutions. The Chair had previously worked on privacy, access, and legislative issues and "having experience with trying to create legislation forces you to be practical and understand compromise." The plan for PharmaSNPs' EAB was to have the right role, structure, process, and level of trust among the members, so that the group could offer independent advice that recognized the business realities of this early-stage company.

ROLE OF THE EAB

PharmaSNPs' EAB was set up to be, true to its title, strictly advisory, with a focus on the area of research and clinical practice, not on ethical questions related strictly to business activities. It was up to the company leaders to decide how to act on this advice, according to the head of the EAB:

> We are very clearly an advisory board. We do not help the company make final decisions. We work with the company at the staff level to come up with answers to daily decisions. We would be happy to meet with the company's Board of Directors and interact with them, but we're not set up to deal with liability issues. We don't want to create the perception that we are.

The company could disagree, and choose not to move ahead with the EAB's recommendations. This helped shield the EAB members from liability, both for the advice they gave and the decisions the company ultimately took, making it much easier to attract members.

PharmaSNPs' executives clearly communicated the company's expectations of the EAB. One member recalled, "At the first meeting, the company told us that it wanted us to be a sounding board to guide it on ethical issues in genetics and international research." Such a role resonated particularly with this member, who was conducting health policy research in developing countries. This strictly advisory role was both an opportunity and a limitation for the company. Because the EAB's recommendations were not binding, PharmaSNPs could ignore advice it considered too constraining or costly or impractical at the time. But choosing to ignore it could increase the firm's risk: "If the firm is going to go against the EAB's advice, then it better have a very good reason and argument for why. If the EAB unanimously disagrees, the firm could face enormous liability if it went against this advice and then something later went wrong," said the Chair.

FORMING THE EAB

The Chair drafted a constitution for the EAB. It was reviewed by another PharmaSNPs' ethics consultant, whom the Chair invited to join the EAB to tap her wealth of international and legal experience. Working together with leaders from the company, they decided to create an EAB whose members were, said the Chair, "explicitly intended to represent the views of different groups and act accordingly." They identified six areas for representation on the EAB: philosophy, law, medicine, public health, bioethics, and the public/consumers, with two members for each area. The Chair, whose past experience suggested that groups were most effective with four to six people, was concerned that it might result in an unwieldy group, but saw the benefits of having multiple perspectives well represented. This outweighed the potential benefits of having a smaller Board, and the Chair and the consultant together identified and invited 10 more Board members. "It was planned in pairs," said EAB members. "The idea was that if someone couldn't come, there [would be someone else] representing the same approach." The Chair "pre-screened them on what was needed by the firm so they'd be comfortable with it," making sure that everyone understood that "we want to be practical as well as ethical" to create value for the company by combining analytical reasoning with workable solutions. They also made sure that the EAB membership contained sufficient international expertise and experience for the global scope of PharmaSNPs' work.

Of the 12 EAB members, the Chair had previously worked with 10. This familiarity facilitated the leadership of the diversity of views and expertise and, ultimately, the development of a balanced perspective for each dilemma. It also started the EAB with a foundation of underlying trust between the Chair and the members that would, in turn, spread among the members, and helped build support for the overall project.

Even so "it takes a board a long time to meld," observed one member. The process of getting acquainted with one another was facilitated by PharmaSNPs, which hosted an informal dinner before each meeting. Social bonds among EAB members and between the EAB and the company developed. "PharmaSNPs was a gracious host," recalled the Chair. "The company showed a lot of respect for EAB members. It made people feel important."

PAYMENT

For a company forming an EAB, an early decision to make is how to balance the desire to create an independent Board with fair compensation. For some, paying for ethical advice means a conflict of interest that could compromise the objectivity of an EAB's members. The Chair explained how they viewed this issue:

> There is no provable way to demonstrate your independence unless you're not paid. I'm a firm believer that those doing ethics work aren't paid enough. We have to pay for their time. Their credentials and reputations are based on the opinions they give. We pay $1,000 per meeting, and $200/hour for any additional work, so no one is going to get rich off of doing this.

This amount excluded travel time to and from meetings, which was not paid.

DISCLOSURE

Another decision that PharmaSNPs had to make was whether it would disclose the existence of the EAB and name its members. The company eventually decided not to publicize either for two reasons: first, it wanted to protect the confidentiality of Board members, who were left free to identify themselves as members of PharmaSNPs' EAB if they chose; second, the company wanted to avoid the perception that the EAB served a public relations purpose. Giving the example of a well-known bioscience company who had publicly announced its newly formed EAB, but "subsequently didn't make use of it so it was seen as a front," the CEO explained that the company "didn't want to do it that way. We see it as a practical internal tool, rather than a marketing device."

THE EAB PROCESS

BUILDING A SHARED MINDSET

In addition to building trust among EAB members, the Chair felt that it was important to build their collective knowledge of the issues facing the firm. Part of this process was providing EAB members with a wealth of information prior to the first meeting about PharmaSNPs' business and its technology. Although this was generally well received, the volumes of material, "with lots of history and terminology, was a little overwhelming" at first, said the lay person representing the consumer perspective.

A next step toward building collective understanding was to begin EAB meetings with formal educational presentations by members of the EAB or the company. "At each meeting, a board member presented current issues from their research to help the decision making, to enlighten us on ethical dilemmas, to get us all on the same footing," a member recounted. At the first meeting, held in August 2002, EAB members were introduced to PharmaSNPs' business and projects by top company executives, who outlined the key ethical issues for advice. Subsequent meetings featured talks by EAB members on ethics in clinical research after data collection, and on privacy, property and informed consent. Interviewees found them "…clear and informative. They were excellent talks. I learned something regarding rules on privacy, the storage of materials and other issues [and judged that] we learned a lot from these presentations. They were very helpful to make informed decisions."

After these presentations, which themselves spurred lively discussion among EAB members, the group focused on specific issues confronting the company. The EAB Chair explained, "I asked the firm to articulate a problem and tell us why they wanted to address it with the EAB." It was important that company executives raise these issues, thought the Chair, because it prompted members to start discussions with problems rather than with principles, which would be instrumental in balancing ethical considerations with practical realities.

The Chair's interpreted the role as "...focusing members on coming up with practical solutions. It tends to make the discussions more productive." Because EAB members were charged with the responsibility of representing a constituency, discussions could have revolved endlessly around different perspectives. The Chair's objective was to get these different viewpoints aired, and then drive the discussion toward key points of agreement, mindful of the firm's resource constraints: "The firm has hard decisions to make with regard to what's do-able on the ethics side given the resources they have... There are limits. [But while] we try like hell to solve a problem, our job is to come down on the ethical side of the issue and not always find a way to go ahead with something."

Applying ethical analysis to practical situations is a process the Chair refers to as reflective criticism: "Typically, we put an issue out on the table, debate it, form opinions, then debate those and raise other issues. Because my job as Chair is to steer us toward decisions, I would say, 'Now that we understand an issue, let's come to a practical solution, and then look at testing these solutions.'"

A Board member also reflected on the process:

> It's very difficult to reach consensus. One thing is that there is always some positioning at the start with any group. It's inevitable. But we started to get beyond that as meetings progressed... The group was finding ways to communicate together and to arrive at decisions across disciplines. We were able to reach consensus on most issues. [The Chair] was very instrumental in this. Discussions would go around and around, [the chair]would stop to recapitulate, saying: 'Have we said this or that? Can we agree on this?' [This] contribution was very valuable.

The CEO confirmed this combination of "...intellectual rigor with extreme practicality. This is an unusual combination. [The Chair] understands the personalities of the people on the EAB [and] put together a group that is very challenging. This is not a group of 'yes' people." Although it made it more challenging to reach a consensus, everyone felt that the diversity of expertise on the EAB added great value to the discussions. Said one member:

> There was a wide variety of people to balance out varying viewpoints and make sure we had a balanced perspective. It especially helped to look at issues from other perspectives... Each member was confident in his or her own area and brought that confidence with them. At the same time, there was mutual respect among EAB members. We had very good discussions and came to consensus on most issues.

Another member expressed the view that "the diversity of perspectives was excellent. Members were able to bring in other cases of how things had been done or are being done. It was very good to have that." The Chair concluded: "We discuss what's possible and come up with three options that are good if not ideal."

After each meeting, the Chair would prepare a report that summed up the main points of consensus and contention around each of the issues brought to the EAB by the company. EAB members responded to the e-mailed report. After incorporating the feedback from members, the report went to the company, including consensus points and contention issues. The Chair also worked with PharmaSNPs' clinical staff on a regular basis to assist in the implementation of ethical processes: "I do a lot of work with the firm between meetings to help implement in great detail the suggestions of the EAB. I'll call on others—members of the EAB or other experts—to help, but the work is mostly done by me."

There was little communication among EAB members between meetings and virtually no interaction between the company and EAB members outside of meetings. This was in fact reflective of the specific function of the EAB within PharmaSNPs: to act as an independent advisor on clinical issues.

ROLE OF THE COMPANY

PharmaSNPs' representatives, especially clinical staff, who attend meetings appreciated the diversity of views on the EAB and gained insights into the numerous issues surrounding company activity. Members observed: "[Company representatives] were more interested in listening than in blocking or limiting the discussion, or countering with the argument that we have to make money," and "They wanted to hear what we put forward." The Chair concurred: "They would bring questions and then watch deliberations."

One Clinical Project Manager had previous work experience in the highly regulated area of Phase III clinical trials where, "because of tight restrictions, we saw uniform practices across all sites." This new area at PharmaSNPs was without those clear regulatory guidelines and fraught with questions. The manager had also worked as a genetic counselor and was thus cognizant of the complexity of ethical issues surrounding genetic research and of the need to put in place ethical processes. But PharmaSNPs did not have the internal capabilities to address all the ethical issues. "We felt the need [to address these] but didn't have the resources in-house to answer all the questions." Attending the EAB meetings gave some guidance in managing the PharmaSNPs' project.

EAB-COMPANY RELATIONSHIP

EAB members reported that PharmaSNPs was genuinely interested in hearing their views, giving them the flexibility to analyze and openly debate issues with each other. As important was that they felt free to disagree with the company. This freedom created an atmosphere of trust between the EAB and the company that grew over time. Explained the Chair:

> From a game theory perspective, this is a multi-round game so individuals [EAB members] are wary about going public with disagreements over an issue. There isn't a level of conspiracy where it's clear that this action is wrong but the company is covering it up. This is much more a case of trying to resolve shades of gray. People were allowed to file clear documentation on areas of disagreement.

There were, in fact, instances where the EAB had concerns about the company's activities. "They had some things that they just couldn't do. There were cases where we said, 'You can't do this going forward, you have to change your practice,'" recalled the Chair. The EAB, however, did not have the power to stop whatever action the company chose to take. One member reiterated: "Our role was to act as advisor. Making decisions was not our task."

One of the most interesting discussions occurred when the EAB disagreed with the judgment of the external Institutional Review Board (IRB) the company was using to approve its clinical-trial protocols.[3] One member explained the problem identified by the EAB: "In this ethical arena, decisions are made from legal perspectives. Instead of talking about what is ethical or the right thing to do, people ask what the legal requirement is." But because what is legal is not necessarily considered ethical, "it was a rude awakening here to say 'the IRB may have approved something but you still shouldn't do it.'" The Chair observed: "Not everyone accepted this." The CEO recalled that this discussion helped spur an internal debate: "We had disagreements between our clinical group and the EAB on one side, and [Eastern Cooperative Oncology Group] ECOG's IRB around how much [informed consent] is needed for approval. Our scientists question why we need extra standards asked for by our EAB if they're good enough for ECOG."

Throughout the discussion, the Chief Executive Officer (CEO) felt strongly that the company should aim to do what was right and not rely solely on the IRB to determine what actions are acceptable: "The standard I use is, if our decision were to be reported factually correctly in *The New York Times*, would it be damaging to the company? We have to take a stand." He was willing to engage the EAB on contentious issues. The EAB contributed a great deal to this debate about what standards should be followed to approve clinical trial protocols.

After several meetings, however, the matter had not been settled, in part because the EAB was still defining its role within the organization. "We are still in the

---

[3]The Eastern Cooperative Oncology Group (ECOG) was established in 1955 as one of the first cooperative groups launched to perform multicenter cancer clinical trials. A cooperative group is a large network of researchers, physicians, and healthcare professionals at public and private institutions across the country who are members of the group. Funded primarily by the National Cancer Institute (NCI), ECOG has evolved from a five-member consortium of institutions on the East Coast to one of the largest clinical cancer research organizations in the US with almost 6000 physicians, nurses, pharmacists, statisticians, and clinical research associates (CRAs) from the United States, Canada, and South Africa. Information available at http://www.ecog.org/general/intro.html.

process of integrating the EAB into the activities of the firm," explained the CEO. According to the Chair, this was a dynamic process that should evolve over time; it was important not to have preconceived notions of how the EAB should fit into the firm. Although it was necessary to set parameters at the outset, it was also necessary "to feel your way through…see how it goes, how the two sides interact, how many recommendations are internalized," and this was a fairly ad hoc process. This allowed both the EAB and the company to manage expectations of each other and to create value for each other.

EXAMPLES OF ETHICAL ISSUES

One of the first issues presented to the EAB by PharmaSNPs concerned a study of polymorphisms from genetic tissue obtained from complete hydatidiform moles, a nuclear egg fertilized by sperm. Tissue samples were to be obtained with consent from women who had these molar pregnancies. Collecting this tissue allowed the company to obtain a double set of genetic material (the paternal genetic material duplicates itself in its attempt to make a baby) that would naturally give higher yields of genetic information than regular tissue. The study, however, raised two potential ethical dilemmas, the first of which prompted PharmaSNPs to ask the EAB to review the study, whereas the second surfaced during the Board's discussion of the issue.

The first ethical issue concerned how to balance the patient's right to withdraw from a trial with the protection of privacy achieved through data anonymization. Because consent was obtained from women experiencing a stressful medical situation, it was felt appropriate to provide a 30-day period to withdraw from the study. This meant that samples needed to be linked to individual patients for at least 30 days, while the privacy of patients was safeguarded during that time. PharmaSNPs hired an external CRO to oversee the study. The role of the CRO was threefold: first, to ensure that the study was conducted properly and informed consent was obtained from participants; second, to anonymize samples while retaining separate records of the identity of the tissue donor; third, to maintain a link between samples and donors for 30 days, after which the link was broken (a process called de-linking). However, hiring a separate entity to oversee the project as well as collect the samples was costly for the company. PharmaSNPs thus wondered if it would be ethical for company employees to manage these programs. Specifically, PharmaSNPs' Clinical Project Manager wondered:

> Does it need to be an outside person in that role? Would we be violating guidelines if we use someone internally? If so, who should it be? How should they operate? Could they be involved in other projects at the same time? Would it be ethical to have paid employees oversee tissue collection, anonymize samples, and act as links? And finally, how do we balance the need to create distance to ensure privacy protection with the need to make sure clinicians are close enough to advise patients if needed?

There was extensive EAB debate on this topic. The questions included: How independent could PharmaSNPs staff be? What protections would they have if they came under internal pressure to reveal the identity of a sample? Would staff be loyal to the company? Who would do a better job, an external observer or a company employee? The EAB agreed that it would be appropriate to have company employees act as monitors, to be called *clinical research privacy officers,* under certain strict conditions.

The first condition was to maximize the independence of employees or, as the Chair explained, "to sequester clinical staff who were handling samples with identifiable information." To protect those who managed the data and could identify donors, the EAB recommended that the company set up an internal firewall, a safety precaution that does not allow information to penetrate into another area, such as data analysis. The second condition was that employees who were managing research data collection should not be involved in statistical analysis or any other project. A Board member explained: "We felt that if staff were involved in other projects, there was a risk of compromising projects, the risk of trading information to get projects through." The EAB's third condition was to use an escrow agent or a trustee to store the links between donors and samples, a strategy one of the EAB members had learned from sitting on another company's EAB. Finally, the EAB suggested that PharmaSNPs put a policy in place to guarantee clinical staff that the company would not fire them for protecting the confidentiality of a subject.

In the course of discussing how to handle these samples, another issue emerged that the firm had not anticipated. The EAB asked whether it was sufficient to obtain only the woman's consent for the tissue sample "if the tissue from a complete hydatidiform mole is 100% paternally derived. Should the consent process not also involve the father?" The debate revolved around the following questions:

1. Who owned the sample? As the EAB Chair explained, "The history of ethical discourse around informed consent is battery law, meaning that you can't touch the person without their permission. An emerging view is seeing it as property claim. If this is true, then you need his [paternal] consent as well," because it could be argued, in this case, that he also owns the tissue sample.
2. What was the nature of the tissue? Some EAB members argued that although "the tissue is 100% paternally derived, it is sufficiently biotransformed in the female," explained a Board member, in which case paternal consent would not be required. Moreover, "if paternal consent is not required for fetal tissue even though it is 50% paternal," it should not be required in this case.

The EAB was not able to reach a consensus on this matter and so unable to provide specific guidance to PharmaSNPs.

Another issue presented to the EAB by PharmaSNPs involved its relationship with a European clinical diagnostic laboratory that collected and supplied tissue samples from all over Europe to the company. A Board member explained that because this collaborator's activity was "borderline between clinical treatment and research [since it involved collection of samples for diagnostic use that could also be used for genetic studies] and it was mostly samples from people who later died," the company wondered whether it was necessary to obtain informed consent for the use of these samples in research outside of Europe. Up to that point, consent had not even been obtained to use these samples for research within Europe. The EAB unanimously agreed that it was imperative to obtain consent from patients. "We told the company," recalled the Chair, "that the current practice was not acceptable going forward." A member continued:

> The company was shocked [when the EAB asserted that] as a matter of prudence, it needed documentation showing that it obtained samples with informed consent and the knowledge that tissue samples would be used for research in the US. No one at the firm realized that in Europe, dead people have rights. They didn't know that the deceased still owns his or her material. It never occurred to them that they would run into problems [for not obtaining consent from dead people].

Because the company could not go back to get consent for those tissue samples already collected, the EAB devised different standards for both retrospective and prospective treatment of samples: retrospectively, the EAB advised the company to anonymize past samples; prospectively, the EAB advised the company to obtain informed consent for all newly formed studies.

This discussion also highlighted the importance of understanding different cultural contexts when collecting samples around the globe. A specific example concerned obtaining and using tissue specimens from indigenous populations in the Netherlands. A Board member explained: "Among [certain] citizens of the Netherlands, there is a belief that tissue samples should not go outside their country and exporting their tissue shouldn't be done without the knowledge of relatives. Yet, specimens were being exported to the US without their knowledge or consent."

Projects in other countries raised further questions related to international research, such as which research standards apply and how to obtain informed consent. Because the nature of PharmaSNPs' research required an ethnically and geographically diverse panel of individuals, the company gathered samples in a number of countries, including the Philippines, Brazil, and Poland. "The countries we explored," explained the Clinical Project Manager, "were driven by the need for ethnic diversity, not any effort to escape US. standards." The CEO corroborated this: "The reason we went to other countries for trials was not to get different standards, but to get access to samples that are not available here."

Conducting clinical trials in other countries, however, proved to be challenging for the company. "We've tried all the way through to apply US standards of ethics," said the CEO, "but it has been quite a problem sometimes to get them to understand why we want to do it." Poland, for example, did not require obtaining consent for participation in this kind of research, although PharmaSNPs continued to enforce this requirement. The firm relied on its CROs to educate their staff on PharmaSNPs' guidelines. The Clinical Project Manager said:

> We translated consent forms into local languages and checked that they conformed to our intent. We had to make sure the patients understand, but we weren't there to monitor on site. [The problem was that] we don't generally do it well here in

the US, so why would we expect it to be different elsewhere? But investigators, especially in Brazil, really wanted to do a great job and were top of things.

The CEO described the example of Brazil where genetic counselors found creative ways to obtain informed consent:

> I've been to both [the Philippines and Brazil], and what surprised me was that even in the center of Brazil, where there are a lot of illiterate patients, [CROs] found an elegant way to get consent totally verbally. There were five counselors with genetic counseling training who [worked with patients] to get consent. They explained to patients what the nature of the clinical problem was, the type of surgery and the outcome. They got patients to draw a picture to see if they understood. This method was not developed for our study but we piggybacked on their routine practice.

Perhaps the biggest ethical issue facing PharmaSNPs related to the company's business model of commercializing human genetic material. There are fundamentally different views on the appropriateness of commercializing human genetic material, and PharmaSNPs grappled with the tension between pressures to produce financial returns based on its research and the ethical debate regarding commercial ownership of genes derived from medical research. The company was confronted with this dilemma in Europe, where unresolved tension between US and European attitudes proved to be limiting for PharmaSNPs. Recalled the CEO:

> We had to abandon an attempt to do our program in the UK because [our academic partner]'s guidelines excluded any commercialization of human genetic products. There are altruistic arguments for this, but it caused real problems for us. For a biotech looking for ROI [return on investment], we need to generate some returns... In the US, there is a pragmatic acceptance of patenting anything you can to make money. Europeans see a role for the state in debating that.

LESSONS LEARNED

EAB members reflected on key lessons for how to establish and run an EAB in a small bioscience company. These lessons included the necessary role of senior management; the value of broad representation; identifying the right issues for discussion; bridging the recommendation–implementation gap; and finding a balance between independence and institutionalization.

THE ROLE OF SENIOR MANAGEMENT

Senior management played the most significant role in the development of the company EAB. The ongoing commitment of three senior leaders drove the initiative and resulted in a mechanism to address ethical dilemmas and concerns. In the Chair's view, the buy-in of senior management was absolutely essential:

> One of the things I've learned is that it is extremely hard to manage a process without an internal champion... When we started, we were under supportive directors. But they both left the organization and, after that, the main senior backing was gone. The clinical staff really wanted the EAB to continue and to keep giving

them advice. They had an extremely personal affinity for it, but it wasn't enough when other senior executives weren't going to come to meetings.

THE VALUE OF BROAD REPRESENTATION

As discussed above, there was a consensus among EAB members that the diversity of perspectives and of areas of expertise proved to be an effective and constructive means to achieve the EAB's objectives, which were to "harmonize ethical goals with practical modalities." The Chair, who had been skeptical of the ability of a larger EAB to function effectively, found that having two persons for each discipline worked exceptionally well: "It tends to solidify the opinion structure of people from each discipline, especially when views are challenged by other perspectives." That is, with a broader representation, the EAB was able to articulate more robust opinions.

Members of the EAB agreed. The size was manageable: "There were ten to fifteen members, which is not too big and not too few. With too few, you don't have enough perspectives. With too many, you have too many conflicts," said one. "The diversity of perspectives, especially global perspectives, absolutely added value," said another.

IDENTIFYING THE RIGHT ISSUES FOR DISCUSSION

In the parlance of risk management, some of the greatest dangers exist because company managers "don't know what they don't know." A potential weakness of the PharmaSNPs approach to its EAB was that the company identified which aspects of its operations might pose ethical issues. Though the firm learned from the questions raised in EAB discussions, it was often unaware of other relevant ethical issues. The Chair recognized this but felt it was minimized by the connections among the different research projects and broad-ranging discussion at meetings: "If the management doesn't know to ask us about an issue, then it is not raised at the EAB. But it is very rare that they are doing a one-off project that is isolated from others. As we deal with one project, it raises other issues." Thus, many issues that management didn't intentionally bring to the EAB ended up being discussed.

BRIDGING THE RECOMMENDATION– IMPLEMENTATION GAP

Some EAB members felt that there should have been a feedback mechanism for how the company acted on the EAB recommendations. Members suggested feedback at the beginning of each meeting from a company representative acting as a conduit between the EAB and the company, to convey to the EAB what, if anything, had been implemented and the effects of Board advice on general firm policy. Another suggestion was that there should be a company employee "clearly in charge of actions," in addition to feeding back information to the EAB.

The Chair of the EAB did work with the company between meetings to implement EAB recommendations, but these actions were not discussed at the meetings.

FINDING A BALANCE BETWEEN INDEPENDENCE AND INSTITUTIONALIZATION

As discussed above, the primary role of PharmaSNPs' EAB was to act as advisor to the clinical side of the business. This was judged to be an effective strategy for guiding the company's clinicians in setting up ethical processes and in making decisions based on an ethical understanding of the issues. The EAB, however, was formally separated from the firm's operational structure. Moreover, although company representatives attended EAB meetings, they generally did not participate actively in the discussions. The advantage was that it gave the EAB greater freedom to discuss dilemmas with no pressure from the company to incorporate company views into the debate; the disadvantage was that the EAB was not always sufficiently integrated within the company to affect its practice. The Chair observed: "We probably went to the extreme of independence. In the process, it hurt our ability to be an integral part of the organization... With twenty-twenty hindsight, independence of mind and spirit may be more important than a formal organizational separation."

CONCLUSION

As biotech firms push back the scientific frontiers, they face unprecedented ethical dilemmas, and many have already suffered as a result. According to PharmaSNPs' EAB chair, it is in the best interest of these companies to deal with ethical issues proactively. "It is easy to paint [companies] with a Frankenstein brush. Firms should get ahead to address ethics before it becomes a problem." In addition, companies can benefit from setting standards for ethically difficult corporate activities. EAB members recognized that standard-setting can forestall external regulation and that their input could help PharmaSNPs create a self-regulatory environment. For example, by collecting samples in a responsible way to avoid any ethical controversies, PharmaSNPs and other pharmacogenomics firms could try to avoid the creation of laws that would restrict their ability to obtain the samples their business models require. As the CEO put it, "No one wants legislation if we can avoid it."

CASE (B): POSTSCRIPT

When PharmaSNPs Inc. went public during the biotech boom surrounding the sequencing of the human genome, investors were tantalized by the biotech company's promising future in pharmacogenetics and personalized medicine. Soon afterwards, however, the initial excitement over the genomics revolution faded and the stock market bubble burst, causing the firm's stock to lose more than 90% of its value in 2 years. The share price fell so much that the firm, like many biotech companies that went public at that time, was being valued below the level of its cash reserves, making it an attractive acquisition target for other cash-starved bioscience companies. This is just what happened to PharmaSNPs, as the company announced a merger agreement with a company with a more developed product pipeline that was running low on cash. The newly merged company, as part of its restructuring, announced plans to cut expenses by reducing combined employment by close to 50% and throttling back commitment to various research and development (R&D) projects. One of the cost-cutting initiatives was the disbanding of PharmaSNPs' EAB, which was deemed unnecessary by the new management team, as the company operated under the auspices of an IRB.

PharmaSNPs' EAB Chair concluded that one reason the EAB was shut down was because there was no longer someone—*a champion*—within the company to support it:

> My sense is that [the] bottom-line business folks won out. When it came down to what is [the] bottom-line legal requirement versus what is gravy and nice to have, they went with for the former... It's hard in the face of extreme business necessity for some executives to see the benefits of an EAB. In general, they want to avoid anything illegal, but can't see the benefits of going further than that.

At the time of the merger, the company was still trying to resolve many of the questions the EAB had discussed: how to handle the cross-cultural differences in the business use of human tissue samples, for example. The CEO had intended at the EAB's next meeting to raise the issue of how to reconcile what "appears generally acceptable to us" with the lack of acceptance of the commercialization of tissue-derived genetic information, especially in Europe. The EAB was disbanded before it could address this question. The Chair concluded that some of the "studies we were discussing were never done because of the merger, so I'm not sure how it would have turned out."

13

MONSANTO COMPANY: BIO-AGRICULTURE PIONEER[1]

A LIGHTNING ROD FOR ETHICAL CONTROVERSIES

Hendrik Verfaillie, who had lived through the genetically modified (GM) food controversy as head of Monsanto's agricultural operations from 1993 to 2000, recognized that his firm—a world leader in agricultural biotechnology—was in a state of crisis. This insight came during a meeting of Monsanto executives in Lyon, France. Here he learned that employees were now removing the company logo so that they would not be identified as Monsanto employees, because the company had become the "personification of evil" in Europe.[2]

Verfaillie realized that a radical change was needed. How had a company that had set out *"to improve the world through biotechnology"*—Monsanto's underlying philosophy—come to be vilified in this way? And what could Monsanto do to restore the company's image and the viability of its bioagricultural business?

The answers to these questions are provided in the following two-part case. Case A summarizes the transformation of Monsanto into a life-sciences company and the ethical debates surrounding its pioneering work in bioagriculture. Readers familiar with this now well-documented controversy can skip directly to Case B,

[1]The case study was written by Cécile M. Bensimon under the supervision of Peter Singer, following interviews they conducted with Monsanto executives in December 2002 at company headquarters in St. Louis, Mo. Abdallah Daar, Margaret Eaton and David Finegold, provided constructive comments and edits on earlier drafts. This case study is one in a series produced under grants funded by The Seaver Institute and Genome Canada to study models for ensuring ethical decision making in bioscience firms. Unless otherwise specified, all attributed quotes were obtained during the case study interviews. We would like to thank all the Monsanto executives who participated in the case, especially Gerard Barry, whose invaluable assistance enabled us to complete this study.

[2]Verfaillie interview.

which analyzes how the firm responded in 2000 with the *New Monsanto Pledge*, inspired by the company's first Pledge instituted 10 years earlier. The case describes how the company developed the Pledge and its subsequent work to embed it in its decision making processes, and to create tools to monitor its effectiveness.

FROM CHEMICALS TO ENVIRONMENTAL SUSTAINABILITY: A FIRST PLEDGE

In the late 1970s and early 1980s, Monsanto was the fourth-largest chemical company in the United States. At a time when it was beginning to face substantial setbacks in its chemical business, Monsanto began exploring revenue growth opportunities in agricultural biotechnology. The company had been hit hard by the OPEC oil crises, and executives at Monsanto were challenged to re-evaluate Monsanto's long-term strategy. Realizing that the company's dependence on petroleum made it extremely vulnerable, they sought new research directions.

As early as 1972, Monsanto executives agreed to invest significant resources into basic plant biotechnology research at the suggestion of one of its scientists, Dr. Ernest Jaworski, who was "betting on the future of biotechnology." Chief Executive Officer (CEO) John Hanley, who had begun "speculating that the chemical business was not the best option for the long-term success of Monsanto,"[3] was willing to take Jaworski's bet, sensing as well that "genetic engineering was an exciting and possibly profitable venture."[4]

By 1981, with Richard Mahoney at Monsanto's helm, biotechnology was firmly established as the strategic research focus.[5] Monsanto entered into a collaborative research agreement to pursue biomedical research with Washington University in St. Louis, setting a precedent for the company's future research collaborations. Monsanto invested heavily in plant biotechnology research and development (R&D) for the next decade before bringing its first genetically engineered product to market.

In 1986, Monsanto, like other chemical companies, was further challenged to re-evaluate its business strategy when the United States enacted the Emergency Planning and Community Right to Know Act (EPCRA). Also known as Title III,[6] this Act was a reaction to public outrage and greater environmental awareness over major chemical spills that had caused thousands of deaths and substantial environmental damage.[7] Title III required companies to inform communities of

---

[3]Hertz M, et al. Monsanto and the Development of Genetically Modified Seeds. *Darden School of Business Administration*, University of Virginia Darden School Foundation, Charlottesville, VA, UVA-E-0220, 2001.

[4]Ibid.

[5]Available from www.monsanto.com/monsanto/layout/about_us/timeline/timeline5.asp. Accessed December 2002.

[6]U.S. Environmental Protection Agency, 1986. Available at www.epa.gov/swercepp/factsheets/epcra.pdf. Accessed March 2003.

[7]For example, Bhopal, India hit the world headlines in December 1984 when tons of a highly toxic chemical, methyl isocyanate (MIC) used in the production of pesticides, spilled from a tank at a Union Carbide plant into the city, killing thousands of local residents and injuring tens of thousands.

chemical hazards in their area and of the impact of company operations on the environment. Title III led Monsanto to conduct a study of its toxic air emissions. Scarlett Foster, current Director of Investor Relations, stated, "We were comfortable with the environmental impact of our technology. Then we saw our numbers and we were all shocked." Foster, who was at the time with Monsanto's Public Affairs department, further explained that CEO Mahoney felt he could not, in good conscience, release such "unacceptable" numbers to his community and neighbors without presenting, at the same time, a contingent plan that would improve Monsanto's environmental performance. The CEO evaluated the situation with Monsanto's scientists, who told him that the company could reduce toxic air emissions by 50%. Adamant that this was not enough, he unilaterally pledged in 1988 to reduce Monsanto's toxic air emissions from all its manufacturing facilities by 90% in the next 5 years, with a final goal of zero emissions. Because Mahoney made this decision unilaterally, his announcement shook both the industry and the company. Foster recalled:

> Externally, it forced the industry to lead. Internally, there had been very little knowledge of Mahoney's announcement, which led to major internal fighting for a long time. He knew that a CEO has a major influence on public policy and in setting the direction internally, and that an announcement of that magnitude would have a major ripple effect.

Within two years Monsanto was well on its way to meeting its 90% emission-reduction goal (eventually achieved in 1992), prompting the CEO to go further and address other environmental problem areas. In 1990, he boldly announced the Monsanto Pledge, a seven-point statement of environmental sustainability principles that expanded the company's commitment to sustainable development in the areas of waste reduction, safe operations, groundwater safety, new technologies, openness to communities, and managing all corporate real estate to benefit nature.[8]

EMERGING AS A LEADER IN AGRICULTURAL BIOTECHNOLOGY

In 1993, Robert Shapiro became Monsanto's CEO and continued a dual strategy to make the company a leader in the life sciences and in corporate social responsibility (CSR). Starting in 1985, Monsanto had begun to sell off its traditional, lower-margin chemical businesses, such as AstroTurf stadium surfaces, polyethylene film, and sorbate food preservative. In 1997, the company spun off its remaining chemicals businesses into a separate firm, Solutia Inc. By 2000, it had divested all non-agricultural biotechnology businesses, raising a total of $1.6 billion. It invested these proceeds, and more, into biotechnology research and acquisitions, with a view to engineering proprietary seeds that would become the

[8]Available at www.monsanto.com/monsanto/layout/about_us/timeline/timeline6.asp. Accessed December 2002.

equivalent of Microsoft's operating systems in information technology.[9] In 1997 and 1998, Monsanto acquired a series of seed companies for a total of $5.8 billion. Together these moves represented "a process of transforming the company into a life-sciences business that would have a leading role in advancing new technologies and business paradigms in food and agriculture."[10]

At the same time, Shapiro adopted a sustainability strategy framework to seek "growth through global sustainability,"[11] leveraging Monsanto's early lead in pollution prevention and environmental clean-up, and capitalizing on the strategic implications of corporate environmentalism and sustainable development.[12] These initiatives were "part of a sustainability strategy, even if we didn't call it that," recalled Shapiro.[13] The goal of establishing such a framework was to add value to Monsanto's day-to-day operations and reinforce Monsanto's biotechnology strategy. First, the framework prompted Monsanto's employees to engage in a dynamic and creative process of corporate reflection and redefinition that was organized into seven teams, known as Monsanto's "sustainability teams" (see Exhibit 13.1). Each team focused on developing tools to provide direction for internal management and to identify global sustainability needs that Monsanto might, or should, address.[14]

The new strategy began to pay off in the 1990s with a stream of new products. In 1991, the NatureMark unit was formed to sell a genetically engineered, insect-resistant potato, the NewLeaf Superior Potato, which was ultimately approved for sale in 1995 and introduced to agricultural markets in 1996. In 1993, Monsanto's

[9]The metaphor is Monsanto's. This analogy would eventually find another level of comparison. In 2003, the Department of Justice (DOJ) launched an inquiry into whether Monsanto engaged in antitrust practices in the herbicide industry, especially with respect to Roundup. Analysts, however, seemed unconcerned by antitrust allegations. In a research report on Monsanto, UBS Warburg stated, "We find it curious that DOJ is looking into this matter given that Monsanto has been aggressively cutting prices over the last few years in a market where it has almost 80% share... We hold Monsanto in the highest regard when it comes to customer pricing and fair play in the market; and have absolutely no reason to believe that they have been acting in an anti-competitive way as defined by U.S. law." *Global Equity Research Report on Monsanto authored by UBS Warburg*, March 17, 2003, p. 2.

[10]Austin J, Barrett D. Monsanto: Technology Cooperation and Small-Holder Farmer Projects. *Harvard Business School.* 13 June 2002:9-302-068.

[11]Margetta, J. Growth through Global Sustainability: An Interview with Monsanto's CEO, Robert B. Shapiro. *Harvard Business Review.* January-February 1997:81; Monsanto Company, *1997 Annual Report.*

[12]*Sustainable development* was first coined in the 1987 World Commission on Environment and Development (WCED) report, "Our Common Future: the Bruntland Report," Oxford University Press, from the World Commission on Environment and Development, New York. It was the subject of the UN Conference on Environment and Development in 1992, also known as the Rio Earth Summit. The primary objective of the effort was to rethink economic development and come to an understanding of "development" that would support economic growth, socioeconomic development, and environmental sustainability.

[13]Margetta J. Growth through Global Sustainability: An interview with Monsanto's CEO, Robert B. Shapiro. *Harvard Business Review.* January-February 1997:85.

[14]Ibid, p. 86.

EXHIBIT 13.1   **Monsanto's Seven Sustainability Teams[15]**

The Eco-efficiency Team

Because you can't manage what you don't measure, this team is mapping and measuring the ecologic efficiency of Monsanto's processes. Team members must ask: In relation to the value produced, what inputs are consumed and what outputs are generated? Managers have historically optimized raw material inputs, for example, but they have tended to take energy and water for granted because there is little financial incentive today to do otherwise. And although companies such as Monsanto have focused on toxic waste in the past, true eco-efficiency will require better measures of all waste. Carbon dioxide, for instance, may not be toxic, but it can produce negative environmental effects. Ultimately, Monsanto's goal is to pursue eco-efficiency in all its interactions with suppliers and customers.

The Full-Cost Accounting Team

This team is developing a methodology to account for the total cost of making and using a product during the product's lifecycle, including the true environmental costs associated with producing, using, recycling, and disposing of it. The goal is to keep score in a way that doesn't eliminate from consideration all the environmental costs of what the company does. With better data, it will be possible to make smarter decisions today and as the underlying economics change in the future.

The Index Team

This team is developing criteria by which business units can measure whether they're moving toward sustainability. They are working on a set of metrics that balance economic, social, and environmental factors. Units will be able to track the sustainability of both individual products and whole businesses. These sustainability metrics will, in turn, be integrated into Monsanto's balanced-scorecard approach to the management of its businesses. The scorecard links and sets objectives for financial targets, customer satisfaction, internal processes, and organizational learning.

The New Business/New Products Team

This team is examining what will be valued in a marketplace that increasingly selects products and services that support sustainability. It is looking at areas of stress in natural ecosystems and imagining how Monsanto's technologic skills could meet human needs with new products that don't aggravate, and perhaps even repair, ecologic damage.

The Water Team

The water team is looking at global water needs—a huge and growing problem. Many people don't have access to clean drinking water, and there is a worsening shortage of water for irrigation as well.

The Global Hunger Team

This team is studying how Monsanto might develop and deliver technologies to alleviate world hunger. That goal has been a core focus for the company for a number of years. For example, Monsanto had been studying how it might use its agricultural skills to meet people's nutritional needs in developing countries.

The Communication and Education Team

This team's contribution is to develop the training to give Monsanto's 29,000 employees a common perspective. It offers a framework for understanding what sustainability means, how employees can play a role, and how they can take their knowledge to key audiences outside the company.

---

[15]Margetta J. Growth through Global Sustainability: An Interview with Monsanto's CEO, Robert B. Shapiro. *Harvard Business Review.* January-February 1997:86.

first biotech product, Posilac® bovine somatotropin (BST), a product to boost milk-production in cows, was approved for sale.[16]

The main thrust of its ag–biotech research, however, was the creation of transgenic crops that fall into two main categories: herbicide-tolerant crops, and insect-resistant crops. Herbicide-tolerant crops are genetically engineered to resist the effects of traditional herbicides. The most widely known and successful examples of this approach are Monsanto's Roundup Ready soybeans, corn, and cotton, which contain specific built-in resistance to its popular herbicide. By the mid-1990s, Monsanto was marketing Roundup with its proprietary, genetically engineered crops, so that farmers, for example, who planted Roundup Ready soybeans could spray their field with Roundup after the soybean plant sprouted. This reduced the amount of herbicide used, because farmers could time herbicide applications with more precision. Before Monsanto developed Roundup Ready seeds, farmers had to apply Roundup before the plant sprouted: if used after sprouting, it would kill both weeds and crop plants. Monsanto has greatly benefited from this integrated herbicide/seed approach by extracting multiple revenue and margin streams from each acre.

The second category of GM crops is insect-tolerant crops, YieldGard® and Bollgard®, marketed under well-known seed brands it had acquired such as DeKalb® and Asgrow®. These seeds are genetically engineered to incorporate the *Bacillus thuringiensis* (Bt), the safe and naturally occurring pesticide that has been used by organic farmers for the last century, into large-volume crops (cotton, corn, and potato), so that each produces its own specific toxin to kill its most common pest.[17]

Alongside these new biotechnology businesses, Monsanto's most profitable product continues to be Roundup (generically known as "glyphosate"), a nonselective herbicide used for agricultural, industrial, and residential weed control in over 80 countries worldwide. Global sales of the company's Roundup herbicide exceed those of the next six herbicides combined. In 2000, Roundup and other glyphosate-based herbicides accounted for half of Monsanto's total sales and about 40% of sales in 2002. When Roundup came off patent in September 2000, Monsanto protected its leadership position and staved off generic competitors by entering into arrangements to supply glyphosate to other agricultural–chemical companies for use in their own branded products.[18] Roundup is most widely used around the world for application in "conservation tillage," a process through which a grower reduces or eliminates mechanical tillage from farming operations. The results are increased retention of soil moisture, reduction in loss of topsoil from wind or rain, and reduced use of tractors, diesel fuel, and farm equipment.

---

[16]Available from www.monsanto.com/monsanto/layout/about_us/timeline/timeline6.asp. Accessed January 2003.

[17]Finegold D, et al. To Bt or Not to Bt: A Case Study of Genetically-Modified Corn. In: Rosa H, ed. *Bioetica Para as Ciencias Naturais* (in Portugese). Lisbon: Fundacao Luso-Americana para o Desenvolvimento; 2004.

[18]Monsanto remains the lowest-cost producer of glyphosate, due to economies of scale. Its cost position enables the company to supply its competitors at prices below their competitors' cost of production.

Other attractive features of Roundup are its broad range of applications and the lack of residual activity (it decomposes into harmless products within a short time of application). Monsanto's new products and its extensive biotech pipeline of GM crops clearly delineated the company as the leader in ag–biotech. Monsanto wrote in an internal document: "In 1997, we were on a high; Wall Street loved us; our stock price was rocketing; our people were energetic and enrolled; there was tremendous, positive media coverage; and the demand exceeded the supply of biotech products."[19] The high was not to last, however. As the leader in this industry, Monsanto also became a lightning rod for protests by those opposed to the use of biotechnology in agriculture.

THE POLARIZED DEBATE

Arguments for and against GM crops and foods are diametrically opposed. Opponents claim that GM foods are dangerous and unnecessary; proponents argue that they are safe and indispensable. Because many of these issues originate from fundamentally different views of science and its benefits and risks for society, there is an inherent tension between proponents and opponents of biotechnology that is difficult to resolve. It is a polarized debate, centered on the different perceptions of the risks and benefits of the technology and the ethical issues related to those perceptions.

Proponents argue that agricultural biotechnology is critical to the future of farming. In the early 1990s, researchers estimated that most of the sustainable farmland, or arable land, was already under cultivation and that growing urbanization and other activities were likely to reduce its availability further. At the same time, the world population was growing at a rapid rate, creating a need to increase food production. Moreover, misuse of chemical pesticides and herbicides could cause environmental damage, and many argued that their use was unsustainable. Finegold et al. recount such concerns:

> In the early 1990s, polls showed that 91% of U.S. consumers were concerned about pesticides used to grow food products and 95% were concerned about pesticides and fertilizers getting into their water supply. Pesticides have been linked with human pesticide poisoning, reduction of fish and wildlife populations, livestock losses, destruction of susceptible crops and natural vegetation, honeybee losses, destruction of natural enemies, evolved pesticide resistance and creation of secondary pest problems.[20]

So there was both growing pressure to improve crop yields to increase food production and to find an alternative to chemical pesticides and herbicides to

---

[19]Monsanto Company. "Anticipating, Understanding, and Responding to Societal Expectations." Presentation prepared for internal use only, March 2001. A copy was given to the authors.

[20]Finegold D, *et al.* Ibid.

protect the environment. With the introduction of genetically engineered crops, farmers could apply significantly less pesticide, thus reducing pesticide costs while increasing their yields. In general, proponents of agricultural biotechnology maintained that GM crops could prove essential to world food-production, environmental protection, and agricultural sustainability. In an interview with the *Harvard Business Review* that same year, Robert Shapiro expressed his vision of sustainable development and the future of biotechnology:

> Today there are about 5.8 billion people in the world. About 1.5 billion of them live in conditions of abject poverty–a subsistence life that simply can't be romanticized as some form of simpler, pre-industrial lifestyle. These people spend their days trying to get food and firewood so that they can make it to the next day. As many as 800 million people are so severely malnourished that they can neither work nor participate in family life... Without radical change, the kind of world implied by those numbers is unthinkable. It's a world of mass migrations and environmental degradation on an unimaginable scale. At best, it means the preservation of a few islands of privilege and prosperity in a sea of misery and violence... The whole system has to change; there's a huge opportunity for re-invention... Current agricultural practice isn't sustainable... We don't have 100 years to figure that out; at best, we have decades. In that time-frame, I know of only two viable candidates: biotechnology and information technology.[21]

Proponents of biotechnology also believed that, in addition to the benefits for the environment and farmers, subsequent generations of genetically engineered agricultural products could, because of their improved nutritional characteristics, potentially play significant roles in reducing malnutrition, as well as promoting public health in both developed and developing countries.

However, although their introduction into the United States and Canada has been relatively smooth and product approval has been relatively rapid, there has been considerable, at times virulent, opposition to GM crops in Europe. Rahul Dhanda, in his analysis of the ethical controversy over genetically modified organisms (GMOs), explains that European resistance to GM crops largely derives from existing farming and agricultural concerns and from food scares in farming and agriculture in Europe:

> Unlike many of the genetic technologies that have been introduced over the past few decades, GM foods fit within an already established industry: agriculture [that had already faced substantial criticisms caused by] regulatory failings regarding European livestock practices.[22] In a series of events in Europe that occurred without influence or interference from companies selling GM foods, the agencies that were responsible for ensuring the safety of food products systematically failed at their responsibilities. The outbreak of mad cow disease, scientifically known as 'bovine spongiform encephalopathy' (BSE), proceeded to worsen due to the failure of regulatory bodies. Although the disease had been recognized

[21]Margetta J. Growth through Global Sustainability: An Interview with Monsanto's CEO, Robert B. Shapiro. *Harvard Business Review*. January-February 1997:81–82.
[22]Dhanda R. Genetically Modified Foods. In: *Guiding Icarus: Merging Bioethics with Corporate Interests*. New York: Wiley-Liss; 2002: p. 41.

as a problem as early as 1986 in England, the restrictions put in place to limit the spread of BSE were poorly enforced. In 1996, the scope of the problem increased when a new variant of the human neuro-degenerative disorder, Creutzfeldt-Jakob disease, was discovered and linked to the spread of BSE. Mad cow disease had spread largely due to the lack of intervention on behalf of the Ministry of Agriculture, Fisheries and Food (MAFF), which did not enforce its regulations halting the re-use of slaughtered cattle as animal feed. The recycling of animals included the recycling of BSE until today when the practice is banned, but the damage is done.[23]

On the whole, there was a general suspicion in Europe about foods produced by non-traditional technologies, placed within the context of a universal suspicion of the global regulatory system and of biotechnology in general. These circumstances laid the foundation for the unprecedented anti-GM foods movement in Europe. Opposition swelled to include government officials, environmental activists, scientists, and consumers.

Opponents of agricultural biotechnology argue that biotech seeds and GM foods, and their introduction by large multinational corporations, raise a number of concerns. The main argument against agricultural biotechnology is that genetically modified crops do not confer many, if any, of the claimed benefits, and that they introduce too many unknown risks and unforeseen ecological effects that, given the complex interdependent nature of our ecosystem, would be impossible to remedy. Another argument against agricultural biotechnology stems from opposition to the science itself. Opponents roundly reject the idea of altering organisms, which they view as gene tampering. Those who oppose the technology are often heard quoting Prince Charles who, in 1998, expressed the view, "Genetic engineering takes mankind into realms that belong to God and God alone."[24]

Countering this view, proponents of biotechnology argue that genetic manipulation is not new and that all modern plants are genetically engineered: over hundreds of years, through a process of trial, adaptation, and selection, farmers identified sets of seeds suited for local conditions and that expressed desired properties. Plant-breeding techniques evolved slowly to select for desired properties and eliminate unwanted qualities, so that the quality and yield of plants has steadily improved. Biotechnology proponents claim that plant breeding and genetic manipulation are one and the same thing, the only difference being that the introduction of genetic traits from other species through recombinant DNA techniques enables genetic enhancement in a faster, more targeted manner than traditional plant breeding. They further argue that genetic engineering offers a valuable source of new diversity for the gene pool in a particular region, because it improves the characteristics of the gene source.

But opponents argue that insects would eventually adapt and develop resistance to genetically engineered crops, curtailing the crops' long-term effectiveness. Resistance would also threaten the most efficient tool for organic farmers who use

[23]Ibid, p. 43.
[24]http://www.biotech-info.net/industry_response.html; http://news.bbc.co.uk/1/hi/sci/tech/108630.stm.

Bt as an alternative to chemical pesticides. They thus argue that this "version of sustainable agriculture may threaten precisely those farmers who pioneered sustainable farming."[25] In response, companies that produce Bt crops, including Monsanto, have advised farmers to practice insect-resistance management, an approach that requires farmers to create "refuge" areas by planting non-Bt crops near the Bt crops, so that insects who develop resistance to Bt are less likely to breed with each other.[26]

Another concern is the question of "gene flow," the transfer of genetic material between crops. Opponents worry that the transfer of traits, including resistance to herbicides and insects, would eventually create "super-weeds" that would be practically impossible to eradicate. This concern was dramatized by a senior staff scientist with The Union of Concerned Scientists, who cautioned:

> Where Bt crops are grown near wild relatives, it is highly likely that the Bt gene will transfer to the wild populations – as a result of movement of pollen from the crop to the relatives. Some of the resultant wild plants may produce enough Bt to ward off insects that normally feed on them. What this may mean in natural ecosystems we don't yet know – there has been little research in this area.[27]

Furthermore, in addition to concerns over biodiversity and potentially devastating ecologic effects, opponents feared that eating foods made from GM crops exposes people to unknown, potentially significant, health risks. Although proponents of biotechnology agreed that, like any pioneering technology, the consequences of its applications may not be known to science for many years, they firmly asserted that the development of these products involved intense safety assessments for each new crop, and projects and products found to have unacceptable risks were always abandoned.

MONSANTO UNDER ATTACK

Monsanto quickly became the focal point for protesters when the company introduced BST in the early 1990s, and these protests were recapitulated when the company introduced its genetically modified Roundup Ready Soybeans. European consumers became outraged when they learned that Monsanto's GM soybeans were mixed with ordinary soya and viewed such practices "as an attempt to infiltrate the market undetected."[28] This raised issues of choice, labeling,

[25]Pollan M. (October 25, 1999). Playing God in the Garden. *The New York Times*. Section 6, Page 44, Column 2.

[26]Others companies include Aventis SA and Pioneer Hi-Bred, acquired by DuPont in 1999, and Novartis Seeds, which merged with AstraZeneca's agribusiness operations in 2000 to form Syngenta.

[27]Rissler J. Bt Cotton—Another Magic Bullet? *Global Pesticide Campaigner*. March 1997;7(1). Available at www.ucsusa.org/food_and_environment/biotechnology/page.cfm?pageID=359. Accessed April 2003.

[28]Lilliston B. The Ecologist. *The Progressive*. February 1999;63(1):39.

and transparency.[29] Consumers demanded that GM foods be labeled so that they could exercise the right to choose whether to buy them and to know what the products they were buying and eating contained.

In Rahul Dhanda's view, Monsanto made "*the* paradigmatic error in corporate biotechnology marketing, which was to think of their product simply as a product." He continues:

> As such, GM foods were positioned by the same type of strategies used for all general products: choose the target market and get that market to buy as much of the product as possible. In this case, the target market was the farming industry, and this is the critical mistake that Monsanto made. They pushed the technology into the marketplace while identifying only a single stakeholder to whom they would market, the farmer.[30]

Monsanto's main initial focus was on persuading farmers to buy its products and meeting the formal requirements of government regulators, without taking into account the final, retail consumer. As Verfaillie later acknowledged, with the benefit of hindsight, "We parachuted products on to society without showing that they are safe." Monsanto thought that it would be enough to extol the benefits of their products.

When the company introduced the genetically engineered, insect-resistant NewLeaf Superior potato to agricultural markets, opposition increased and pro-testers began to engage in tactics such as the destruction of company field tests. To add insult to injury, Phillip Angell, then Director of Corporate Communications, was quoted as saying in a *New York Times Magazine* article, "Playing God in the Garden," "Monsanto should not have to vouchsafe the safety of biotech food. Our interest is in selling as much of it as possible. Assuring its safety is the FDA's job."[31] In a letter to the editor published the following month, Angell retorted that the story Pollan didn't tell is that biotechnology is the single most promising approach to feeding a growing world population while reducing damage to the environment.[32] But by then, Angell's comment had been widely published, giving opponents fuel for their fire.

The controversy over the Roundup Ready Soybeans and the New Leaf Superior potato led to influential calls for moratoria (that eventually led to a European

[29]In 1997, the European Commission agreed to a general policy on labeling of GM products that would eventually apply to seeds, as well as animal feed and food, causing other countries to rethink their policies toward GM seeds and foods. In July 1999, Japan announced that it would require GM products to be labeled and in August 1999, Australia and New Zealand announced the same policy in a joint agreement.

[30]Dhanda R. Genetically Modified Foods. In: *Guiding Icarus: Merging Bioethics with Corporate Interests*. New York: Wiley-Liss;2002:44.

[31]Pollan M. (October 25, 1998). Playing God in the Garden. *The New York Times*. Section 6, Page 44, Column 2.

[32]Angell P. (November 15, 1998). Playing God in the Garden: Letter to the Editor. *The New York Times*. Section 6, Page 26, Column 4.

Union [EU] moratorium on all genetically modified organisms in early 1998).[33]
Retailers then began withdrawing their support and stopped buying Monsanto's
products from farmers—not necessarily because they were against ag–biotech or
Monsanto, but because they feared losing customer support, just like politicians
feared losing voter support.

In May 1998, Monsanto's problems worsened when it agreed to buy Delta &
Pine, which had just patented, along with the U.S. Department of Agriculture, the
Technology Protection System—bio-engineered crops that did not produce fertile
seeds. The benefit of the technology for Monsanto was that it could increase the
value of proprietary seeds and open new markets in developing countries. Dubbed
the "Terminator Gene,"[34] the technology was highly criticized by opponents who
believed that it would harm subsistence farmers who depend on seed saving
because they cannot afford to buy new seeds each year.

That summer, Monsanto developed a biotech-acceptance strategy in Europe to
calm consumer fears and to increase public acceptance of GM foods. Part of this
strategy was a sophisticated, multimillion-dollar, European advertising blitz that
sought to persuade European consumers and decision makers of the safety of
Monsanto products and of the benefits of agricultural biotechnology. The ad campaign featured messages such as "Biodiversity matters to Monsanto," "Food
biotechnology is a matter of opinions. Monsanto believes you should hear all of
them," and "Biotechnology will help feed the hungry of the world." Meanwhile,
the company continued pursuing its acquisition strategy to strengthen its presence
in ag–biotech. In June 1998, the company agreed to a merger with American
Home Products. It would have created the world's largest agricultural company
and would have given Monsanto 85% control of the cotton seed market, 33% of
the soybean market, and 15% of the maize seed business in the United States. But
plans for the merger suddenly collapsed in October, causing Monsanto's share
price to drop 27% the day after the announcement of the cancellation. Despite this
setback, CEO Shapiro stressed, "Monsanto remained committed to its goal of
building a life-sciences company, even though the opportunity to move years
ahead in one fell swoop has been lost."[35]

At this time, company executives were beginning to acknowledge that
Monsanto was in trouble. "We could see that Europe was melting down and questioned what we should do. We knew we had to be proactive, and make a public
statement, but we weren't ready to do that then," recalled Jerry Steiner, Director

---

[33]The EU has since lifted its moratorium on GM foods by approving a new GM corn, developed
by the Swiss firm, Syngenta.

[34]The "terminator gene" is named after an Arnold Schwarzenegger movie, *The Terminator*, in which
he plays a robotic killer. The term was introduced by the Rural Advancement Foundation International
(RAFI), recently renamed The *Etc* Group.

[35]Morrow D, Feder B. (14 October 1998). 'American Home Products' Deal with Monsanto
Collapses. *The New York Times.* Section C, Page 1, Column 2.

of Global Strategy. They continued grappling with the controversy that was engulfing them and debated what to do. Meanwhile, things only got worse. By November 1998, worldwide protest was ignited over "terminator technology," and opposition against Monsanto began spreading wildly and rapidly, leading to much negative media attention and massive organized protests around the world. In marches against the company, protesters carried banners with slogans such as "Monsanto is a monster," "Monsanto is anti-farmer," and "Monsanto is sowing the seeds of destruction." That same month, two internal memos concluding that the ad campaign had been an unmitigated failure were leaked to Greenpeace.[36] Stan Greenberg,[37] who conducted this research, wrote in one of the memos:

> The broad climate is extremely inhospitable to biotechnology acceptance and, absent political support in government, Monsanto would surely face unfavorable decisions on its key products... At each point in this project, we keep thinking that we have reached the low point and that public thinking will stabilize, but we apparently have not reached that point... Over the past year, the situation has deteriorated steadily and perhaps at an accelerating pace.[38]

In May 1999, protests began a new crescendo when a laboratory study at Cornell University concluded that transgenic crops kill monarch butterflies, prompting United States regulators to re-evaluate Bt crops and the European Commission to delay approval of pending requests to sell Bt corn.[39] Scarlett Foster, Director of Investor Relations, recalled, "In 1999, all hell broke loose. There were claims that we were killing the monarch butterfly, controversy over biotech corn, it goes on and on..." In an article published in *Canadian Business* magazine, Reynolds described exactly what Foster meant:

> In the past year, genetically modified (GM) seeds and the food they produce have become the focus of an unprecedented international consumer backlash. What began as a fringe movement, focused on fears about GM technology's effects on human health and the environment, is growing, month by month, into a costly, far-reaching disruption of global food markets... All [companies developing GM seeds] have come under fire, but Monsanto—which had $8.6 billion in net sales in 1998 and makes 88% of the GM seed sold in the US—has managed to make itself a lightning rod for the whole controversy. How badly has Monsanto blown it? Even pro-GM farmers like Daynard [a Monsanto client] feel alienated. "I don't think they know what's going on," he says with rare frankness. "Every time I talk to someone at Monsanto, I just keep telling them to keep their

---

[36]Available at http://archive.greenpeace.org/pressreleases/geneng/1998nov18.html. Accessed April 2003.

[37]Stan Greenberg, Chairman and Chief Executive of Greenberg Research, had served in the past as polling advisor to President Clinton, Prime Minister Tony Blair, and Chancellor Gerhard Schroeder.

[38]Available at http://archive.greenpeace.org/pressreleases/geneng/1998nov18.html. Accessed April 2003.

[39]Marshall C. Corn Study Another Hot Potato for Biotech. *Forbes.com*, May 25, 1999. See B Case for more details on scientific debate surrounding this research.

mouths shut,"…referring to the way Monsanto confronted the [earlier] backlash with an aggressive, unyielding approach that gave the impression it intended to force-feed GM foods down European throats.[40]

Exacerbating the controversy, the debate on GM foods spread through the United States and Canada that summer. American and Canadian activist opponents launched domestic anti-GM campaigns, contending that large multinational firms, particularly Monsanto, had too much market power, and that farmers required to buy biotech seeds every year would become dependent on Monsanto. These concerns were echoed by American and Canadian farmers who felt that Monsanto was trying to stifle competition in the seed business through licensing agreements with other producers and restrictive contracts with farmers.[41]

In a significant departure from previous practice, Monsanto's contract with farmers, the Technology Use Agreement (TUA), sought to prevent farmers from saving seeds from plants grown 1 year and planting them the following spring. Farmers purchasing Monsanto's GE seeds had to agree that they would not resell, store, or collect seeds from the plants that they grew. The TUA gave Monsanto the right to conduct inspections of farmers' crops by seed company representatives, and to take samples from any part of their fields. Farmers found this practice to be highly objectionable.[42] Others recognized that, from Monsanto's standpoint, the TUA was a legitimate tool to recover the company's R&D investments, to protect its patent rights, as well as to be fair to farmers who respected their contract with the company. In addition to the backlash against the TUAs, Monsanto was also embroiled in a number of patent infringement disputes, both offensive and defensive. Together, Monsanto's efforts to protect its ag–biotech intellectual property and its lead in the industry sector reinforced consumer distrust of GM crops, and of Monsanto especially.

The collapse of public support for biotechnology and GM foods began to have a profound negative effect on the company. By the summer of 1999, when Monsanto's stock price was plummeting, and public opinion of Monsanto had grown intensely negative, Monsanto "hit a wall."[43] Yet no one at Monsanto quite understood why the company evoked so much hostility, furor and fear. Everyone was left wondering what went wrong.

---

[40]Reynold C. Frankenstein's Harvest. *Canadian Business*. October 1999;72(8):64–90.

[41]This would set in motion a suit filed against Monsanto at the end of that year, on December 14, 1999, by several farmers.

[42]Reinhardt F. Agricultural Biotechnology and Its Regulations. *Harvard Business School*. Boston, Mass: Harvard Business School Publishing; 9-701-004, April 30, 2001:3–4.

[43]Steiner interview.

MONSANTO COMPANY (B)[1]

THE NEW MONSANTO PLEDGE

Institutionalizing Values in the Face of Ethical Controversy

Visitors walking into Monsanto's head offices in St. Louis are greeted by large, colorful placards featuring the *New Monsanto Pledge*. They see it on company walls, in corporate literature, and just about everywhere that Monsanto operates. They hear, in conversations with company executives, how the Pledge permeates their consciousness. They inquire into the company's business processes and find out that the *New Monsanto Pledge* has changed the way Monsanto does business. After years of controversy and criticism (see Case A), and in the wake of an intense anti-biotech, anti–GM-foods and anti-Monsanto environment, especially in Europe, the company has used the Pledge to reformulate the way it manages ethical dilemmas and refocus its biotech-acceptance strategy. It is a series of commitments that have become the building blocks defining Monsanto's business and ethical strategies.

"We turned a disadvantage into an advantage," said Hendrik Verfaillie, Monsanto's recently retired CEO. The company had learned that even a great business model would fail if it didn't account for the realities of public opinion and consumer acceptance. "Our business model had no room for consumer biotech-acceptance," said Bob Echols, Director of Global Compliance. Monsanto had, in fact, underestimated the resistance and latent fear of European consumers, a fear reinforced and exploited by activist opponents. "It became apparent that regurgitation of science was not enough to alleviate concerns. Science alone won't convince. It's more than science," he added. Verfaillie concurred. "We could see the benefits of our technology. We knew we had products that would improve the quality of food and its nutritional value; that we had the ability to deliver products at more affordable prices to developing countries. Yet, there was *no* acceptance."

Following a 1999 meeting with Monsanto employees in Lyon, when he was the company's ag–bio head, Verfaillie:

> came to the realization that we were not sensitive to our stakeholders and we had to change our behavior. Society did not understand Monsanto... Here there were people, us, who believed we could make a difference, who really cared about doing good things, but were seen as the devil. It didn't make sense to any of us.

[1]This case study is designed to follow Monsanto (A), which describes the evolution of the firm's business model. It was written by Cécile M. Bensimon under the supervision of Peter Singer, following interviews that they conducted with Monsanto executives on September 11–12, 2002 at company headquarters in St. Louis, Mo. Abdallah Daar, Margaret Eaton and David Finegold, provided constructive comments and edits on earlier drafts. This case study is one in a series produced under grants funded by The Seaver Institute and Genome Canada to study models for ensuring ethical decision making in bioscience firms.

Jerry Steiner, Director of Strategy, concurred: "We had no concept that not talking would incite people to oppose and reject us. It's through adversity that you find out the track you're on is not working." And Robert Horsch, Vice-President of Product and Technology Cooperation, concluded, "For the most part, we didn't get it." So in the summer of 1999, the leadership team took the bold step of turning to those who best understood these concerns–their opponents. This was the beginning of what would become a carefully planned stakeholder dialogue intended to seek and listen to stakeholder views and concerns.

STAKEHOLDER DIALOGUE

Monsanto began a process of identifying stakeholder groups, including interest groups, non-governmental organizations, and government and regulatory officials. The company also began conducting customer and agricultural trade surveys, and initiated a 10-country tracking-study of consumers and what the company calls "influentials."[2] It undertook these initiatives to broach issues within communities, to listen to their views and concerns, and to garner support and rebuild trust. Steiner explained, "We wanted to have a network of friends who would act not because we were asking them to but because they believed it was right." Said Horsch: "We had shepherded a lot of projects, for example, sweet potato, rice genome,[3] but there was so much negativity. So we launched a large dialogue and asked stakeholders to tell us what they were worried about." Echols added: "We showed that we were willing to engage."

Both then-CEO Shapiro and soon-to-be CEO Verfaillie, as well as other top executives, met personally with European activist opponents, such as Greenpeace and Friends of the Earth, to listen to their concerns. Diane Herndon, Director of Public Policy, said, "We went through an extensive process to learn how to listen, not be defensive and not be aggressive." Verfaillie concurred. "I had been counseled not to talk but to listen." CEO Shapiro addressed the June 1999 Greenpeace Business Conference held in London through videoconference, an address that reverberated in all corners of the world because he candidly conceded, "We had got things wrong... Our enthusiasm for it [biotechnology] has, I think, widely been seen, and understandably so, as condescension or indeed arrogance, because we thought it was our job to persuade. Too often we've forgotten to listen."[4]

Meanwhile, Verfaillie also met with other companies that had dealt with, and overcome, public outrage, including Shell and BP. He wanted to gain insight into the way they addressed the controversy that surrounded them and work through his own dilemmas with executives who had faced exactly the same circumstances. As part of its stakeholder dialogue, Monsanto also hosted speakers

---

[2]Monsanto Company. Anticipating, Understanding, and Responding to Societal Expectations. Presentation prepared for internal use, March 2001, given to the authors.

[3]Such projects were designed to help developing countries.

[4]Available at http://archive.greenpeace.org/report99/index2.html. Accessed March 2003.

from the Public Policy Institute. Engaging with both allies and critics allowed Monsanto to hear the views and concerns of stakeholders. Echols said, "We were willing to engage fairly and honestly with all with whom we do business, and say, 'You have some concerns. It is legitimate to raise concerns. Let us address them.'"

But Monsanto's management team realized that even these concerted efforts would not resolve the ethical debates raised by agricultural biotechnology. For instance, the public opposition to plant-sterility research, wrongly attributed to Monsanto, or so-called "terminator technology," had not been limited to opponents of biotechnology. Even strong supporters of ag–biotech advised Monsanto not to proceed with the "terminator seeds." Gordon Conway, President of the Rockefeller Foundation, which had invested more than $100 million in ag–biotech research, expressed this view: "There may be arguments for 'terminator technology' in the industrial world, to secure [Monsanto's] proprietary rights. But in terms of the public debate, it's important that Monsanto make a number of concessions to the public." In October 1999, under intense public pressure and in a show of good faith, CEO Shapiro pledged in an open letter to Conway that Monsanto would not use systems such as the one it had acquired from the Delta and Pine Land Company's, and the U.S. Department of Agriculture's, Technology Protector System. In December 1999, Monsanto learned that it had not yet addressed farmer concerns when several farmers filed suit against the company, arguing that the company had deceived them about the risks of biotechnologies, and that its practices violated United States antitrust laws.[5]

It was the end of a long year of attempts to address the controversy surrounding the company, yet the situation seemed to be deteriorating. "In Europe, every time we thought it couldn't get worse, it did," said Steiner. Amid ever-increasing controversy, Monsanto's management was in search of a new strategy. Its aggressive biotech-acceptance strategy, including public education programs and a sophisticated advertising blitz in Europe, had visibly failed. The company had pushed its case too hard, too fast, and had failed to deflate protests and criticism. The management team faced the enormous challenge of coming up with a strategy that would inspire trust and acceptance.

In late 1999, Scarlett Foster, who had spent almost two decades in Monsanto's Public Affairs department, worked with a team "to get back to neutral media coverage, effective crisis management and a less aggressive stance on biotech." In January 2000, her team began formulating a new strategy. The intent was to unveil it in conjunction with the initial public offering (IPO) that would result from the spin-off from Pharmacia, re-establishing Monsanto Company as an independently traded company (see section below on "The New Monsanto"). "In a sense we were a new business, so we felt it was the right forum to make new commitments and

[5]Lieff Cabraser, Heimann & Bernstein LLP. Lieff Cabraser Takes Leading Role in Litigation on Behalf of Small Farmers and Consumers Against the Monsanto Company. *Press Release*, December 15, 1999. Available at www.lieffcabraser.com/monsanto_press.htm. Accessed April 2003; Barboza D. (December 15, 1999). Monsanto Sued Over Use of Biotechnology in Developing Seeds. *The New York Times*, Section C, Page 1, Column 2.

disclose a new strategy. We also felt that it was appropriate to do it in front of our customers first," explained Foster. Meanwhile, the team pondered how to devise a strategy that would profit from past successes and avoid past failures. Lessons from the stakeholder dialogue provided a framework for understanding and diagnosing the controversy. From this analysis emerged the new Pledge.

WHAT MONSANTO LEARNED

While he was Monsanto's Head of Agricultural Operations, Verfaillie had gained many insights from his discussions with activist opponents, but two were most prominent. First, he realized, "most [activists] sincerely believed that their objectives were not helped by our technologies. Most of them were not being negative just to be negative... I realized that there had been no communication. We were like two trains passing by each other with no connection." Secondly, it struck him that both activists and his own company "had similar objectives, for example, bio-diversity, sustainability, protection of wildlife, wetlands [but that] while our objectives were the same, there was a great divergence on how to get there... I asked: what information would you need to have to accept that our tactics could fit with your objectives?" Though few, if any, answered Verfaillie's question, Monsanto executives continued seeking the concerns of stakeholders. "It was a gradual process, negotiative. People were able to say 'I can agree with three of your points but I don't like your stance.'"

After hundreds of interviews, Monsanto began identifying stakeholder issues that were at the root of the debate over the risks and benefits of biotechnology and GM foods. The leadership team delineated the following twelve concerns:

1. sufficiency of testing/long-term effects unknown
2. transparency/sharing data
3. biodiversity/environmental safety
4. speed of introductions
5. lack of confidence in regulators
6. labeling/choice
7. no consumer benefits
8. corporate arrogance
9. power/corporate control
10. need for corporate principles
11. need for public education
12. need for public-sector investment

Herndon summed up some of the main lessons the company learned:

> We realized that our science and research were not publicly available. People couldn't see our extensive studies and results, and that approvals were based on extensive studies. For example, people questioned the environmental impact of our products because they had no way of knowing that we had looked at that extensively already. We also heard that people felt that we had no parameters.

This scared people. Without parameters, nothing seemed off limits. People wondered what we would not do to make money.

Monsanto also learned that excellent science and regulatory approval were not sufficient to market a product successfully. Said Verfaillie, "With the old business model, a good product was all that mattered. With biotechnology, having a good product is not enough. We have to be sensitive to stakeholders. Society wants to have a company they can trust. However, we were not trusted... Monsanto's strategies had been viewed as a power play. This attitude is typically non-European."

Monsanto further learned that consumers questioned the benefits of GM foods; that these had been forced on them without their consent or knowledge; that new technologies, especially biotechnology, trigger ethical concerns that cannot be ignored. Monsanto's leaders concluded that they must listen to the concerns and expectations of people and groups who have a stake in their activities and consider those concerns in their decision making.

Monsanto's management team also reviewed the underlying assumptions that drove their business strategy. The first was that people would agree with the company that biotechnology promises more sustainable agriculture and improved well-being. According to Steiner, "It was naïve arrogance. We had a *naïve* sense that our view would be unanimously accepted. We were *arrogant* by making decisions about what people would or should want." The second was that regulatory approval would lead to public approval, which proved incorrect in the face of public distrust of European regulatory agencies; third, that addressing questions about the science would silence most legitimate concerns; and fourth, that negative public reaction was "irrational" and would wane with public understanding of the virtues of GM foods.[6] Steiner said, "We believed that our products would have a highly positive influence and thought that only people who didn't understand the technology would oppose it, in which case it was really only a matter of education."

CHANGING THE FRAMEWORK

The lessons that Monsanto was able to distill from its stakeholders served as guideposts for Monsanto's new ethics-management strategy. The internal team strategically reconceived Monsanto's biotech-acceptance strategy. "We wanted to encapsulate things Monsanto was already doing and to incorporate lessons we learned," described Foster. They packaged this in the form of a new Pledge. Foster and other key members of her team had been involved in the formulation of the first Pledge (instituted in 1990 by then CEO Richard Mahoney (see Case A) and had seen how effective it had been both internally and externally. "Although the first Pledge had been a top-down initiative that stunned a lot of people, it was nevertheless inspiring and exciting," she recalled, citing an example in which

---

[6]Monsanto Company. Anticipating, Understanding, and Responding to Societal Expectations. Presentation prepared for internal use, March 2001.

employees had been motivated by Monsanto's Pledge to care for the environment by taking the initiative to restore the land and native plants on an abandoned property in front of their Texas plant. She wanted to build on that success by formulating a new Pledge relevant to the ag–biotech industry and reinvigorating the spirit of the Pledge within the company. The main objective for this new Pledge was to move away from a "trust us" to a "show me" stance, meaning that Monsanto would no longer just *say* that their products are safe but would *show* why and how they are safe.

THE NEW MONSANTO

In March 2000, Pharmacia took advantage of the strong negative impact that the ag–biotech controversies had had on Monsanto's share price to merge with Monsanto, acquiring its G.D. Searle pharmaceutical unit (with its prized arthritis blockbuster drug Celebrex, which made $1.8 billion in 1998). The new company, Pharmacia Corporation, announced that it would spin off the agricultural unit, which would keep the Monsanto name, with plans for an IPO in October of that year. Verfaillie was picked to move from agricultural operations to replace Shapiro as CEO of the new Monsanto.

Verfaillie knew that he had to confront the skeptics who opposed the concept of a new Pledge. Verfaillie recalled, "At first, there was a lot of resistance, a lot of barriers to overcome. If people didn't see a direct benefit, they challenged the initiative. It was hard to convince people to spend on something that didn't lead to increased sales, especially in an economic downturn." Because the company was still mired in controversy and had crash-landed under its acquisition-debt burden, there was significant internal debate and reflection on Monsanto's short- and long-term business strategies. The leadership team was apprehensive about making any commitments to investors who had already grown weary of the life-science business model. With the new Monsanto's future inexorably tied to ag–biotech, however, they concluded that the Pledge was needed to gain acceptance for its products.

THE NEW MONSANTO PLEDGE

In November 2000, Verfaillie announced the *New Monsanto Pledge for a New Company.* Echoing Shapiro in the face of earlier protests, he candidly expressed what Monsanto didn't do right: "We were blinded by our enthusiasm… We missed the fact that this technology raises major issues for people—issues of ethics, of choice, of trust, even of democracy and globalization."[7] Foster, recalled: "The announcement of the [new] Monsanto Pledge was a sort of *mea culpa*. It was a way to say we didn't do this right and this is what we're going to do."

---

[7]Monsanto Chief Executive Outlines Commitments on New Agricultural Technologies in the "New Monsanto Pledge." November 27, 2000. Available from www.monsanto.com/monsanto/layout/media/00/11-27-00.asp. Accessed November 2002.

The *New Monsanto Pledge for a New Company* embraced five (now six) principles—(integrity), dialogue, transparency, respect, sharing, and benefits— that aim to articulate the company's commitments and guide its operations, in an effort to incorporate sensitivity to public concerns and address ethical dilemmas related to biotechnology and GM foods.[8] The new Pledge is distinct from its predecessor in three ways. First, it is rooted in stakeholder views and concerns, or, as Herndon said, "What we understand people want from us," and is thus designed to "compel Monsanto to listen more, consider our actions and their impact broadly, and lead responsibly."[9] Second, it is based on dynamic principles that translate into a systematic approach to ethics management, rather than a set of tangible goals focused on specific issues. Third, it permeates all levels of the organization, thus introducing a new corporate dynamic that sparks and fuels action both internally and externally. "It's process and behavior-based. It's not a checklist or a list of activities. It's a mandate to change the company's behavior."[10]

Monsanto's pledge commitments are as follows:

Dialogue

The purpose of this commitment is to listen to ethical concerns and share ethical dilemmas. To address this purpose, Monsanto pledged:

We will listen carefully to diverse points of view and engage in thoughtful dialogue to broaden our understanding of issues in order to better address the needs and concerns of society and each other.

As part of its commitment to dialogue, Monsanto formed two advisory councils, the Biotechnology Advisory Council and the Grower Advisory Council, to help company executives better understand and respond to public issues that surround the development and commercialization of new technologies, as well as to guide the leadership team to make decisions based on the input of stakeholders. For example, Monsanto requested that its Biotechnology Advisory Council address dialogue issues associated with launching Roundup Ready wheat products. The company then formed a Wheat Biotechnology Industry Advisory Committee, the intent of which is:

> to facilitate an effective dialogue between Monsanto and participants in the wheat industry. This would include advising Monsanto in a number of areas including the feasibility, strategy, and standards for market acceptance; the development and review of plans for biotechnology spring wheat grain handling protocols; and the stewardship of biotechnology spring wheat and the process for commercialization.[11]

---

[8]Available from www.monsanto.com/monsanto/layout/our_pledge/default.asp. Accessed April 2003.
[9]Available from www.monsanto.com/monsanto/layout/media/00/11-27-00.asp. Accessed December 2002.
[10]Ibid.
[11]Available from www.monsanto.com/monsanto/layout/our_pledge/dialogue.asp. Accessed December 2002.

Transparency

The purpose of this commitment is to show that Monsanto products are safe, as well as to respond to demands for labeling or greater visibility. Monsanto pledged: *We will ensure that information is available, accessible, and understandable.* This goal, which represented a radical departure from Monsanto's traditional strategy and that of its competitors, spurred intense debate both within Monsanto and the industry. Horsch recalled, "At first, the idea of transparency—making available our research and data—was *really* scary. Once we grappled with it, we saw the benefits." Roy Fuchs, Director of Regulatory Science and leader of the team working on transparency, said, "We believe making this information more accessible has significant value for the scientific community and the public, because access to comprehensive and scientifically sound information is essential for advancing an informed discussion about new technologies."[12] In addition, transparency allows the public to make an informed decision on the acceptability of new technologies. To support transparency, Monsanto posted on its website the safety data summaries for all its biotech products. Monsanto's transparency was an important factor in the debate on the monarch butterfly,[13] which began when laboratory researchers at Cornell University found that large amounts of Bt corn pollen fed to monarch larvae affected larval weight and survival. As a result of these findings, which were published in *Nature* in May 1999,[14] comprehensive research was initiated by scientists in the US and Canada. It appears to have shown that Bt crops do not pose a threat to the monarchs. Environmental interest groups and the public became engaged when lay press stories were published claiming that bio-engineered Bt crops were poisonous to this popular butterfly. Monsanto participated constructively in the ensuing debates about the effects on the monarch butterfly of Bt exposure, and was supported by scientists who were able to access Monsanto's data. These data, combined with those of a new series of studies, showed that the initial speculations on the effects of Bt maize pollen on the monarch butterfly were not supported by the biology of the interactions between the crop and the insect under actual field conditions. In fact, one report showed that populations of monarch butterflies were increasing in the U.S. corn-belt, a fact that could be attributed to the reduction in pesticide use.

Respect

The purpose of this commitment was to show what Monsanto was already doing to take into account diverse religious and cultural beliefs. This component of the Pledge is especially relevant in the context of Monsanto's multinational activities. Monsanto pledged:

[12]Available from www.monsanto.com/monsanto/layout/our_pledge/transparency/default.asp. Accessed December 2002.

[13]Ibid.

[14]Losey JE, Rayor LS, Carter ME. Transgenic Pollen Harms Monarch Larvae. *Nature*. 1999;399: 214–215.

We will respect the religious, cultural, and ethical concerns of people through-out the world. The safety of our employees, the communities where we operate, our customers, consumers, and the environment will take highest priority.

While respect serves as the foundation for the other four Pledge commitments, Verfaillie explained why it was given its own section in the Pledge:

> As we were working on the development of the Monsanto Pledge, the concept of respect kept coming up again and again. Regardless of what section of the Pledge we were talking about, we always seemed to come back to the central concept of respect. So it's no surprise that we felt respect was important enough to be included as one of the five elements of the Pledge. But what was difficult to express in the Pledge was how important we felt respect was to the achievement of all the other elements. A simple, honest respect for others emerged as the essential beginning of everything else the Pledge guides us to do.

The commitment to Respect includes business practices that Monsanto had already established, such as the decisions not to use genes from animals or humans in products intended for food or feed, and to make certain that products intended for commercialization do not contain known allergens.

Sharing

The purpose of this commitment is to share proprietary genomic-sequencing data and other technology with public researchers, non-profit organizations, and local farmers around the world. Monsanto pledged:

We will share knowledge and technology to advance scientific understanding, to improve agriculture and the environment, to improve crops, and to help farmers in developing countries.

Monsanto had had a long tradition of sharing. In the early 1980s, the company began to share knowledge with public-sector researchers, and in 1990 the company began to share commercial application rights for GM crops. But sharing technology and research was new. It led to new strategies, such as small-holder farmer projects in developing countries,[15] which do not undercut commercial market-opportunities for Monsanto's products. Horsch explained: "There are two very different strategies with basic science; one is to be secretive and the other is to share, discuss, publish, teach… Sharing our research and data has proven to diffuse a lot of benefits." Steiner enumerated the purposes of Monsanto's sharing strategy:

> First, with broad-based technology sharing, there is more impact, and the broader the impact, the broader the support; second, by working with developing countries, it can motivate them to set up regulatory systems; and thirdly, it's a privilege to help. I'm a dad; I have two young kids at home. I think all the time about people in need and I can't imagine what it's like not to be able to put food on the table for my kids.

[15]Austin J, Barrett D. Monsanto: Technology Cooperation and Small-Holder Farmer Projects. *Harvard Business School*. Boston, Mass: Harvard Business School Publishing; 9-302-068, June 13, 2002.

Similarly, Verfaillie explained that Monsanto's involvement in developing countries is partly philanthropic and partly driven by a long-term return on investment. He said, "It's a means to demonstrate the benefits of biotechnology, which impacts biotech acceptance in Europe." Said Horsch, "There are no scalable costs to sharing resources. It doesn't cost us more to share with 879 scientists than with 1 or to benefit 500 farmers than 1."

Benefits

The purpose of this commitment is to show the environmental, agricultural, and economic benefits of agricultural biotechnology. Monsanto pledged:

We will deliver high-quality products that are beneficial to our customers and for the environment, through sound and innovative science, and thoughtful and effective stewardship.

Showing the benefits of agricultural biotechnology is central to Monsanto's "biotechnology acceptance strategy." To promote a deeper understanding of biotechnology, and to further bolster its acceptance strategy, Monsanto set up a website, Biotech Knowledge Center.[16] It answers questions on genetic modification, the safety of GM foods, and labeling; it lists the agricultural, environmental, and economic benefits of biotechnology; and it notes food quality improvements.[17]

Monsanto recently added two additional areas to the Pledge:

1. Act as Owners to Achieve Results

We will create clarity of direction, roles, and accountability; build strong relationships with our customers and external partners; make wise decisions; steward our company resources; and take responsibility for achieving agreed-upon results.

2. Create a Great Place to Work

We will ensure diversity of people and thought; foster innovation, creativity, and learning; practice inclusive teamwork; and reward and recognize our people.

Both reflect and reinforce internal principles and processes. It may be a reflection of the strengths and dynamism of the new Pledge that it created enough new issues and perceptions to necessitate the creation of two new Pledge commitments.

INSTITUTIONALIZING THE PLEDGE

The Pledge was designed to bring together and institutionalize ethics management strategies and processes that the company was already implementing, along with lessons that the company had learned from its stakeholder dialogue. Existing practices, for example, included the Monsanto Fund (see next sections), the stakeholder dialogue, and the Biotechnology Advisory Council. In a sense, the Pledge embodied both continuity and innovation. Monsanto had always had pockets of value-driven projects and had undertaken major policy initiatives to address

---

[16]Available at www.biotechknowledge.com. Accessed March 2003.
[17]Available at www.biotechknowledge.com/biotech/bbasics. Accessed March 2003.

ethical concerns, such as issues in manufacturing, safety, and corporate citizenship. But at no point in the company's history had any such projects and initiatives been organized into a company-wide strategy, which is what was most innovative about the Pledge. Said Echols, "There were multiple formats, structures, and sets of values. The Pledge was a way to focus them, centralize and manage ethical processes." It was a strategy for change and it was systematically implemented to reach every level of, and everyone involved in, the organization—employees and shareholders, customers, and stakeholders.

The Role of Senior Management

Senior management was the single most important driving force in re-managing the company's ethical dilemmas, developing and implementing the Pledge, and in mobilizing everyone toward a new vision. "If the Pledge was not led from the top, if it wasn't strongly supported by the executive team, there would be little, if any, commitment to the Pledge," stated Verfaillie. Echols added, "If senior leadership had not been willing to walk the walk, to embrace this, it would have been a horrendous task to make it successful." The management team's ongoing commitment to the Pledge has also been a powerful force in inspiring and guiding Monsanto's 14,000 employees, who have embraced the challenge of the five Pledge commitments through various mechanisms.

Foster also emphasized the strategic role of senior management in making new commitments, especially at a time of crisis. She described how, although the Pledge was not entirely a top-down initiative (like the first Pledge had been), it remained largely covert until it was announced. She said, "I took a lot of flack after the announcement of the Pledge for not telling a lot of people. This time, we had shared it with more people but still, not many knew about this… If you do something radical, and try to build consensus, you get a watered-down consensus. You get the lowest common denominator." She added that, at a time of crisis, there is no time to build consensus. Monsanto needed leadership and change, which is precisely what the management team set out to do when they formulated, communicated and implemented the Pledge.

Pledge Teams

The role of senior management was not only instrumental in setting a new direction and creating the context for change, it also laid the foundation for sustaining long-term change through the creation of Pledge Teams. Building on a Monsanto tradition that had been established with Shapiro's seven Sustainability Teams, the management team created five teams, one for each Pledge commitment, to chart the course of, and provide direction for, each commitment. The ongoing support of senior management for the Pledge is reflected in their active participation in these teams, which are each chaired by an executive member of the company. Teams meet once a month to lead activities and monitor the progress of their respective pledge. Each team seeks stakeholder feedback related to their respective commitment and continues to listen to stakeholder views, needs, and expectations.

In essence, the Pledge teams embody the underlying action-value of the Pledge, which Horsch emphasized. The value of the Pledge, he said, "is not from *making* the Pledge but in *doing* it."

Excellence Awards Program

One of the most important distinctions between the new Pledge and previous commitments is that the new Pledge does not prescribe outcomes, but guides processes and behaviors. Echols explained that one of the biggest factors in implementing behavior-based commitments is to find ways to align values and behaviors with business practices. One way to do that is to create mechanisms to reward "desirable" behaviors. Monsanto did this by setting up an Excellence Awards Program. The Awards Program is intended to promote and reward behaviors encouraged by the Pledge found in, for example, community- and dialogue-based programs. It also serves to measure deliverables against each Pledge commitment.

Town Halls

In addition to the Excellence Awards Program, Monsanto organizes Town Halls, which create a forum in which to seek the views of employees, to listen to their concerns, and to answer their questions. Monsanto employees from around the world can connect to Town-Hall meetings, hosted by Monsanto's head offices. Both Steiner and Herndon said that this mechanism has proven to be constructive. In particular, it has given Monsanto employees an opportunity "to talk about things, explore what we can and cannot do, what we *pledged to* do and not do."

Monsanto Fund

The Monsanto Fund was originally created in the mid-1960s to provide charitable grants in communities with a Monsanto presence. Today, the Fund's objectives and allocated resources have been aligned with the Pledge in five commitments. One of the Fund's missions, for example, is to promote science education. "We want to change the dynamics of science education and make sure that schools teach science on an inquiry-based model," said Deborah Patterson, the President of the Monsanto Fund. To this end, the Fund invested $675,000 in the "Vashon Compact" project to provide, in collaboration with Boeing Co., professional development for science education for 10 inner-city schools in the St. Louis area, where Monsanto is headquartered.[18] The Monsanto Fund's role serves the dual purpose of providing funding and education opportunities to local communities and, at the same time, applying and furthering Monsanto's pledge commitments.

Biotechnology Advisory Council

The Biotechnology Advisory Council (BAC) is an implementation strategy, born out of the stakeholder dialogue, to incorporate and institutionalize what

---

[18]Patterson interview.

Monsanto learned, in particular the need to be more inclusive and to seek external feedback. The role of the BAC is to give feedback and provide insights into critical issues that emerge through various outreach efforts, including:

1. Industry and food chain outreach
2. Scientific outreach
3. Government outreach/support
4. Stakeholder dialogue
5. Biotech industry coalition
6. Visibility of benefits and education

The BAC gives Monsanto guidance on ethical dilemmas that the company faces. Herndon explained that the BAC has been an effective and useful resource, providing advice on questions such as corporate responsibility, global regulatory frameworks, poverty alleviation, and issues related to sustainable agriculture. The BAC has also proven to be an effective means to engage stakeholders. Another Council, the Grower Advisory Council, allows Monsanto to seek feedback from what Verfaillie calls the food chain growers, distributors, and retailers. "Businesses see the value in being included," said Herndon; she finds they respond positively to companies that enlist their active participation in making decisions.

Kathryn Kissam, Director of Public Affairs, summed up the role of the BAC as encapsulating what all of Monsanto's implementation strategies achieve: "These tools are effective for three things: 1) they hold us accountable; 2) they provide us with feedback and criticism, from which we grow; and 3) they force us to have conversations about our processes, policies and infrastructure."

One issue facing Monsanto is to find ways to integrate the Pledge, as an ethical strategy specific to Monsanto, with the more general business ethics framework. Indeed, efforts are currently underway to reconcile the Pledge with two other frameworks of action, CSR and a Code of Conduct, which both reinforce and expand Monsanto's Pledge commitments.

Corporate Social Responsibility Integration Team

Just as the Pledge had its roots in the stakeholder dialogue, so conversations with stakeholders have led Monsanto to explore the concept of CSR and what it means for an agricultural company to be a responsible corporate citizen. Patterson explained, "CSR is an assessment tool to survey functions within Monsanto" and to develop standards and indicators of good business practices and corporate behaviors. To achieve this, Monsanto is working with its Business Social Responsibility (BSR) team to align Monsanto's business conduct program with corporate social responsibility.

The challenge is to "come to terms with how the Pledge and CSR fit," explained Echols. "We've had discussions on how to merge them, but right now there is no model." Corroborating this, Patterson explained: "We want to bring CSR and the Pledge together but haven't yet figured out how they fit. The Pledge gives us a set of values. CSR would give us standards. Then the question would be: 'How do we implement standards that are reflective of our values?'"

Code of Conduct

Echols explained that, although the Pledge provides frameworks and boundaries to guide the company's operations and decisions, it does not give values-based tools to deal with situations in which there is a conflict of values. For example, if a client makes a request that violates the company's beliefs or convictions, should the company stand for what it believes in, staying true to its corporate values? Or should Monsanto respect the client's request, which is one of the pillars of the Pledge? According to Echols, "There has to be something guiding decisions, something that gives us room to achieve greater good." Therefore, his team is focusing on putting together a Code of Conduct that:

- is solely designed for Monsanto and intended to replace Pharmacia's code of conduct, which had been largely designed for a pharmaceutical and chemical company
- provides underlying values to achieve a greater good
- will weave corporate values, the Pledge, and the Corporate Code together.

THE IMPACT OF THE PLEDGE ON MONSANTO

There is a consensus among company executives that the new Pledge has transformed Monsanto, at both the senior management and at employee levels. Foster observed that the announcement of the Pledge was "immediately impactful," especially at the senior executive level. She recalled, "It changed our conversations. We could talk about the future. It brought things together, things that had been disjointed. It became a tool to work with and to measure against."

Internally, the impact of the Pledge is apparent in three distinct ways. First, it created clarity. "It's a line in the sand," said Steiner. For Kissam, "Implicitly, the Pledge was a catalyst to be clear about internal policies." Second, the Pledge has created new ways of operating. Kissam continued: "Explicitly, it was a catalyst to seek feedback [from stakeholders] that directly impacts business decisions such as defining transparency, sharing data and sharing dilemmas." The Pledge has integrated ethical and stakeholder issues directly into the firm's strategic decision-making process. "The real impact is that to make decisions, we go through each Pledge commitment to make sure we are honouring all of them," said Horsch. Similarly, Steiner said, "The Pledge is used as the main criterion to answer questions: Is this going to work? Do we have a successful business plan? Do our business decisions have good common sense? The Pledge contributes to all our business decisions because we understand why we made these commitments." Third, the Pledge is building a unified corporate culture. "We have gotten to the point that there are people here because of the kind of company we are," said Verfaillie. The Pledge has had a distinct effect on the working atmosphere of the company. "It's creating a great place to work. People are starting to get pride again," added Verfaillie. In Steiner's view, "It has created a culture of safety within the company. For example, it's safe to express opposition." Herndon also believes that the Pledge is becoming part of the company's social fabric, as evidenced by this year's applications by 308 teams to the Excellence Awards Program, compared to 188 in 2001. "308 applications to the Awards Program shows engagement and alignment," she said.

THE IMPACT OF THE PLEDGE EXTERNALLY

Externally, the Pledge has had some effect on the way the company is perceived, although Monsanto's executives feel that it is still too early to assess this impact. To Verfaillie, one measure of success is that Monsanto had recently been rated first in a survey on how the company was perceived by farmers and retailers, up from seventh place in the 1999 survey. Echoing the perspective that the tide is turning and the opposition has somewhat subsided, Herndon said: "Critics have lost traction because people can see what we do. Without the pledge, we'd be fighting every battle on ag-biotech."

From an investor perspective, the announcement of the Pledge has not yet had a direct impact on the company's share price, although the new strategy was well received by Wall Street analysts. Other, more fundamental, factors may have contributed to the lack of a market response to the Pledge. According to UBS Warburg, the new Monsanto was, at that time, "spectacularly undervalued," in large part because of the market's deep concern over the future of agricultural biotechnology, and biotechnology more generally, in the depressed market of 2001–2003. Analysts viewed the ag–biotech sector as unattractive, and are likely to continue to view it as such so long as there is uncertainty about the public acceptance of GM foods and the other challenges and costs of developing this new field of science. Given this situation, Wall Street had been urging Monsanto to find ways to deal with the controversy and acknowledged Monsanto's efforts to address public concerns. After the launch of the Pledge, Salomon Smith Barney saw the firm as moving in the right direction:

> We believe there is no better-qualified group of individuals available in this business to run Monsanto. It has begun a program to educate the public better on the benefits of biotechnology through the Biotechnology Industry Association and in general has become more sensitive to the needs of the ultimate end-users in the food chain. In short, the arrogant Monsanto has become more accommodating.[19]

In a similar report, Lehman Brothers acknowledged the Pledge when it wrote, "Monsanto's challenge [to overcome controversy] is to be carried out through a number of initiatives to increase public-safety awareness and the benefits of plant biotechnology."[20]

[19]Salomon Smith Barney. Monsanto. *Equity Research: United States: Ag Biotech/Fertilizers*, November 30, 2000:27.
[20]Lehman Brothers. Monsanto: The reincarnation of the MONstar or MONster? *Equity Research: Major Chemicals*, October 20, 2000.

From a shareholder perspective, the ultimate test of the Pledge's utility will be whether it leads to greater sales of ag–biotech products. Both Foster and Verfaillie explained the hope for a causal chain: "Values serve as an enabler," said Foster, to help create a climate for greater biotech acceptance; this, in turn, should help ease regulatory approval and market acceptance that will, according to Verfaillie, be how the Pledge generates "explicit business benefits."

Perhaps the biggest external impact of the Pledge is at the industry level. The announcement of the Monsanto Pledge initially took the industry by surprise, but Monsanto's implementation of its Pledge commitments has forced industry peers to follow suit. For instance, Monsanto's decision to share the rice genome-sequence data, made public in August 2000, surprised companies that were also sequencing rice but had chosen not to make the data public. "Twenty months later," Verfaillie noted, "Syngenta followed."

A MORE NUANCED PUBLIC DEBATE

Coinciding with Monsanto's decision to make public the rice genome, Golden Rice[21] emerged as an important topic for discussion in the debate on GM foods. In a statement released in February 2001, Greenpeace stated, "Genetically engineered rice does not address the underlying causes of vitamin A deficiency (VAD), which are mainly poverty and lack of access to a more diverse diet."[22] Greenpeace remains opposed to GM crops and foods. The organization did, however, announce at the 2001 BioVision conference on biotechnology in Lyon, France, that it would not disrupt field trials of Golden Rice. Many observers thought that this announcement signaled a significant softening in Greenpeace's trenchant opposition to GM food.[23]

In March 2001, Dr. Patrick Moore,[24] an ecologist and cofounder of Greenpeace, reversed his opinion on agricultural biotechnology:

> The campaign of fear now being waged against genetic modification is based largely on fantasy and a complete lack of respect for science and logic. In the balance it is clear that the real benefits of genetic modification far outweigh the hypothetical and sometimes contrived risks claimed by its detractors. Genetic modification can reduce the chemical load in the environment, reduce the impact on non-target

[21]Golden Rice is genetically modified rice that contains a provitamin A substance called beta-carotene. The intent of the genetic modification was to enhance the nutritional value of the rice and, specifically, to prevent vitamin A deficiency-related childhood death and blindness, which is prevalent in some developing countries where rice is a staple.

[22]Greenpeace. Genetically Engineered "Golden Rice" is Fool's Gold. Greenpeace Statement, February 2001. www.biotech-info.net/fools_gold.html.

[23]Moral Maze. February 2001. www.newscientist.com/news/news.jsp?id=ns9999408.

[24]Moore, who is now an environmental consultant, was a founding member of Greenpeace. He served for 9 years as President of Greenpeace Canada and 7 years as Director of Greenpeace International. In 2001, however, he broke with Greenpeace, accusing it of abandoning science and following agendas that have little to do with saving the Earth.

species, and reduce the amount of land required for food crops. There are so many real benefits from genetic modification compared to the largely hypothetical and contrived risks that it would be foolish to ban genetic modification.[25]

There have also been signs that the tide of regulatory and public opinion could begin to turn in favor of GM seeds and foods. At a debate on GM crops in July 2000, the European Commission stated that it wanted to resume the approval process "in the near future," while addressing public concerns. The pressure from governments against GM crops and foods has also lessened, and recent initiatives suggest that there is a more constructive debate on these matters. Today, governments are mainly focusing on drawing up a regulatory framework to address labeling, traceability, and monitoring of GM foods.[26] Although the EU and many other countries and non-governmental organizations (NGOs) around the world continue to resist GM foods and to protest against agricultural biotechnology products, the difference today is that the biotech industry, along with Monsanto, is engaging, listening to, and addressing the concerns of all stakeholders in the GM debate. The biotech industry has recently created an organized pro-GM alliance that will be organizing a campaign through industry-funded lobby groups, think tanks and websites aimed at regulators, legislators, retailers, and consumer groups to approve, and approve of, GM crops.[27]

MEASURING THE EFFECTIVENESS OF THE PLEDGE

Just as Monsanto designed processes to implement the Pledge, company executives designed tools to measure the effectiveness of the Pledge. "We want to continue delivering against the Pledge, so we continue to revisit our performance against the Pledge," said Herndon. To do that, and to show stakeholders what the company has done in support of the Pledge, Monsanto came up with the idea of a Pledge Progress Report.

The first Pledge Progress Report, *Fulfilling Our Pledge: 2000–2001,* was a collection of Monsanto's accomplishments against each pledge commitment. There were three parts to this report, described in its Letter from the President and CEO:

> The first reports specifically on the progress we've made toward fulfilling the five elements of the Pledge. The second called "Seeds of Change" presents other projects and activities that were already underway when the Pledge was launched.

[25]Greenpeace Founder Supports Biotechnology: Moore Criticizes Colleagues for Opposing Golden Rice. March 2001. www.agbioworld.org/biotech_info/pr/moore.html.

[26]The EU has since lifted its moratorium on GM foods by approving a new GM corn developed by the Swiss firm, Syngenta. The EU has also mandated that all forms of GM corn sold in Europe have to be clearly labeled.

[27]Rowell A. (March 26, 2003). The Alliance of Science: Independent Groups Share Pro-GM Common Ground. *The Guardian*, p. 9.

The third section continues our long-standing commitment to report on our environmental, safety, and health performance.[28]

This first report was received with mixed emotions from stakeholders. Herndon said that stakeholders reported that "this book seems to have a lot of good news," but wondered "why you still seem to be having problems." She concluded that what was lacking from this first report was stakeholder feedback. This was part of the second Pledge Progress Report. This report is titled *Commitments to Our Stakeholders: 2001–2002*. Like the first, it chronicles the ways in which Monsanto is applying the Pledge to the company's business activities around the world. There are detailed examples for each Pledge commitment, showing what and how Monsanto is doing to advance and reinforce the spirit of the Pledge, both within and outside the company. There have been additions to improve and broaden the Pledge, as Hugh Grant, Executive Vice President and Chief Operating Officer, informs stakeholders: "We have also added two additional areas to our Pledge: Create a Great Place to Work and Act as Owners to Achieve Results. These two areas are important for how Monsanto people work together as a company and we decided that they should become an explicit part of our corporate commitments in the future.[29]

The second Pledge Progress-Report, like the first, highlights activities related to Monsanto's commitments that preceded, and continue to be an integral part of, the Pledge. These commitments include safety, community involvement, environmental stewardship, and diversity. To show what the company has done in support of these commitments, Monsanto included eco-efficiency data on resource use efficiency and other environmental data, using a format established by the World Business Council for Sustainable Development. As the title suggests, this second report acts on stakeholders' suggestions to take heed of their views and feedback. In it, Monsanto included discussions about some of the dilemmas the company continues to face, such as challenges around labeling, adventitious presence, intellectual property, and plant-made pharmaceuticals. Each dilemma features outside perspectives as well as Monsanto's approach to the issues at stake.

The objective of the Pledge Progress Reports, therefore, is to reflect periodically on whether the company is doing the right things to deliver on its Pledge commitments and whether they are making a difference to people. But measuring how the Pledge is making a difference remains a challenge. This is because the Pledge is not quantifiable. Because the seven Pledge commitments are process goals, emphasizing whole processes rather than narrow tasks, company executives find it hard to know where Monsanto makes progress. "When we ask the questions, How are we doing? Are we living up to our standards? Are we living up to expectations of stakeholders?," said Patterson, "our answers start with conversations." So although there is a consensus at Monsanto that the key to measuring the

---

[28]Monsanto. *Fulfilling Our Pledge*:2000-2001.
[29]http://www.monsanto.com/monsanto/layout/our_pledge/default.asp. (Accessed April 2003.)

effectiveness of the Pledge is external feedback, Verfaillie said the question is, "How can we measure something that is not quantifiable? We have to define objective measures." Patterson echoed this perspective. She said, "A standard is something you can measure and measure against, which we don't have. We have a values-based program that has focused our thinking. But we have not yet fleshed out standards against values. So there's no evaluation of how we're meeting these values."

Even so, company executives believe that the Progress Report is an effective internal audit mechanism. "It allows us to know what we've accomplished in relationship to Pledge concepts, what's important to outside groups, which also allows us to see if their criteria overlap with our commitments," said Herndon. "We want to know what's important to society, because if this is what's important to society, then this is what's important to us," continued Herndon. "Because the biggest factor is trust," said Verfaillie. "We can't sell our products if there's no acceptance, and there can be no acceptance if there is no trust." Perhaps this has been the Pledge's most daunting task: repairing, building, and maintaining trust.

TOWARD THE FUTURE

Today, the management team feels that it has become more humble as a result of the controversy that engulfed it and the company. Its members have a greater appreciation for the ethical concerns and dilemmas that biotechnology raises. In addition, Verfaillie believes that Monsanto has a greater understanding of public perceptions and expectations, as well as what it takes to be successful in today's society. He described four basic ingredients for success. First, the company has to offer products and services that excite the customer. Second, the employees need to understand the company's vision and perceive that they can make a difference as Monsanto employees. Third is to satisfy investors, which means to have a lucrative business. The fourth ingredient is to make a difference to society by showing that the company does not exist merely to create economic benefit for itself, but to do well as a company doing good things to benefit society.

Monsanto's leaders still have a lot of questions to which they seek answers as they continue to address their corporate responsibilities: Are we doing the right thing? Are we living up to the expectations of stakeholders? Are we doing something that can hurt people? What does it mean to be a responsible corporate citizen? What behaviors are productive? What is the relationship between the Pledge, CSR, and a values-based Code of Conduct? Monsanto executives also ask questions about the societal implications of their business: Who benefits from our new technologies? Does the developing world benefit if we share our proprietary research? How much intellectual property should a single company control? How do we deal with questions of intellectual property rights, fairness, equity, and distribution.

Finally, the management team still wonders whether the Pledge will have a long-term impact on the company and, if so, what that impact will be. The company still grapples with two necessary goals that can sometimes pull the company in separate directions: maximize shareholder value to satisfy investors and comply with their Pledge commitments and Code of Conduct to satisfy stakeholders. It is still Monsanto's challenge, and that of the wider ag–biotech industry, to try to satisfy the needs of both shareholders and stakeholders simultaneously.

14

NOVO NORDISK: THE
TRIPLE BOTTOM LINE[1]

INTRODUCTION

Novo Nordisk, the Danish pharmaceutical company focused on diabetes care, has a new clause in its Articles of Association. It was added by the Chief Executive Officer (CEO) and Board of Directors, and passed by the shareholders at the annual general meeting of March 15, 2004. The article holds the company's management responsible for delivering not only on its financial targets, but also on its environmental and social mandates.

This three-pronged approach (Finance, Environment, Social) is what Novo Nordisk calls the Triple Bottom Line (TBL), and it influences decisions throughout Novo Nordisk on how to maintain what the company calls its "societal license to operate," while maximizing long-term profitability. Elements of the TBL are:

- The Novo Nordisk Way of Management
- Noticeably passionate and dedicated leaders
- Human resource policies with a focus on human rights
- An initiative that holds the company's suppliers accountable to the TBL
- A search for a cure for diabetes—the disease their products help control and on which the current success of the company depends
- Global corporate social responsibility focused on building sustainable local health infrastructures
- A concentrated effort to engage and reconcile dilemmas with stakeholders

[1]This case study was written by Jocelyn E. Mackie under the supervision of Abdallah S. Daar. All data for this case come from interviews performed by Daar and Mackie on March 15–17, 2004, at Novo Nordisk, unless otherwise noted. This case is one in a series produced under grants funded by The Seaver Institute and Genome Canada to study models for ensuring ethical decision making in bioscience firms. We would like to thank all the employees of Novo Nordisk who took part in this study.

- A commitment to find investors interested in socially responsible investing (SRI)
- A bioethics department with a director who has won an animal welfare award from the Danish Animal Welfare Society

All of this is supported by in-depth reporting mechanisms, a Balanced Scorecard (BSC) that holds employees and the company accountable to the TBL, and company facilitators who ensure that the Novo Nordisk culture is lived and practiced. As CEO Lars Rebien Sørensen sums up:

> Before, I had to get approval from shareholders to give money to the Diabetes Foundation, whereas now that it has been passed and it's in the company's mandate, it is just part of what the company does. Now that it's actually in our charter—now there's no way for us not to follow the Triple Bottom Line.

THE COMPANY

THE NOVO AND NORDISK HISTORY

Novo Nordisk's story began in 1922 when Danish Nobel Prize winner August Krogh and his wife, Marie, a doctor and researcher in metabolic diseases who suffered from type 2 diabetes, traveled to America. The Kroghs heard of two Canadian researchers, Frederick Banting and Charles Best, who were treating diabetics with insulin extracted from bovine pancreases. Because of Marie's disease, the Kroghs traveled to Toronto to meet with Banting and Best and were ultimately granted permission to produce insulin and treat diabetic patients in Denmark. Together with Dr. H. C. Hagedorn, a specialist in the regulation of blood sugar, Krogh started Nordisk Insulinlaboratorium (Nordisk) in 1923.

The Pedersen brothers (Harald was an engineer and Thorvald was a chemist) joined Nordisk to build the machines and analyze the chemical processes for insulin production. Thorvald did not get along with a company executive and was fired, so both brothers left. By 1924 the Pedersens had created Novo Terapeutisk Laboratorium (Novo) and were also successfully producing insulin in Denmark.

Over the next 65 years, Novo and Nordisk expanded rapidly, and both created large research units and hospitals focused on diabetes care. Both companies also diversified: Novo became the world's largest producer of industrial enzymes, and Nordisk developed drugs for hemophilia and growth disorders.

Having competed for many years, the two companies decided to merge in 1989 to form Novo Nordisk. Together they could now focus on establishing a presence in international markets. Just over 10 years later, in 2000, Novo Nordisk split into two main businesses: Novo Nordisk for healthcare and Novozymes for enzymes (Novo A/S is the holding company).

Employees of Novo Nordisk feel that the company's history has helped create the current corporate culture. Lise Kingo, Executive Vice President of Stakeholder Relations, says: "Novo Nordisk is founded on the passion for diabetes, for people and for values—it is this seed that the company, as you see it today, has grown from."

Vernon Jennings, a Sustainable Development Consultant to Novo Nordisk, and former Vice President of Sustainable Development at the company, adds, "They were not purely about making money; they were about producing insulin to treat diabetes. They were led by science and a sense of purpose." Stig Pramming, Vice President of Global Health Strategy and Innovation, adds that the fact that these two companies were "always trying to outdo each other was enormously healthy for the company. It kept both companies in the front seat of innovation."

NOVO NORDISK TODAY

Novo Nordisk is currently focused on four therapeutic areas: diabetes care, hemostasis management (blood-clotting agents), growth hormone therapy, and hormone-replacement therapy. The company is a world leader in diabetes care,[2] with a broad range of products that include pharmaceuticals, injection devices, and needles. Mads Krogsgaard Thomsen, Executive Vice President of Research and Development, explains part of the company's present research focus:

> We are looking for better insulin. Early in the pipeline are improved versions for delivery. We are looking into an inhaled version which is in phase III trials—and we're also looking into tablets for type 2 diabetes. We are looking into proteins for cancer and chronic inflammation as well. We are using our technical skill-base from biotechnology and recombinant technology.

Novo Nordisk had an operating profit of 857 million Euros in 2003, an increase of 7% from 2002; total sales increased by 15% from 2002; and the net profit was 653 million Euros, which was an increase of 19% from 2002. The company has a 48% global market share in insulin.[3] Much of the company's success is due to patents it holds for insulin-delivery-methods, as there is no patent for insulin itself. The firm has 19,000 employees working in 69 countries around the world. Novo Nordisk maintains its strong presence in Europe—the eighth-largest Europe-based pharmaceutical company in 2003[4]—and continues to grow in new markets. For example, the company is adding approximately 150 new sales representatives in the United States over the next few years and has just acquired a Brazilian pharmaceutical company, Biobrás, to continue expansion in Latin America.

DEFINING EVENTS

Although Novo Nordisk has always been focused on delivering quality products to their consumers around the world, two defining events broadened the company's stakeholder focus and helped mobilize a more institutionalized approach to corporate ethics and responsibility. Spurred by a new culture of consumer awareness

[2]http://uk.us.biz.yahoo.com/cnw/040419/nj_novo_nordisk_appt_1.html.
[3]Danske Bank–Danske Equities. Company Report: Novo Nordisk, p. 3.
[4]Pharmaceutical Executive, June 2004.

and activism in the late 1960s and 1970s, society, mainly in the United States, became fearful of the use of enzymes in detergents. Ralph Nader, a leading consumer-activist, was warning the public not to buy detergents containing enzymes. Kingo explains what happened next:

> It was a big surprise when Novo's Industry was hit by a campaign created by the American consumer-activist, Ralph Nader, in 1969 in the US. At that time, detergent enzymes was one of the biggest products in the company, and the campaign warned consumers against buying detergent with enzymes due to allergy risks. The issue of allergy risks to consumers was beyond the imagination of the company, as enzymes made up only 2–3% of the detergent. But it took the Investigation Commission that was set up in the States more than 2 years to conclude that there were no risks, but only benefits from adding enzymes to detergents. But meanwhile, an undocumented claim from an NGO had spread to the media, to our customers, and finally hit our employees in the sense that we had to let go a group of employees due to the decline in turnover from detergent enzymes. By listening to my colleagues who were in the company at that time, I think that this incident prepared the ground for defining the company's stakeholders more broadly, for being alert to critical voices and addressing them proactively, and for understanding new signals and trends from stakeholders in society.

The second event began in the late 1990s, when more than 40 pharmaceutical companies started a court case against the South African government. They were trying to stop the government from enacting legislation that would reduce the price of drugs by allowing parallel importing of cheaper generics as substitutions for brand-name drugs.[5] A Novo Nordisk subsidiary was one of the 40 companies bringing the suit. These events were in relation to fighting the AIDS epidemic that had reached catastrophic levels in South Africa by this time. In 2001, Oxfam started a campaign against the companies to raise awareness about public health, and Novo Nordisk was forcefully targeted by the Danish press. Kingo explains Novo Nordisk's involvement and how it was handled:

> In 2001 we were hit by the South African court-case, which was a debate over patents of AIDS drugs. It came as a surprise. One reason was that it was a debate over AIDS that we, as a diabetes company, suddenly were dragged into. It was a debate over patents and we do not have any patents on human insulin. Another reason might have been that we had just come out of a de-merger process that had demanded significant internal focus. It did, however, make us update our trend-spotting activities regarding access to health in general. Lars Rebien Sørensen had just taken up his new position as Novo Nordisk's CEO and handled the situation beautifully. He was very open and direct and didn't shy back. Even when there was a public demonstration in front of one of our plants, Lars was there, speaking to the protesters, debating the patent issue on the television, having dialogues with NGOs. He managed to explain that Novo Nordisk was not in the AIDS business, that the company did not have patents on insulin sold in the developing world and that Novo Nordisk, as always, would take a responsible business approach. It was a tough process, but I think that Lars, as well as the company, stood stronger than ever after the waves had calmed. Novo Nordisk was awarded the most admired company in Denmark one month later.

[5]Sidley P. Drug Companies Sue South African Government Over Generics. *BMJ*. 2001;322(7284):447.

In company meetings with concerned non-governmental organizations (NGOs), the CEO initially tried to explain the technical issues, but realized these details were not important to the NGOs. Novo Nordisk then decided to work with the World Health Organization (WHO) to try and resolve the concerns of their stakeholders. These defining events were an impetus for the company to broaden its stakeholder model and engage more proactively with yet another set of stakeholders' concerns: health in the Third World.

MAKING THE CASE FOR ACTING RESPONSIBLY

When Novo Nordisk employees were asked why the company emphasizes acting ethically, some said it was partly because of the history of the company, others that it had something to do with the Danish culture, and others that it was their duty as world citizens.

Sørensen feels that it's "part of the DNA or history of the company and it is society that actually decides the framework within which we can operate." Pramming explains his perspective:

> Do we, as a pharmaceutical company, have more responsibility to health than you do, or, say, journalists do? I would say no. But we have an equal responsibility. We must all try to find ways to contribute to society. You should do what you are good at; we are good at diabetes. We have an obligation to help here. An obligation to the person with diabetes wherever they are in the world. We have an obligation to the societies who have people with diabetes and we all have an obligation to the poor with diabetes as well as a responsibility to fight poverty in the world.

Companies are often trying to make a business case for acting ethically, whether it is to justify it internally to employees or externally to shareholders. Sørensen sees it like this: "To be honest, I'm mixed about trying to make the business case for ethics. There is a proportion of it that will never be proven. And neither should you try. Many of the benefits are intangible and if you didn't do them you'd be somewhere else. Those benefits are really important and there is a linkage between staff and performance."

Novo Nordisk is confident that the benefits of behaving in a sustainable way are important for the long-term success of the company. The question is whether this model is transferable to other contexts outside of the company and Danish culture. Jennings' answer is, "I think the model is transferable—but it is easier to do it in a Scandinavian context."

"THE NOVO NORDISK WAY OF MANAGEMENT"

The company's model for acting ethically begins with "The Novo Nordisk Way of Management," a framework that is intended to guide the way that every company employee and manager works around the world (Exhibit 14.1). It consists of three elements: The Vision, The Charter, and Policies. The first element, *The Vision* (Exhibit 14.2), is meant to explain what the company wants to achieve and what the

EXHIBIT 14.1   **The Novo Nordisk Way of Management**

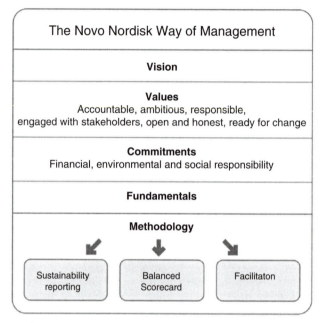

The Novo Nordisk Way of Management serves as the solid footing from which innovative ideas can take off. Its immediate strengths lie in its consistency, coherence and systematic follow-up methods. To people working at Novo Nordisk, it simply is the way do things.

overall goals are relative to patients, society, and their employees. The second element, *The Charter,* includes a declaration of the company's values, which are to be accountable, ambitious, responsible, engaged with stakeholders, open and honest, and ready for change. The Charter also includes the company's commitments to being financially responsible, environmentally responsible, and socially responsible (the TBL). The Charter's final component is a list of fundamentals (Exhibit 14.3) that each unit and team must follow. The Charter guides all companies in the Novo Group (including affiliates), and the employees commit to working within this framework. The third aspect is the *Policies*, which give good-practice and ethical guidance on specific operational issues within these areas: communication, environment, finance, health and safety, information technology, legal, people, purchasing, quality, risk management, global health, and bioethics. Within each policy, Novo Nordisk explains what it will do and how it will do it. For example, the bioethics policy, which is written in The Novo Nordisk Way of Management's handbook, says:

> In Novo Nordisk we will continuously improve our bioethical performance. This means we will:
>
> • Promote bioethical awareness throughout the company

Exhibit 14.2   **Our Vision**

Our Vision paints a picture of what Novo Nordisk wants to achieve as a company – our place in the market and our overall goals in relation to patients, society, and employees.

We will be the world's leading diabetes care company.
Our aspiration is to defeat diabetes by finding better methods of diabetes prevention, detection, and treatment. We will work actively to promote collaboration between all parties in the healthcare system in order to achieve our common goals.

We will offer products and services in other areas where we can make a difference.
Our research will lead to the discovery of new, innovative products also outside diabetes. We will develop and market such products ourselves whenever we can do it as well as or better than others.

We will achieve competitive business results.
Our focus is our strength. We will stay independent and form alliances whenever they serve our business purpose and the cause we stand for.

A job here is never just a job.
We are committed to being there for our customers whenever they need us. We will be innovative and effective in everything we do. We will attract and retain the best people by making our company a challenging place to work.

Our values are expressed in all our actions.
Decency is what counts. Every day we strive to find the right balance between compassion and competitiveness, the short and the long term, self and commitment to colleagues and society, work, and family life.

Our history tells us, it can be done.

Exhibit 14.3   **The Charter Fundamentals**

1. Each unit must share and use better practices.
2. Each unit must have a clear definition of where accountabilities and decision powers reside.
3. Each unit must have an action plan to ensure improvement of its business performance and working climate.
4. Every team and employee must have updated business and competency targets and receive timely feedback on performance against these targets.
5. Each unit must have an action plan to ensure the development of teams and individuals based on business requirements and employee input.
6. Every manager must establish and maintain procedures in the unit for living up to relevant laws, regulations, and group commitments.
7. Each unit and every employee must know how they create value for their customers.
8. Every manager requiring reporting from others must explain the actual use of the reports and the added value.
9. Every manager must continuously make it easier for the employees to liberate energy for customer-related issues.
10. Every manager and unit must actively support cross-unit projects and working relationships of relevance to the business.
11. Everyone must continuously improve the quality of their work.

- Establish and ensure high ethical standards for:
  - Experiments on living animals
  - Clinical trials and use of human material
  - Gene technology
- Set high ethical standards for our external partners, contract research organizations and suppliers and monitor their performance
- Engage in stakeholder dialogue and partnerships, and report on our performance
- Live up to the spirit, values, principles and content of relevant conventions, laws and requirements

LEADERSHIP

The employees at Novo Nordisk feel that the Way of Management is not just a piece of paper or a plaque on the wall; they say that it is really the way they work. Michala Fischer, a controller in International Marketing, says:

> We are very serious about what we say and we really mean it. Lars really means what he says to us. When you have the management actually building on these values internally it has a very strong impact—it's not just in our annual report. I would be surprised if any one [of the employees you interview] were to say that it's just a marketing tool or a big game.

Jennings says this about the Novo Nordisk management culture: "Some people are more pragmatic than others, but they have a very clear set of values and a very clear vision. Management really drove the development of the Way of Management—and this really drove the implementation of the values into the company."

One particularly important leader for the company was the previous CEO, Mads Øvlisen. As Kingo explains:

> Mads Øvlisen, the chairman of the Board and former CEO of Novo Nordisk, is an icon. The 'holistic' business principle that makes the company so special today was developed under his leadership. In the Executive Management team we are striving to embed and develop this principle, and anchor it throughout the entire business. We call it the Triple Bottom Line–balancing economy, social and environmental responsibility in everything we do.

Corporate Social Responsibility Manager, Annette Stube, adds, "It's also a personality thing; he is such an ethical person. He and his values are still part of the company as the Chairman of the Board."

As Chairman of the Board, Øvlisen acts as a mentor to the current CEO, Sørensen. Sørensen greatly admires Øvlisen and has visibly followed his example as he now leads the company dedicated to continuing the company's successes and ethical orientation. Under the direction of these two leaders, Novo Nordisk has broadened its stakeholder model, infused the Novo Nordisk Way of Management with various reinforcement mechanisms (discussed below), and institutionalized the TBL way of doing business.

THE TRIPLE BOTTOM LINE

The Novo Nordisk Way of Management outlines how decisions are made at the company, and the TBL directs how decisions are executed. The company has an entire strategic unit, Corporate Stakeholder Relations, which "drives, challenges and monitors the TBL strategy in Novo Nordisk and helps the [employees] implement new activities to pursue sustainable business goals."[6] Jennings explains that "many will say in the company that the social and the environmental goals are just as important as the financial. It's about optimizing each aspect of the TBL, not trading one off against the other." The specific elements under each of the three parts of the TBL, as listed in their Stakeholder Relations booklet, are:

Socially responsible towards:
• Employees
• People whose healthcare needs we serve
• Local communities and the global community

Financially and economically viable regarding:
• Corporate growth
• Investors
• National growth

Environmentally sound regarding:
• Environment
• Animal welfare
• Bioethics

Within each of these categories, Novo Nordisk has implemented projects, commitments, and initiatives to meet their social, environmental and financial goals. Examples of the initiatives that fall under these three focuses are described below.

SOCIALLY RESPONSIBLE

One of the three aspects of Novo Nordisk's Triple Bottom Line is to be socially responsible. This commitment is with their employees, with their customers, and with the local and global communities in which they operate.

Socially Responsible: Employees

As of January 2004, Novo Nordisk restructured its Human Resources department (called People & Organization) to be part of Corporate Stakeholder Relations. This way, internal stakeholders (employees) are aligned organizationally with external stakeholders (consumers and the public). People & Organization has several mechanisms that support the culture of the TBL among employees, including

---

[6]Company document: *Introducing Stakeholder Relations.*

recruiting, training, monitoring, salary levels, and an ombudsman program with a focus on respecting human rights.

Recruiting

During the recruiting process, Novo Nordisk pays particular attention to whether the candidate will fit in with the company's culture and not just be able to perform well technically. Human Resources partner, Ove Munch Ovesen, explains:

> During the interview, I explain and communicate to the candidate about the Novo Nordisk Way of Management. Plus, we have our 'Fundamentals' (Exhibit 14.2) beyond the 'Values' that they will have to follow. My task, when I am doing the recruitment, is that I have to measure the person not only on the business perspective but also to make sure the person will fit in with the company as someone who will live our values and perform well.

Training

The People & Organization department continues to reinforce the culture with regular training. CSO Thomsen says:

> When we have seminars, we talk about the business targets and the softer issues also. Part of training as a new employee is learning how things are done the Novo way... We also have an in-house training program for those conducting the clinical trials—about being ethical and professional, about informed consent, and making sure the protocols are followed.

Monitoring

Beyond training, the other mechanisms used to reinforce and hold employees accountable to the Novo Nordisk values and fundamentals include:

- Detailed reporting mechanisms
- A Balanced-Scorecard method that measures employees on goals in four areas: Finance, Customers & Society, Business Processes, and People & Organization
- In-house facilitators

These mechanisms work to monitor and reinforce the goals of The Novo Nordisk Way of Management and of the TBL and will be discussed in more detail below.

Salary Levels

A policy at Novo Nordisk is its strategy to pay slightly lower salary-levels, given the success of the company. Although they still pay above average, they are not at the top, even though they are so successful. CEO Sørensen says, "I don't want people to come to work here just to maximize their personal financial situation. I want them to come here for other reasons."

Ombudsman

An interesting role found at Novo Nordisk is an employee Ombudsman. Peter Bonne Eriksen is Senior Vice President of Regulatory Affairs and also has the role of company Ombudsman. He explains:

> When there's a conflict in the company between people that cannot be brought to an agreement, then it is open for the employee to bring it to me for final resolution.

> It is known by the employees that my position is here... I listen to the person who comes with the complaint, I get access to as much written documentation as possible and I talk to other parties to try and get a neutral picture... And my decision is binding on the company... The employee isn't always happy with my decision, but glad that there is a neutral person that has looked into it.

Eriksen says he gets about 15 to 20 cases a year; most cases are from the Danish sites, but there are also several from the international organizations. He says the majority of cases are when someone is being terminated and they feel they have been treated unfairly. He adds that the "most interesting part is whether I was able to help right their perception of being wronged."

Human Rights

Novo Nordisk has been particularly focused on human-rights issues of equal opportunity and diversity. Ovesen says:

> We are measuring the organization's ability to put more women into management. But it goes very slowly. We are seeing more and more women taking up management positions, but still the majority are men. A position will never be filled based on gender; it is always based on the person's ability. And, for example, a woman's year-off for maternity is still counted as a year for experience.

He adds:

> In Denmark we have an increasing problem with what we call third-generation immigrants who have been here for three generations but still feel alienated. We are encouraging people who have a background other than Danish to apply to Novo Nordisk. We published little booklets [about the company] in other languages to be handed out in high schools.

Vice President of Corporate Responsibility Management, Elin Schmidt, explains about the Equal Opportunity Program that has been implemented to address human rights and employee diversity issues:

> It is a very important, and at the same time very challenging program. We have employed the program throughout the company, also reaching all affiliates. It is important both because everything that falls into this program is totally in line with our values and beliefs, and at the same time it creates future business opportunities; making Novo Nordisk even more international and innovative... In the program you will not find any kind of quotas. We do not believe in affirmative action so we do not want to set up any quotas—people don't want to be a part of a quota! So instead we asked all our Senior VPs to develop an action plan covering their organization for the next five years, and we support them in this work. We asked them to tell us about all the diversity issues they identify today and the issues for the future, and how they see themselves reaching compliance with these challenges. We understand the many challenges they face on this agenda, working in local cultures where issues can be very different compared to a headquarters situation. We meet with them twice a year to discuss their action plans, and the challenges and progress they have made. The Equal Opportunities targets are part of the Balanced Scorecard.

Within these action plans, VPs for instance, have decided to run courses for their staff about gender issues.

An external project that Novo Nordisk has been asked to help lead is the Business Leaders Initiative on Human Rights (BLIHR)[7] run by Mary Robinson,[8] the former United Nations High Commissioner for Human Rights. Schmidt explains that together with six other companies, Novo Nordisk will be working with Robinson "over a four-year period to further develop the human-rights agenda and to try to bring in other companies that are not so experienced in this area. Also, to try to show them how behaving ethically is not only the right thing to do but also creates a more sound business."

For Sørensen, one measure of the success of the People & Organization initiatives (which is part of the TBL) is the happiness expressed by Novo Nordisk employees. For example, he says:

> Our employees are not members of an employee association here in Denmark. They deal directly with us. The workers are unionized in various unions but then they form a local club to negotiate with us as an employer. We're at the forefront with employee issues—we have one-year, paid maternity leave. Other companies got upset at us for that. We have average salary positions, given our superior position, but our employees look at the total package, such as security of employment.

Jennings says, "I have worked as a consultant for many other companies. There were no other companies at that time that I would have gone to work for." Also, in a Danish reputation survey done by an external agency who asks students which company they would most like to work for, Novo Nordisk has always ranked either first, second, or third.

Supplier Project

Novo Nordisk has decided that it is not enough for the company and its affiliates to follow the TBL, but that its way of doing business should also extend to their suppliers. Bisgaard explains what the Supplier Project is all about:

> Three years ago we started the project about the social and environmental responsibility of our suppliers... We made a questionnaire and did a study. We spent a day with some of our key suppliers in Mexico, India, Scandinavia, and Central Europe. We asked them: 'If you were to receive a questionnaire from us, how would you perceive it?' From this, we built a questionnaire that we use today to evaluate our suppliers and contractors. (Exhibit 14.4)

Bisgaard says that they are not using the questionnaire as a way to terminate business, but as a way to improve human rights and environmental standards. He says that Novo Nordisk is trying to create an atmosphere in which the company's suppliers will come and talk to them to help resolve these types of issues. Sometimes, however, the supplier is out of line with the Novo Nordisk way of doing business and something has to be done. He gives an example of when one of their suppliers was not following internationally accepted labor practices and

---

[7]http://www.business-and-human-rights-seminar.org/blihr.htm.
[8]http://www.unhchr.ch/html/hchr/unhc.htm.

Exhibit 14.4   **Supplier Questionnaire**

Please observe the following before completing the questionnaire

- If your company is part of a major group, you should answer on behalf of your own operations, not the entire group.
- If the product/s we buy are not produced in your country, please answer for the country where the product/s or the majority of the product/s are produced.
- Please answer each question as fully as possible. This gives you the opportunity to tell us about your initiatives and thereby obtain a higher score. A blank space will be rated "nonsatisfactory." The questions are not intended to include apprentices/trainees.
- No documentation is required, but it is recommended that the data be thoroughly commented to increase understanding on our part. The data must be verifiable through an audit.

Environmental information

1. **Environmental impact**

   We expect our suppliers and contractors to identify and monitor the environmental impacts of their company's activities. In this context, environmental impact covers changes to the environment, wholly or partially resulting from the company's activities, products, or services. We consider it a fact that any activity causes an environmental impact.

   a. What are the environmental impacts of your company's activities?
   b. How have these environmental impacts been identified or assessed?
   c. Has your company established a continuous monitoring of these impacts? If yes, please indicate how.

2. **Environmental management**

   Active management of the environmental impacts related to the company's activities means, at a minimum, ensuring compliance with local environmental legislation and having an environmental policy or statement. More progressive companies may have an environmental management system (EMS), which may be externally certified under ISO 14001 or EMAS.

   a. Do you ensure compliance with local environmental regulation?
      How is this performed?
      Does your company have a documented Environmental Policy?
      If yes, when was it introduced?
   b. Does your company have an environmental management system or programme?
      If yes, which main environmental targets has your company worked toward during the last year?
   c. Is your company planning to achieve an ISO14001 certification or EMAS registration?

3. **Environmental actions**

   We expect our suppliers and contractors to take action to improve their environmental performance. Actions can include, e.g., environmental improvements of production or products, initiatives to increase the employees' environmental awareness, improvements of the environmental organization, sharing of better practices.

   a. Which environmental actions has your company introduced or supported within the last year, if any?

4. **Environmental dialogue**

   We expect our suppliers and contractors to provide appropriate environmental information to interested parties including authorities, employees, shareholders, neighbours, and other stakeholders. This can be a voluntary initiative or part of a legal requirement.

Continued

EXHIBIT 14.4 **Supplier Questionnaire—cont'd**

a. Does your company publish an environmental account or statement?

b. What environmental training has your company given your employees?

c. Please give details of any breaches of environmental regulations within the last year.

d. If you have received enquiries during the last year regarding your environmental performance, what practical steps did you take to respond?

Social information

5. Wages and benefits

Everyone who works has the right to just and favourable remuneration, ensuring for himself and his family an existence worthy of human dignity, and supplemented, if necessary, by other means of social protection (UDHR). Wages earned for regular working hours should be sufficient for the worker and his/her dependents to meet their basic needs (UN interpretation).

a. Do you ensure that your employees are adequately protected according to the Universal Declaration of Human Rights regarding wages and benefits?

b. What is the minimum wage as defined by the law in your country or industry average (whichever is higher)?

c. What is the lowest wage paid by you to an employee on a full-time basis?

d. Which benefits do you offer to all employees beyond the legal minimum?

6. Working hours

Everyone has the right to rest and leisure, including reasonable limitation of working hours and periodic holidays with pay (UDHR).

a. Do you ensure that your employees are adequately protected according to the ILO convention regarding working hours, overtime, time off, and holidays?

b. How many hours of normal working time and overtime would a full-time production worker or the like work in an average week?

c. How many days off would a full-time production worker or the like have in an average week?

7. Health & Safety

Everyone has the right to ... just and favorable conditions of work (UDHR). A safe and hygienic working environment shall be provided. Adequate steps must be taken to prevent accidents and injuries to health arising out of work. The causes of hazards in the working environment must be minimized. Workers shall receive regular and recorded health and safety training (ILO convention).

a. Do you ensure that your employees are adequately protected according to the ILO convention regarding health and safety?

b. What health and safety training do you provide for your employees?

c. Is responsibility for the working environment clearly defined at all levels in the organization? Kindly describe.

d. Please give details of any breaches of health and safety regulations during the last year.

8. Child labour

Everyone has the right to education .... Elementary education shall be compulsory ... and childhood [children] are entitled to special care and assistance (UDHR). The minimum age of an employee in developing countries is 14, and 15 in developed countries. However, in companies with hazardous working conditions, the minimum age must be 18 (ILO convention).

a. Do you ensure that your employees are adequately protected according to the ILO convention on child labour?

b. What is the date of birth (year) of your youngest employee?

EXHIBIT 14.4   **Supplier Questionnaire—cont'd**

c. Do you have special procedures for employees younger than age 18?

d. If child workers are identified, how would you ensure that the negative consequences of laying them off are reduced to a minimum?

9. Forced labour

No one shall be held in slavery or servitude; slavery and the slave trade shall be prohibited in all their forms. Everyone has the right to work, to free choice of employment (UDHR). There is no forced or bonded labour. This covers all work or service that is extracted from any person under the menace of any penalty for which the person has not offered him/herself voluntarily (ILO convention.)

a. Do you ensure that your employees are adequately protected according to the ILO convention regarding forced and bonded labour?

b. Are all your employees working in your company of their own free will?

c. Does everybody have employment contracts?

10. Freedom of association and collective bargaining

Everyone has the right to freedom of peaceful assembly and association. No one may be compelled to belong to an association. Everyone has the right to form and to join trade unions for the protection of his interests (UDHR). Employees have the right to join or form trade unions of their own choosing and to bargain collectively (ILO convention).

a. Do you ensure that your employees are adequately protected according to the UDHR and ILO convention regarding freedom of association and to collective bargaining?

b. Is there a union or alternative representative body with the right to bargain collectively on behalf of production workers?

11. Nondiscrimination and Equal Opportunities

Everyone is entitled to all the rights and freedoms set forth in this Declaration, without distinction of any kind (UDHR). There is no discrimination in hiring, compensation, access to training, promotion, termination, or retirement based on race, caste, national origin, religion, age, disability, gender, marital status, sexual orientation, union membership. or political affiliation (ILO convention).

a. Do you ensure that your employees are adequately protected according to the ILO convention regarding discrimination?

b. What policies and procedures do you have on nondiscrimination and equal opportunities?

c. What improvements have been made as a result of the policy?

12. Disciplinary measures

No one shall be subjected to torture or to cruel, inhuman, or degrading treatment or punishment (UDHR). Physical abuse, the threat of physical abuse, sexual or other harassment, and verbal abuse or other forms of intimidation shall be prohibited (UN interpretation).

a. Do you ensure that your employees are adequately protected according to the UN Universal Declaration of Human Rights regarding disciplinary practices?

b. What disciplinary and grievance (e.g., anonymous complaint, whistleblower system) procedures do you have?

c. How many times have the grievance procedures been used during the last year?

13. Privacy

No one shall be subjected to arbitrary interference with his privacy, family, home or correspondence (UDHR). Everyone has the right to his/her privacy. This covers among others the monitoring of persons and the handling of personal information. Interference with privacy should be guided by rules and should only be used if necessary to achieve a justified aim. (UN Interpretation)

Continued

EXHIBIT 14.4   **Supplier Questionnaire—cont'd**

---

   a. Do you ensure that your employees are adequately protected according to the UN Universal Declaration of Human Rights regarding the privacy of employees?

   b. Are employees given clear information beforehand when they are being monitored (via computers, telephone, cameras, or other)?

   c. What procedures do you have for keeping employees' personal information confidential?

14. Your suppliers

   1. Does your company evaluate your own suppliers on environmental issues?

   2. Does your company evaluate your own suppliers on social issues?

   3. What environmental and/or social risks do you anticipate in your supply chain?

Novo Nordisk decided to verify a new supplier, but at the same time work closely with the current supplier. He explains:

> We set up an external audit committee who went and did an audit of this supplier once we had received their questionnaire back. Unfortunately the questionnaire did reflect what was going on [with labor violations]. We did an audit of another company in the same area that did the same type of production and we are now trying to improve standards with our current supplier but have at the same time identified a new vendor who meets the standard.

SOCIALLY RESPONSIBLE: PEOPLE WHOSE HEALTHCARE NEEDS WE SERVE

The most important stakeholders for Novo Nordisk are those people suffering from the diseases it treats, primarily those with diabetes. The majority of Novo Nordisk's income comes from the treatment of diabetes, yet company employees say they are actively seeking to prevent and cure the disease their business depends on. Fischer says, "Prevention and curing are in our goals and values. I have no doubt that if we end up finding a cure we will definitely reveal it." Sørensen explains:

> People ask me whether I'm crazy looking for a cure for the disease that is the centre of our business; whether I am cutting the branch on which we stand. The truth of the matter is that the branch is growing faster than I can cut it. And, unless we do work towards a cure, we'll have no credibility.

The growth to which Sørensen refers is the 177 million people worldwide who have diabetes, an increase of 15% in the last 10 years, which is expected to continue to increase drastically.[9] The WHO predicts that the diabetes epidemic will increase further by 2025, with at least 300 million of the world's population suffering from the disease.[10]

---

[9]http://www.insulinfree.org/facts/reasons.htm, http://www.who.int/dietphysicalactivity/publications/facts/diabetes/en/.

[10]http://www.emcc.eurofound.eu.int/publications/2003/dk03001a.pdf.

Finding a cure, if it ever happens, is many years away and so Novo Nordisk is trying to meet the needs of their consumers/patients living with diabetes the best they can. Pramming explains a project Novo Nordisk started in the countries in which they operate to help initiate and continue diabetes programs:

> If we are going to help the person with diabetes, the fundamental thing is to find out what matters to them. So, we conducted a survey in 15 countries where we asked diabetes sufferers about their needs, attitudes and wishes about the disease and what they think is important... Then we started the National Diabetes Programme to make sure our affiliates and subsidiaries in the 69 countries where we operate get involved... What we want these subsidiaries to do is get involved with the national diabetes strategy in their country and if there is no program in the country, they should try to develop one. For example, we are working in China to develop a program with the health authorities using some of our money and skills. We have had a program with indigenous people in Australia and with the elderly in UK.

SOCIALLY RESPONSIBLE: LOCAL COMMUNITIES AND GLOBAL COMMUNITY

To reach out beyond their current customers, Novo Nordisk's Corporate Social Responsibility (CSR) programs have been focused on developing infrastructure and knowledge in the world's poorest countries to help treat people suffering from diabetes. Novo Nordisk has three main approaches for their CSR initiatives in these countries: their National Diabetes Programme (formerly termed the World Partner Programme), their pricing policy and their donations to the World Diabetes Foundation.

National Diabetes Programme

Pramming explains the first initiative:

> Our guidance here comes from the recommendation of the WHO, which is: these countries need to have capacity—not just the drugs but the infrastructure also. We set up a program called the World Partner Program, which was a comparison looking at the diabetes treatment in poor countries to find out why it was a success in one and a failure in another. We did that in six countries [Bangladesh, Malaysia, Tanzania, Zambia, El Salvador, and Costa Rica]. Having been there and discussed with the authorities in these countries, we then felt obliged to do something as well. So we started working to build infrastructure and capacity in those six countries and added India and China as well.

Dr. Mapoko Ilondo is the project manager for Novo Nordisk's National Diabetes Programmes in developing countries, which was the outcome of the World Partner Programme, and he explains how he works with these countries:

> What management decided here three years ago, when we had the problem about the lawsuits in South Africa, was that we needed to do things more proactively coordinated out of headquarters. Lars knew that Novo Nordisk was doing a lot of things in development but it was all very decentralized... So, I first went to talk to local diabetes federations, to talk to hospitals, and officials—just to have an idea of what is going on. We talked with affiliates about what resources are in

the country... We signed partnerships with selected partners in these countries that will last for three years. So in practice what I do is go down and talk to them about how they would like things done. They develop a list of priorities, we discuss these with them. We make decisions together. We provide them with funding. We provide them with experts too if they need them.

Ilondo says that they are in the process of evaluating all of these projects. They are currently collecting baseline information so that in 3 years, they will be able to evaluate the progress made. For some detail about these projects, see Exhibit 14.5.

EXHIBIT 14.5   **Corporate Donations: Country Examples**[11]

Bangladesh
Novo Nordisk has worked closely with the diabetes association in Bangladesh. Ilondo started by asking them how the company could help them, and they expressed two main areas where they needed help:

* Firstly, they wanted a place for foot care. They have a separate center for foot care to perform things like amputations, but they were lacking a place where they could do prevention of foot complications.
* Secondly, they wanted a place to train nurses to take care of feet.

So these are the two main areas that Novo is helping them with. The company funded them to open an area dedicated to foot care and gave them money to buy the necessary equipment for foot care, and for prevention and education on foot care.

Later, they identified that they needed doctors to be trained in diabetes. So, Novo Nordisk is aiming to train 300 doctors in diabetes in Bangladesh. Novo plans to do some practical training in hospitals and to take them off-site for some training as well. On top of this, Novo gives the diabetes association some free insulin to use for the families that cannot afford insulin for their children.

Tanzania
Novo Nordisk has done a lot of work in Tanzania. They have built a clinic there and they are working with the government to find the right way to build diabetes infrastructure. They have learned that it doesn't help to just give a lot of insulin to a country; you have to train the doctors and nurses and build infrastructure.

Tanzania does have a few specialists focused on diabetes, and a relatively good healthcare system, with a referral system that works properly. Ilondo says that the healthcare system, however, is geared to work for infectious diseases (such as malaria, tuberculosis, etc.). A doctor or nurse in Tanzania will react to a fever by looking for malaria. But when someone reports to a hospital with symptoms of diabetes, the nurses and doctors will usually miss it.

Many people will come with foot lesions and other complications. When the patient reports to the hospital, his or her foot will be treated, but the patient will not be tested for diabetes.

Only when things get really bad, perhaps when the patient has to go to see a surgeon to have the foot amputated, will it be recognized as diabetes. This is why Novo came to the conclusion that a lack of insulin is not the main problem and that education has to happen first.

Ilondo explains that they decided to start top-down at the national teaching hospital. They renovated a small building to use as a diabetes center. That center is used to take care of patients with diabetes and as a place to train doctors and nurses in diabetes.

By March 2004, Novo had opened clinics in three district hospitals as well. Ilondo explains that the problem with the district hospitals is that they didn't have space for the clinics, so Novo Nordisk bought and set up container systems to house the clinics. There are now four diabetes centers in Tanzania.

Exhibit 14.5   **Corporate Donations: Country Examples**[11]—cont'd

Novo Nordisk pays for renovating the space (or donates the container), all the furniture, lab equipment and weigh scales, and the IT equipment. They also help install the Internet and other set-ups that are needed. They then pay for some doctors and nurses to attend a 5-day training session to learn about diabetes. They pay for their time to be at the session and their rooms, food, and transportation.

Ilondo adds that they have also engaged with a regional medical officer in the capital, Dar Es Salaam, who has accepted two doctors and two nurses in each of the district hospitals who will deal exclusively with diabetes and hypertension. So now there are people who are more or less permanently working on diabetes. In the past, these doctors would be involved in other activities at that hospital, but now they can focus on diabetes because Novo Nordisk is providing them with some financial help.

Zambia
When Novo started discussions with the central board of health in Zambia, they said, "We don't have a problem with diabetes. Diabetes is for you guys in Europe because you eat too much. Who has diabetes here?" What is happening, though, is that as the population moves from rural to urban areas, the instances of diabetes are becoming much more prevalent. Ilondo says that once they started seeing these types of statistics, they finally started listening. He says this is a good example of how, even if you give them free insulin, if they don't see the problem, it won't help. Novo Nordisk hopes to do a similar thing in Zambia as they did in Tanzania, but the difficulty is that in Zambia there are fewer specialists.

They are currently talking to a hospital where they hope to start training nurses. Ilondo explains that they have data comparing health centers, and the places where nurses were trained in diabetes have had a significant positive impact on care of diabetes. Thus, they are putting an emphasis on nurse training. He adds that the good thing about Zambia is that they have very good, well-maintained hospitals. The problem is that they are no longer receiving funds for training or for medicines, so the doctors and nurses are getting discouraged.

Costa Rica
Diabetes care and medications are free in Costa Rica, but Ilondo says there is a disproportionate incidence of diabetes complications for this type of country. Therefore, the country has asked Novo Nordisk to help with diabetes education.

El Salvador
Ilondo says that the emphasis in El Salvador has been on patient education. The Deputy Minister of Health asked that Novo Nordisk look into the possibility of training doctors in the country's public system. Novo is currently working on that and on renovating their diabetes center in San Salvador.

Foot Clinic in Kilpak, India
Another recent corporate donation initiative is the foot clinic that Novo Nordisk just built and established in Kilpak, India. The funds for this endeavor came from personal donations from Novo Nordisk employees. Sørensen says, "It's crazy that in India there are 40 thousand limbs lost due to diabetes alone. With this clinic we want to help prevent this. The clinic also is about getting education about diabetes and buying people $2 sandals so their feet don't become infected."

[11]Interview with Dr. Mapoko Ilondo, Novo Nordisk's Project Manager for National Diabetes Programs.

Pricing Policy

The second approach, Novo Nordisk's pricing policy, is explained by Pramming:

> For the 49 least-developed countries we will sell the product at 20% of the price that we charge in the developed country. We don't give it away—if we did it would not have any value, and it would be against the sustainability principle. And it doesn't respect the employees of this company who work hard to develop these products. But these poor people shouldn't have to pay for part of new drugs [future research], but they should pay for the cost of that drug. But, even though this is so cheap, there are still some who cannot afford it. That is why we suggested creating a foundation for diabetes that could support improvements in care in the neediest countries in the world.

World Diabetes Foundation

Sørensen explains that this is where the third initiative, the World Diabetes Foundation, comes in:

> The foundation is independent of the company, but the company donates the money for it. I am a member of the board of the foundation. We accept applications from countries and the criterion is that the proposed project has to be linked into existing infrastructure (for example, they can't be building roads with this money) and it has to be appropriate, and it's really important to have a local organization involved too.... The World Diabetes Foundation presently supports 26 projects with a total portfolio of USD 27.3 million, of which USD 11.6 million are donated by the Foundation. There was only one project, in the Congo, where we donated money for buying drugs; all the rest were for capacity building. We are a funding agency and there is always a partnership with people on the ground. We are also monitoring what comes out of the money we put in.... There are lots of pharmas who want to do partnerships but what we do, which I think is different, is that we find a good partner, then we ask them what they think we should do. We do not go in with a concept. We don't pretend to know the conditions and needs of these countries. But, it is a difficult balance for companies to find. As far as pushing your brand, you really don't have to push it. People know where the money is coming from for these projects... we don't have to push it on people.

A question that comes to every investor's mind is how all of this time and money invested in these projects is helping the company. Sørensen explains how he defends Novo Nordisk's CSR initiatives to his shareholders:

> What I'm doing here with our CSR issues is partly for risk mitigation. I had to ask our shareholders for the millions for the Diabetes Foundation. I can't guarantee a return for them but what I did tell them is that the energy this type of investment puts into the firm will increase our internal productivity; it increases our attractiveness as a company to work for by employees. Then the other aspect, that in the shareholders'eyes is only speculation, is that the development of developing countries will benefit the world in the long-run. The problem is that this is out of the scope of investors. But I have been able to show that our CSR initiatives do increase the attraction rate [for new employees], reduce employee turnover, and increase employee satisfaction.

Stakeholder Dialogue

Another important stakeholder for Novo Nordisk is the general public; this includes local communities around company facilities as well as NGOs and activists.

Stube explains that the company occasionally holds "town-hall meetings," in which the public can come and raise their concerns to company representatives. Some Novo Nordisk facilities have newspapers, and most production sites produce their own sustainability reports that they send out to the community to communicate with them.

Thomsen recalls one time when the company had to stand its ground in order to get public feedback: "For instance with stem cells—the government said they would have renewed legislation on it. The politicians wanted to rush it and get it passed in a month's time. We needed time to get public feedback. We needed half a year to discuss the issues with the public."

Kingo describes the stakeholder-engagement process:

> We are very focused on stakeholder dialogue in the company and have been for decades. We work systematically with each stakeholder group and try to tailor the process to each different group. It has been particularly challenging and interesting to develop our dialogue with critical NGOs, which we began to put into a formal process by the end of the 1980s. At that time many NGOs were worried about genetic engineering and environmental issues, and we were approached with many questions as a biotech company... In 1990 we had the first in a series of annual two-day visits to Novo Nordisk for NGOs around the world. We learned a lot from this process, which helped us shape the Triple Bottom Line concept that has made Novo Nordisk a leader in addressing sustainable development. It was not an easy process and it took a while before we learned that in a dialogue it is alright to disagree. But I also know from the NGOs that have visited Novo Nordisk over the years that it was a learning experience for them, too.

Jennings, who has done a lot of work with NGOs, explains what the stakeholder engagement process was about, and why it worked at Novo Nordisk:

> It was successful because of the openness and because we weren't seeking consensus. What we were seeking was to understand each other and to look for areas of commonality. Out of the discussions many issues emerged and they reinforced the need for greater transparency on environmental—and later social—information, and hence voluntary reporting... It also raised issues that weren't purely environmental—like animal welfare... Some think that the dialogue itself is sufficient. But it's not. It also requires action and responsiveness. There has to be a tangible outcome. There are those who feel that engaging alone will do the job; but there has to be learning and there has to be movement.

FINANCIALLY AND ECONOMICALLY VIABLE

The financial component of the TBL is an aspect that is very familiar to every company. At Novo Nordisk, it includes ensuring corporate growth and economic viability for the company, along with providing a fair return to shareholders. The company's economic impact on surrounding communities is also a concern. For example, the Novo Nordisk affiliate in Clayton, North Carolina is assessing whether its company site has had a positive economic impact on that community.

In the Finance aspect of the TBL, Novo Nordisk also includes helping the national growth of the country, thus avoiding a hostile takeover, in order to keep the company headquartered in Denmark. This area also includes taking a long-term perspective while still meeting short-term gains, which are particularly important to many investors.

Socially responsible investings (SRI) is an aspect of the Financial Bottom Line that Novo Nordisk is trying to encourage. SRI is slowly starting to become more important to investors around the world.[12] SRI considers both the investor's financial needs and an investment's impact on society and the environment. Stube explains, "Shareholder interest in CSR started very slowly since [shareholders] are usually mainly interested in the money. But, we have a new group of SRI investors, and part of their portfolio is that they screen companies to look at how companies do on the other aspects. These investors are very interested in our company."

Sørensen adds, "Ethical screening from the institutional investors is still not a priority for them. The individual investors are very concerned with the type of company we are and the work we do, but the big ones are still really only concerned with the returns. We've invited SRI investors to come here and work with us."

The company feels that the interest in SRI is increasing, partly because more investors feel that "socially responsible corporations are seen as being better at managing risks and opportunities on a long-term basis and consumers are increasingly aware of corporations' ethical behavior"[13]

ENVIRONMENTALLY SOUND

The third aspect of the Triple Bottom Line involves the environment, and for Novo Nordisk this includes issues that fall under the headings of bioethics, animal welfare, and the natural environment.

Bioethics

Hanne Gürtler is Novo Nordisk's Director of Bioethics Management. She explains the role of bioethics in the company:

> We work in applied bioethics; ethics in practice; hands on. We are really down in the issues. We're not having theoretical debates here. I'm not an ethicist, and my colleagues are not. But we work with government ethicists and we are putting ethics into practice. I don't think in my position we should be ethicists. We need a natural-science background to understand the issues and to get the respect of your colleagues in science. To initiate the discussion, you need somebody coming from within. Psychologically you need someone from inside.

Issues that fall under the bioethics umbrella at Novo Nordisk include biobanking, cloning, stem-cell research, clinical trials, and animal welfare. The company's general approach is to assemble committees of internal and external experts to

---

[12]http://www.socialinvest.org/Areas/SRIGuide/.
[13]http://www.novonordisk.com/sustainability/socially_responsible_investment/socially_responsible_investment.asp.

discuss these issues and to take a stance on them. Thomsen gives some examples of the issues:

> Genomics and genetics are not things you can avoid having a view on. We have no interest in genotyping people because diabetes is so environmentally influenced and so polygenetic that you can never genotype out of the problem. If there was really only one gene for diabetes, it would have been found by now.

Thomsen also discussed the growing debate around stem-cell research and how this area of research holds both very promising discoveries and many ethical considerations. For Novo Nordisk in particular, stem-cell research is the most promising approach to find a cure for type 1 diabetes. The company recognizes the ethical issues that surround this research and have clear guidelines and approaches concerning stem-cell research listed on their website.[14]

Thomsen explains a recent initiative that the company spearheaded to address another bioethics issue—biobanking:

> We [Denmark] were the first to develop a law regarding the rules to consider in developing a biobank. [The government] wanted to amend it. They asked for [Novo Nordisk's] opinion. We did not want to appear to be bullying them and we wanted to take it even one step further than what academic researchers do. So, we got philosophers and ethicists together and formed a committee in order to make appropriate recommendations.

Clinical Trials

In the area of clinical trials, all preclinical data are reviewed by an internal Safety Committee before deciding whether it is safe to go into humans. The company also has independent Safety Review Boards composed of external experts who evaluate all safety data during trial execution with new chemical entities. Thomsen explains that they are currently undergoing International Standards Organization (ISO) certification in their drug development area to have an even higher standard than GXP[15] for clinical research.

Berit Edsberg, the Director of Clinical Reporting and Clinical Drug Development, says that part of an ethical approach in this area includes publishing academic papers on every study to the extent that that is possible, regardless of whether it has negative or positive results, and only doing drug development in areas that can help people. "We are not developing beauty-enhancing drugs, for example."

Animal Welfare

Animal welfare has been such a focus for Novo Nordisk that while many pharmaceutical companies in Europe are being condemned by animal-rights activists,

---

[14]http://www.novonordisk.com/sustainability/positions/stem_cell_research_uk.asp.

[15]GXP is the standard certification for pharmaceutical companies that is the combination of GMP (Good Manufacturing Practice), GLP (Good Laboratory Practice), and GCP (Good Clinical Practice).

Novo Nordisk's bioethics manager received an "Animal Friend of the Year" award from the Danish Animal Welfare Society. Thomsen explains how it all began:

> We realized 4 years ago that the animals we use will spend 1% of their life in an experiment and the rest of their life being a normal animal. So, we did research with leading ethologists[16] and a Danish animal protection agency, and together with them we found out what makes a happy rabbit, mouse, etc. When we invested in a new animal facility, we developed it based on the findings of the research we did. We invested 25 million DKK into the new animal-facility.

Gürtler explains what they did next:

> First we had a workshop with leading animal behavioral experts and animal welfare advocates to find out the basic needs of the different animals. We made a pilot plan on how to address these needs, and then invited the experts and advocates back to get their feedback on what we had implemented. Ethologists and animal-welfare advocates have actually been the ones promoting our company and facility. We have a new housing system for rabbits, rodents, and improved standards for dogs... When I started my work on improving the welfare of experimental animals I interviewed everyone in the animal experiment area. I asked questions like: Why are the rabbits kept like that? Because I have a long background in research, I know how a project works; you can't just gloss over an answer on me!

A video demonstrates how Novo Nordisk has completely changed the way the animals used in experiments are treated. The small animals have large pens and rabbits, for example, are fed carrots and fresh vegetables (which was unheard of before). The dogs are taken for daily walks in the country and now are trained to jump up onto the veterinarian table to have their blood drawn.

The company decided to be open and transparent about what they were doing from the beginning. Gürtler says:

> Something else we did was to invite the press early on to keep the general population involved from the beginning... We became trustworthy. Once they saw that I truly am a woman who cares about animals, I had the credibility from the beginning. I've been able to say openly in the press that I don't like animal testing. We also have a huge number of external stakeholders, including schoolchildren, who visit our animal facilities to view our housing and care standards... We also have two ethical review committees which are not required by Danish law. In Denmark, all you have to do is apply and get permission to do a certain test on animals, but we felt that wasn't good enough. So, before applying for each type of experiment it has to go through our own ethics committee.

The company, over the last decade, has reduced its use of animals by approximately 70%. This was done partly by using new technologies when possible, and by shifting the research culture in the company to using a "need-to-know" instead of a "nice-to-know" criterion. Now they only use animals for research when they need to know something to help advance their research.

---

[16]An ethologist is a person who studies how animals behave in a laboratory or in the wild.

Gürtler and her Bioethics Department, with the support of the company, have also gone out to try to influence other companies. She says:

> There is the Danish Pharmaceutical Industry Association, and we were trying to get our ideas into that organization. The Association and the Danish Animal Welfare Society made a joint proposal to the Danish Parliament on improved standards within animal experimentation. The results and experiences at Novo Nordisk were used in the legal process for the new Danish legislation on housing and care of experimental animals.

A sign of success for Gürtler in this area is how the work done by Novo Nordisk has affected the general perception of testing on animals in the country. She explains:

> There was a hearing in the parliament in 2002 where I gave a presentation on the work we've done, and it was interesting to see how all the animal-rights groups were very pleased. Now animal experimentation has become the example for animal welfare and other users of animals are now implementing the same things. It is now reflected in the legislation and has resulted in a very positive view by people in Denmark. You don't have the polarization of 'for and against' any more. When I was taking the approach of being critical about animal experimentation, the public was actually defending animal experimentation!

Natural Environment

Novo Nordisk's concentrated focus on the environment started in the early 1990s when management developed a small department called Environment Services to examine the company's environmental performance. This group, led by Kingo, was in charge of interacting with the NGOs. She explains the process:

> When we had the first NGO visit to Novo Nordisk in 1990, it included 20 representatives from different NGOs. John Elkington, CEO of Sustainability, was facilitating the meeting, and I have to admit that we were a bit worried about the outcome as we never had so many critical NGOs at the company at the same time. So the visit was discussed and approved by executive management before the guests were invited. It was a serious event... When the NGOs came they quickly got tired of listening to all the presentations that we had prepared and said: "Why don't you prepare a report on your environmental performance that we can read in advance. This way we might have time for some dialogue—and not a monologue." A few years later, Novo Nordisk published its first Environmental Report and has received numerous awards and recognition for its transparency ever since.

The company's Environmental Report later merged with the Social Report to form the yearly Sustainability Report.

MONITORING AND REINFORCEMENT MECHANISMS

The Novo Nordisk Way of Management not only includes a vision, a set of values, and commitments to the TBL, but also the methodology to monitor and ensure that what the management says and hopes is happening is truly occurring throughout the company. The three methods used by Novo Nordisk for this purpose are sustainability reporting, a Balanced-Scorecard method of measurement, and company facilitators.

Sustainability Reporting

Novo Nordisk produces a yearly Sustainability Report prepared in accordance with Global Reporting Initiative (GRI)[17] guidelines. The report strives to explain Novo Nordisk's approach to doing business and discusses what it is doing to live up to its promise to deliver on the TBL. The report aims to transparently describe Novo Nordisk's "'footprint' in terms of social, environmental and economic impacts on society."[18]

The report also acknowledges the many challenges the company is facing and explains how they are trying to deal with them. Some of the dilemmas raised in the 2003 report are:

- How can we make public–private partnerships succeed and still ensure the independence and integrity of the contribution each partner brings to the table?
- How can we help empower people to take more responsibility for their health and yet respect their personal integrity?
- How can we justify dedicating scarce economic resources to diabetes care and yet recognize that there are other life-threatening epidemics in the world?
- How can we encourage investors to challenge conventional measures of long-term performance and yet maintain comparable reporting standards?
- How can we be focused on investing in the health of society and yet not compromise the need to invest in the global environment?

The report then discusses each question at length. Stube explains that it is not just about producing the physical report:

> It is all the processes around the reporting that is a key factor. It's not just about making a glossy report; it is about accountability and writing about the issues that are really relevant to stakeholders, not just the ones where we are doing well. When you go through that process, it is a mapping exercise; it helps find the gaps that we need to work on.

The Balanced Scorecard

The BSC is a performance measurement system that can be used to monitor and measure individual employees, departments, and entire companies. The concept of the BSC is that it is meant to include financial as well as non-financial targets and goals.[19] Traditionally, the BSC tool includes four areas for measurement: Financial, Customer, Internal Business Process, and Learning and Growth.

---

[17]Global Reporting Initiative, a multistakeholder process and independent institution whose mission is to develop and disseminate globally applicable Sustainability Reporting Guidelines. http://www.globalreporting.org/.

[18]Novo Nordisk: Sustainability Report 2003. What does being there mean to you? p. 61.

[19]Kaplan R, Norton P. (1996) Translating Strategy Into Action: The Balanced Scorecard. President and Fellows of Harvard College: USA. p. 8.

At Novo Nordisk, they are called: Finance, Customer & Society, Business Processes, and People & Organization (Exhibit 14.6). All employees have individual annual targets and development plans, which are linked to the corporate and departmental BSCs. In this way corporate and departmental targets, set in the BSC process, guide the setting of individual performance targets.

Bisgaard explains one example of a goal on his departmental BSC:

> This year it was decided that the employees don't know enough about Novo Nordisk in the market place. This is when the Balance Scorecard comes into play – I have on mine that 80% of my employees should know more about Novo Nordisk in the market place. That will be measured and I will be held accountable for reaching that goal.

Another goal that appeared on the scorecards was that, by the end of the year, 90% of all employees should meet people with diabetes to discuss their condition and gain a better understanding of life with diabetes. To meet this, managers spent a day with a sales representative, meeting with pharmacists and physicians. Employees also had the opportunity to attend information sessions to learn from diabetes patients.

FACILITATORS

The third process used by Novo Nordisk for monitoring is Novo Group facilitators who perform mini-audits of departments throughout the company. The facilitators visit all departments on a 3-year rotational cycle to monitor compliance with the elements of the Novo Nordisk Way of Management and to assist in developing action plans and improvement targets. This occurs as a consultancy

EXHIBIT 14.6   **Company Balanced Scorecard 2004**

Balanced Scorecard 2004

- Establish strong position in global analogue market
- Realise full potential of NovoSeven® indications
- Implementation of global value upgrade program
- Achieve superior customer satisfaction & company reputation
- Ensure sound environmental, social and bioethical performance
- Improve our collaboration with key stakeholders in diabetes care worldwide
- Ensure timely and efficient delivery to market

- Realise growth in Operating Profit
- Ensure competitive ROIC
- Improve Operating Margin
- Ensure competitive Cash to Earnings Ratio

- Improve R&D speed, quality and productivity
- Ensure competitive portfolio of insulin analogues and delivery systems
- Generate new diabetes and biopharmaceutical drug candidates that pass development candidate selection
- Ensure execution and completion of NovoSeven® and phase 3 trials on time
- Launch Levemir successfully
- Improve quality management focus in all business processes
- Improve high level risk management focus
- Ensure successful implementation of IT projects

- Improve customer relations
- Enhance winning culture
- Attract and retain the best
- Ensure development of people
- Embed social responsibility

service, set out in contracts that are negotiated between individual companies in the Novo Group and the holding company Novo A/S. These facilitators are company employees based at Novo A/S who assess whether departments are performing in line with the Triple Bottom Line and the Novo Nordisk Way of Management. Bisgaard explains, "We have facilitators around the world and their only job is to go into the departments and interview the employees to see if they know our values, and if they think of them... They would then report to me as the manager; so we get a feedback. If you're not meeting the standard, you get an action plan to make changes."

The facilitators work with the manager of the department to plan the facilitation process that includes interviewing employees, evaluation and reporting, and a detailed action plan and follow-up. It is meant to be a learning process and a mechanism for improvement.

FUTURE CHALLENGES

With Novo Nordisk's Way of Management and the TBL approach to doing business, the company feels they have laid a good foundation on which to continue growing. However, every employee seems aware that there are still many challenges ahead and continued re-evaluation of strategies and processes will be needed. Sørensen says, "A big challenge for the future is going to be as we start to expand our production internationally, to maintain this culture and family atmosphere."

The company has started scenario planning to estimate how the world will look in 20 years and how they can best situate the company depending on various outcomes. Pramming explains that a possible future is one in which the company finds a cure for diabetes. In their scenario-planning exercises, senior management is trying to figure out how the company can be a part of that and still be successful.

Kingo explains her perception of how certain pressures are going to force Novo Nordisk to stay on its toes:

> The pharmaceutical industry is faced with a number of challenges that have developed in the cross-fire between globalisation and ethics: intellectual-property rights, generics, pricing, clinical trials, marketing ethics are just a few examples. These issues are putting the well-known value-chain business-model under pressure, and we might need to experiment with new models: models that work better in the developing world, models that include partnerships with other organisations, and models that build more on knowledge combined with a product.

Jennings would like to see a model in which, instead of selling a physical product, the company will concern itself with promoting health. He explains the questions in his mind:

> How can we change the focus from selling drugs to promoting health, so drugs are a means not an end? If we're looking to where Novo Nordisk is going in the future right now, it is a relatively small and highly focused company. There are advantages to this, but it makes the company potentially a lot more vulnerable to

a take-over and we want to remain independent. If you want to find a cure or prevent diabetes, how do you make a business out of curing or prevention as your product or service? How will these trends unfold?

These are just some of the considerations for the future into which Novo Nordisk's management is looking.

Internally, the company regularly performs TBL studies at their sites outside Denmark and finds that there is still a lot of room for improvement. Externally, new opinions are coming up all the time. Society has been pressuring large pharmaceutical companies, including Novo Nordisk, to waive their patent protection in developing countries, and some are asking for drugs to be given away for free. In animal welfare, Gürtler gives an example of a recent shift:

> There are new opinions arising all the time; some people feel it is okay to use animals for type 1 diabetes but not with type 2 diabetes research because it has to do with lifestyle—a similar feeling to testing for cosmetics. Industry has been relying on saying 'It's for medicine,' but now people are saying 'Okay, but for what type of medicine?' That is an opinion that is coming up more often. Industry can't rely any more on a broad public acceptance on the use of experimental animals for the development of all drugs.

An unusual concern raised by Thomsen was that Novo Nordisk should not become a victim of its own success by abusing its good standing in Denmark. "The country is so positive towards us as a company that it is almost too easy for us to say something to get our way. Which is a worry... we get worried about railroading. You have power and a voice from being good."

A consideration raised by Pramming is how to integrate responsible business completely into the way the company is run. He asks, "How do you, instead of having 'corporate social programs,' just be the way we run our business? I don't like to see it as a sideline. If this is going to survive, it needs to be the way that people are doing business. That's probably the next way: integrating CSR into the business model. And, that is a tough thing to crack."

Although the management of Novo Nordisk has many challenges ahead, perhaps the first step in the right direction occurred on March 15, 2004 when the CEO and Board of Directors convinced their shareholders to include the TBL in the company's Articles of Association.

15

CONCLUSION: LESSONS FOR COMPANIES AND FUTURE ISSUES

The case studies in this collection demonstrate the complexity of ethical decision making in bioscience companies. Each company is distinct and faces a unique combination of ethical challenges; the management of each has addressed these challenges in different ways, although there are common elements in their approaches. Overall, the message is that these companies, large and small, North American and European, health or agriculture focused, are making good-faith efforts to address the challenges inherent in bioscience business ethics.

The experiences and lessons of these companies are not normally widely available to their peers. But ethics is a pre-competitive issue, and wide sharing of good practice is beneficial for the industry. To distill more general insights from these cases, and to draw lessons for bioscience companies and the industry as a whole, this final chapter identifies cross-case themes: the ethical issues facing bioscience firms, how firms address ethics, and future trends for bioscience firms.

WHAT ETHICAL ISSUES DO BIOSCIENCE COMPANIES FACE?

All 13 companies interviewed faced ethical issues in various aspects of their business. Some are issues that only companies in the bioscience industry face; others could exist in any type of firm. Not every bioscience company will face all these ethical issues all the time, but every bioscience company ought to be aware that ethical issues such as these could be present within their biobusiness.

These 13 companies are all biobusinesses, but they are very different (in size, focus, and location). So the ethical issues also vary. After coding the interview data and case studies, the following 10 recurring themes emerged (see Exhibit 15.1). Each will be described using examples from some of the companies.

EXHIBIT 15.1 **Biobusiness Ethical Issues**

---

1. Financial Pressures
2. Developing New Technologies
3. Research Ethics
4. Working with Regulators
5. Marketing and Delivery of Products ·
6. Value, Pricing, Access to Products
7. Doing Business Globally
8. Managing Conflicts of Interest
9. Corporate and Social Responsibility
10. Business Ethics and Good Goverance

ISSUE 1: FINANCIAL PRESSURES

Most public companies face the challenge of balancing the short-term demands of shareholders and investors with the longer-term investments in research, capital equipment, and people development needed to build sustained success. This challenge is particularly acute for bioscience firms. New products take, on average, more than a decade and $800 million or more to develop; business plans require large investments in research and development with returns materializing many years later (if the product actually makes it that far). During this process, bioscience companies still have to answer to shareholders and investors. These companies are also involved in products that could have a major impact on the health and well-being of very sick segments of the world's population. So the need to be financially profitable and to do good at the same time is an ethical issue with which most of these companies struggle.

Monsanto employees, for example, discussed the major tension involved in trying consistently to meet short-term Wall Street expectations while satisfying long-term societal expectations. The Merck case focuses on the same dilemma: the conflicting pressures to increase both financial returns and corporate donations to address major global health problems. At a time of declining earnings, major setbacks in its development of new drugs, and the largest layoff in the firm's history, Merck has still substantially increased its charitable programs. Merck's Vice President of Marketing and Operations Europe, Middle East, and Africa said, "The real ethical challenge that the pharmaceutical industry faces is juggling these competing forces—staying true to one's corporate mission, not just adopting any practices to compete, [while being] competitive and profitable at the same time." Merck's Board of Directors is coming under growing pressure from the investment community to make changes to the company's leadership and strategic direction. Merck's Senior Vice President and General Council Member said, "From shareholders, we sometimes get questioned why we are giving their money away at all."

One approach to reducing this tension is to seek an investor base that values a longer-term and broader societal focus for the company. Novo Nordisk has been able to attract socially responsible investors (SRI) to buy shares in the company, but as the Chief Executive Officer (CEO) explained, the "majority of shareholders are still only concerned with the financial bottom-line."

ISSUE 2: DEVELOPING CUTTING-EDGE TECHNOLOGIES

All the employees we interviewed believe that their company's science and technology is either currently, or will potentially be, a benefit to society. It is this very confidence in the social benefits of their work, however, that can often blind them to the concerns that the public and stakeholder groups may have about ethical issues raised by these new, often poorly understood, cutting-edge technologies. Monsanto is perhaps the best-known example of this. Focused on developing a new technology that it saw as having a vital role (genetically modified [GM] seeds that would create a sustainable approach to feeding the world's population), the company was blindsided by fierce opposition. This ranged from those for whom any genetic modification is "playing God," to others concerned with the unknown long-term environmental or health effects, such as the threat of gene migration or resistant pests. Diversa has encountered some similar issues with its pioneering work in establishing biodiversity partnerships. "Typically we find that we're pushing the envelope, we're ahead of the curve," said Diversa's Biodiversity Manager. "There are environmental organizations against patenting of life forms—others are just totally against genetic research." The CEO of TGN seemed to agree when he said that there would always be some people who simply would not agree with their efforts to develop transgenic animals. Although many interview participants acknowledged that their firms could do more to engage proactively with stakeholder groups, they also recognized that, even after educating the public and trying to explain the safety guidelines that they have in place, there would always be some who simply would not agree.

ISSUE 3: RESEARCH ETHICS

Along with emerging new technologies, such as stem cells and genetically modified organisms (GMOs), most bioscience companies have to deal with a set of more traditional medical and research ethics questions associated with the development and clinical testing of their products. Participants explained that there are always concerns in clinical trials about safety, the balance of risks and benefits for the trial participants, and proper informed consent. However, they all noted that there are so many protocols and regulatory guidelines in place that the issues are usually more about ensuring proper and consistent execution than dealing with uncharted ethical territory. But there are exceptions. Genzyme, for example, was in an ethical quandary when the Food and Drug Administration (FDA) approved one of its orphan drugs, but only on the condition that it continue a double-blind, placebo-controlled trial. If the company did not complete the trial, the drug would not be as widely adopted and fewer patients would have access; if it did, the placebo group would be delayed access to an approved drug it desperately needed. As Genzyme's Senior Vice President for Clinical Research observed:

> This was the most difficult ethical issue I have faced. The standard I've always used to determine whether what we're doing is ethical is if I would be willing to do it myself or recommend it to a patient. If I wouldn't be willing to participate in the trial, how can I ask someone else to? This is the first time that we've been in a situation that didn't fully pass my personal test.

Similarly, interview participants from companies conducting experiments on animals were aware of the ethical significance and complexity of their proper treatment. The Pipeline case illustrates issues that surround animal rights and the significant consequences they can have for bioscience companies. Explaining how tricky the ethics can be, one interviewee said, "The beautiful and frightening thing about ethics is you can't always know what is right beforehand; you don't know how an animal will respond to treatment."

With the decoding of the human genome and moves toward more personalized medicine, new ethical issues concerning the proper conduct of medical research and treatment are emerging. Among those identified in the cases are:

- How to protect the privacy of the information gathered
- Whether consent is obtained not just for one, but for multiple uses of information in various tests or trials
- The potential for discrimination against certain races or groups of people that can result from genotyping
- Access to the information by the police
- Access to the information by insurance companies who could use the information to disqualify applicants
- Biobanking and what should be the legislation surrounding it

Generally, all the employees who mentioned these issues felt they could be mitigated through open disclosure with research subjects and instituting careful company safeguards, including the anonymization of patient data.

ISSUE 4: REGULATORY GUIDELINES INSUFFICIENT

Many bioscience companies assume that the best way to avoid any ethical issues is to follow the rules laid down by the FDA and other regulators. This not only fails to distinguish compliance (doing what's legal) from ethics (doing what's right), but also, as our cases illustrate, is an insufficient strategy for guiding corporate decision making: with continually evolving science and technology, regulations often do not exist, or are poorly defined, or inconsistent, or subject to change. In the area of pharmacogenomics, for example, there are as yet no regulatory requirements for drug companies to adopt diagnostics linked to drug safety and efficacy, but it would be hard to argue that it is ethical to continue to sell drugs that are ineffective for a high percentage of patients, when a test is available to determine which patients can benefit.

Employees at both Sciona and Interleukin said that the infancy of their science (nutrigenomics) means that there is no regulatory framework to help guide the testing and marketing of their products to the public. An employee at Maxim explained that different regulations can sometimes conflict and are subject to significant changes based on the shifts in leadership of the agency. Each regulating agency has its own rules and objectives, and the individuals within them may have different interpretations of the regulations. So the lack of a central mediator to help the company figure out which rules to follow can be frustrating. Participants from other companies mentioned that the regulations that govern their research are often not enough to guide ethical behavior in their respective industries. For example, Diversa's biodiversity coordinator noted that most countries, including the United States, do not have a regulatory framework for the collection of genetic samples, and thus provide no clear guidance on how to collect samples ethically or with whom the firms should negotiate and share benefits. An Affymetrix participant explained that the firm must make its own ethical judgments regarding its DNA-chips customers, as currently there is insufficient protection against discrimination and invasion of privacy in genetic testing. Likewise, TGN Biotech, as a pioneer in the still-emerging field of transgenic animals, faces a situation in which "the regulations are developing at the same time as the science," according to its Director of Regulatory Affairs.

The case studies identified a number of ways forward in the frequent absence of a clear regulatory framework. All the firms agreed on the importance of a company establishing its own independent ethical standards that put patients' interests first. At the same time, they acknowledged the importance of building a close working relationship with regulatory bodies. In cases in which patients want access to drugs that are still in clinical testing, for example, Genzyme's Vice President of Regulatory Affairs observed: "We work closely with the FDA to allow compassionate use for those who are very sick... The FDA has the final say on the ethics. It is absolutely key to work with the FDA in doing what we do." The efforts at international harmonization of regulatory standards is gradually making the task of reconciling different demands easier. But until these harmonization efforts are complete, and in fields where regulations still don't exist, the cases identified several possible ways forward, including creating an Ethics or Clinical Advisory Board, developing cooperative industry mechanisms for self-regulation, or being recognized by the Canadian Council on Animal Care.

ISSUE 5: MARKETING AND DELIVERY OF PRODUCTS

How products are marketed and sold is one of the most high-profile and persistent sets of ethical issues facing the bioscience industry. What forms of direct-to-consumer advertising are appropriate? What information can be given to doctors relating to off-label use of drugs? What benefits or medical education is it fair to give doctors to encourage use of a particular treatment? Participants from several companies explained the challenge their sales representatives face when competing against other companies that are using lavish inducements, such as sending doctors on expensive vacations. Said one marketing manager, "The reps will come back and say, 'My competitors in that country take doctors out and spend money, and I know it's not legal in the United States. but it is here, and I'm not going to make my quota if I don't do this.' And we just have to say, 'No, that's not how we do things here'."

Industry efforts at self-regulation in this area have not proved very effective: "Generally speaking, product promotion is not a level playing-field," said Merck's Vice President of Economic and Industrial Policy, "since not enough firms adhere to the voluntary codes of conduct and guidelines of the American Medical Association and of industry associations in the United States and Europe."

Sciona and Interleukin were entering less-charted ethical water, attempting to build an effective and ethical business model that entailed analysis of individuals' genetic information on which to base dietary recommendations. Sciona used to market its services directly to the consumer and Interleukin is considering the direct-to-consumer approach through independent sales representatives who will sell the service from their homes (the Amway sales method). Interleukin's in-house ethics experts acknowledged that this form of sales needs to be thought through carefully.

ISSUE 6: VALUE, PRICING, AND ACCESS TO PRODUCTS

Criticisms of the high costs of drugs and other bioscience products are intensifying for bioscience companies, particularly in countries, like the United States, without national healthcare systems, where individuals and companies must pay for their own drugs. The rapid rise in healthcare costs and the diffusion of the Internet, with its ability to reveal price differences across countries and allow individuals to order drugs online, seem likely to sustain the growing focus on this issue. Bioscience firms justify the costs of their products by noting the major R&D investment required, the need to pay for the costs of products that never make it from development to market (more than 90% of them), and the relative cost-effectiveness of many of their new technologies compared with hospitalization or other therapies. Critics counter that the industry is now spending more on marketing and sales than on R&D, and that much of the R&D investment is going to "me-too" drugs that entail lower risk and offer less clear-cut benefits.

Of all the case-study companies, Genzyme faces perhaps the greatest challenge on the pricing issue, because it develops new orphan drugs for very small numbers of patients, and thus must charge a very high price to recover the costs and reinvest in new drug development. Genzyme has decided to adopt a strategy of charging a single global price for its products, which can range from $150,000 to

$200,000 per year per patient, but then to give the drug free to patients around the world who cannot afford the product.

Even in developed countries, many people cannot afford, or do not have access to, the therapies that drug companies produce. Many of the companies have compassionate-use (donation) programs like Genzyme's, but the demand often far surpasses the supply and ability to donate. Another consideration is that donation (at least on a whole-country scale) is not always the best solution, because it is unsustainable and does not encourage the recipients to value the drugs. (Once Genzyme begins supplying patients in a developing country with treatment, the company seeks reimbursement from the government, arguing that even poorer countries can afford it because so few patients are affected by these rare diseases.)

Even in cases in which drug companies are giving their drugs away or selling them at the cost of production, they can run into difficult issues. Merck has been criticized for not providing sufficient access to its discounted acquired immune deficiency syndrome (AIDS) drugs in Africa. But in many instances the countries have been slow to grant regulatory approval, even though the drugs are being widely used in the United States and Europe. In contrast, its partnership with the Bill and Melinda Gates Foundation and the Government of Botswana serves as a model of human immunodeficiency (HIV) drug-service delivery. Novo Nordisk is facing similar issues around the price and accessibility of its insulin that many health systems still cannot afford, even at a heavily discounted price. Donating the insulin would not solve the problem, because an underdeveloped infrastructure and lack of specially trained doctors and nurses can mean that many diabetic patients cannot get diagnosed and treated.

ISSUE 7: DOING BUSINESS GLOBALLY

To cover the huge costs of developing new drugs or bioagricultural products, most bioscience companies must become global, either by expanding their own operations or through partnerships. And going global brings new ethical issues: variations in proper conduct of clinical trials in developing countries and the definition of responsibilities once these are complete; cultural differences over what constitutes acceptable business practices (for example, payments to individuals to enable sale or distribution of a product); how to conduct corporate donation programs; how to treat employees fairly across different countries and healthcare systems; how to ensure informed consent when conducting clinical trials in remote, relatively uneducated populations in developing countries; how to continue offering treatment for these patients on completion of the trial, with little medical infrastructure in place.

Diversa's Biodiversity Manager noted that it could be a challenge to strike a balance between achieving the company's objectives while respecting the host country's decision making processes. Novo Nordisk employees gave the example of trying to work with developing countries on corporate responsibility programs, to arrive at common goals, without excessively imposing the company's values.

A Novo Nordisk international project manager talked about how important it is to help such countries build infrastructure in a sustainable manner, in line with their needs and wants, so that if Novo ever had to withdraw from giving support, the initiative continue independently. "We pay to train doctors in diabetes but we do not supplement their income; that would be disruptive and not sustainable," he said. Merck was concerned that, as it expanded its drug-donation programs into developing nations, its employees and their families were now at greater risk of infectious diseases and instituted policies to ensure that all of them would receive healthcare coverage to try to minimize these risks.

Common to both developed and developing countries were the diversity issues mentioned by a few interview participants. Although most of these companies are trying to ensure equal opportunities for all employees, one Human Resources manager noted "non-verbal cues that are problematic," such as the example in the Merck case where a man gets recognition for a woman's suggestion.

ISSUE 8: CONFLICTS OF INTEREST

Given the large and ever-changing network of partnerships in the bioscience industry, potential conflicts of interest are an almost inherent part of operating in this sector. Our case studies identified conflicts of interest in the following situations: between academic and business goals, between personal goals and business goals, and between business units at the same company.[1]

ISSUE 8.1: SCIENCE PROFESSIONALS' VERSUS BUSINESS NEEDS

All research-driven bioscience firms must decide how to balance the scientific norms—free access to information and rapid publication of new knowledge—with the company's need to protect its intellectual property. Participants mentioned the conflicting personal desires to publish and to be transparent about the company's research (an academic orientation) and to keep information on the products they're developing secret from potential competitors (a business orientation). This is even more complicated when the founders of a company retain their positions as university professors, as is the case at TGN, while holding significant ownership positions in a start-up company based on their university research. Although TGN's founders have transparent contracts with the university, one of them explained that people outside the company have raised concerns about the arrangement. This issue has had more attention recently, as leading medical journals have insisted that they will only publish articles based on clinical trials that have been disclosed from the outset.

---

[1]Another conflict of interest issue, between ethics advisors and the companies they are working with, is discussed in the final section.

ISSUE 8.2: PERSONAL GOALS VERSUS BUSINESS GOALS

Our case studies identified other personal/company conflicts. When Interleukin merged with Alticor, Interleukin's CEO and top management team had to decide between what was right for their own careers and what was right for the company, the employees, and shareholders to whom they were responsible. The CEO explained, "[Alticor] would not invest in our company unless each of us signed a three-year employment contract. Never before had I faced a situation where my decision about a job would directly affect so many other people."

For a firm that focuses on extremely rare diseases, Genzyme has a surprisingly large number of employees who are personally affected (either themselves or a close relative) by these conditions for which no other good treatments are available. In cases in which a drug is still in clinical development and the supply of the drug may be severely limited, the company has had to face difficult choices between looking after the best interests of its employees and doing what is best for the patient population as a whole.

ISSUE 8.3: BETWEEN BUSINESS UNITS

Potential conflicts of interest also can arise within different business units of the same firm. In a well-known example, Novartis' bioagricultural division was developing GM crops while its Gerber division made the decision to ban GMOs in its baby food because of public safety concerns. Within our cases is the example of Genzyme, which has a diagnostics division that tests people for genetic diseases and a therapeutics division that develops and sells the treatment for some of the orphan diseases for which they test. One of Genzyme's Vice Presidents explained, "We have to be very careful not to get accused of generating a market for our products."

ISSUE 9: CORPORATE AND SOCIAL RESPONSIBILITY

Bioscience companies, like most global corporations, are coming under growing pressure not just to generate returns for shareholders, but also to measure and try to improve their wider impact on society. Merck's General Counsel drew a distinction between corporate and social responsibility (CSR), compliance, and ethics:

> Ethics is what we teach our children (about) behaving the same way when their parents are gone as when they are there. Corporate and social responsibility touches on compliance and ethics but goes beyond these concepts. Because of our field, we have the ability to reach out to disadvantaged people around the world and make a difference... We have a responsibility to behave ethically and comply with laws, but that is not enough for us... If we have the capabilities to help and reach out to needy, sick people, we ought to do more.

CSR can take the form of efforts to mitigate the negative impacts of a firm's activities, such as TGN's efforts to minimize the pollution their pig farms create.

And it can involve initiatives to improve the quality of life in communities where they operate, through such things as employee volunteering or local outreach and educational programs, or much more ambitious efforts to make therapies available to individuals who cannot afford them. Although the intrinsic enjoyment of "doing good" is often a prime motivation for these programs, they can also be in the corporation's long-term interest. Diversa explained that one of their rationales for setting up benefit-sharing agreements was to create a sustainable solution to protect the biodiversity from which its discoveries stem. As we note in the next section, some leading large bioscience companies are now systematically tracking their social and environmental, as well as their financial, performance.

ISSUE 10: BUSINESS ETHICS AND GOOD GOVERNANCE

A string of high-profile corporate scandals and ensuing regulations, such as the Sarbanes-Oxley Act and stock-exchange reforms, have encouraged all public companies to focus more on business ethics and corporate governance. Bioscience companies featured prominently in these scandals, for example, Elan Pharmaceuticals and ImClone, causing the industry as a whole, and not just the firms involved, to lose both public trust and tens of billions of dollars in value. Off-balance-sheet accounting and insider trading characterized these scandals. Financial irregularities reported to us were of being asked to pay bribes to get drugs into a country, even when the drugs were being given away for free; dilemmas regarding at what stage to disclose any new research or clinical data to investors; and the need to rescind any loans to executives, even when these were seen by the Board to be in the best interests of the firm.

Our case studies also revealed some of the steps that bioscience companies are taking to ensure good governance, which Merck's general counsel defined as "having the controls and systems in place to run the company in the best interests of the shareholders." Millennium, like most public firms, has completed a thorough review of its audit procedures, tax practices, and executive compensation. Maxim Pharmaceuticals has instituted an annual strategic retreat that includes several outside experts who sign a confidentiality agreement and are able to provide fresh, unbiased perspectives on the firm's plans and activities. And Merck has underwritten the creation of independent, non-profit Ethics Resource Centers, in the United Arab Emirates, South Africa, Turkey, and Columbia, to promote broad use of ethical business practices across these developing economies.

BUILDING AN ETHICAL DECISION MAKING PROCESS

Having identified the wide range of ethical challenges that bioscience firms face, we now distill 10 key steps that bioscience companies can take to enhance

EXHIBIT 15.2   **10 Steps to Ethical Decision Making in Biobusiness**

---

1. Put People First
2. Start Early
3. Lead by Example
4. Build Ethics Capabilities
5. Integrate Ethics with Business Strategy
6. Communicate: Create an Ongoing Dialogue
7. Build Structural Protections for Ethical Behavior
8. Treat Ethics as a Process not a Plaque
9. Extend ethics to partners
10. Measure Effectiveness

---

their ethical decision making process (see Figure 15.2). It is crucial to recognize, however, that although these lessons appear to have broad applicability, each company faces a unique set of circumstances, and the particular approach that is best suited to it is likely to depend on factors such as stage of growth, the nature of its technology and business model, as well as the surrounding cultural and institutional context in which the company is operating.

1. PUT PEOPLE FIRST

One of the simplest, yet most powerful, maxims that several of our case-study firms have used to help make ethical business decisions is consistently to ask the question: Will the end-users of our product benefit from this choice? At Merck, this has been the central part of the firm's credo, from the time George W. Merck first articulated it more than 50 years ago, and has influenced many key decisions, such as opting to provide millions of free doses of Mectizan to cure river blindness when no payer could be identified.[2] Genzyme has similarly made "Putting Patients First" its motto since the firm was founded in 1980. This has led both companies to focus their research and product-development efforts on first-in-class drugs for severe, unmet medical needs, rather than investing in less risky, but arguably less socially beneficial, "me-too" or lifestyle drugs.

It is important to note the distinction between this maxim and the one commonly taught in business schools to "put the customer first." In the bioscience sector, the two are often not synonymous, with governments, insurance companies, or hospitals (whose interests may not always align with those of patients) often paying the bioscience firms for the drug or medical device a patient receives. Monsanto learned this to its cost with the marketing of the first generation of GMOs. Monsanto did an excellent job of meeting the needs of its primary customer, the large, industrial farmer, but failed to take into account the views of the end-consumers of crops around the world.

---

[2] Merck's controversial decision to pull the pain-killing drug Vioxx off the market because of the evidence of increased heart attack risk for long-term users of the drug occurred after this case study was completed, and is thus not covered in this volume.

2. START EARLY

As noted in Chapter 1, early-stage bioscience companies, often a decade or more from reaching the market, and confronted with the twin challenges of developing a viable product and raising enough money to do so, often feel it is premature for them to worry about ethics. Our case studies, however, revealed a number of key reasons why it is important for them to develop a strong ethical culture early in the firm's life.

- Early decisions, such as which products to develop for what markets, and which partner to choose, can have lasting implications.
- The ethical challenges a company faces grow significantly once it has a product approved. Then it has to deal with issues of drug pricing and access, marketing practices, and pressure from Wall Street to meet quarterly earnings growth targets, rather than being assessed just on its potential. If a strong ethics foundation is not in place before it is confronted with these issues, the firm may find it much harder to emphasize ethics in business decision making.
- Once a culture and way of making decisions is imprinted in an organization, it is very difficult to change. This can be a problem, if a firm is trying to reverse a pattern of unethical behavior, or just a lack of attention to these issues, but conversely can be an asset if it has already built a strong ethical base.

3. LEAD BY EXAMPLE

All the companies we studied agreed that having a strong ethics focus starts from the top. Senior management must not only talk frequently about ethical issues, but must demonstrate through their actions the importance attached to it, to send a strong signal about the need for ethical behavior throughout the company. Participants explained that, with ethical leadership, they feel empowered and fully supported to make ethical decisions. The Vice President of Ethics and Corporate Responsibility at Millennium described the role of ethical leadership in a company:

> In the business ethics arena it is particularly important, notwithstanding any programs you do, to have an identifiable leader with charisma and a deep sense of commitment to the institution and who models ethical behavior for his or her employees. Most stuff I've read suggests that a good business ethics program needs as its anchor an ethical and deeply committed CEO.

In some cases this leadership came from the company founder(s); in others, from a later chief executive or a group of managers who encouraged the ethical culture and the implementation of ethical approaches. In different firms, the leadership embedded values such as honesty, putting patients first, openness to criticism, integrity, and trust. In Maxim, for example, CEO Larry Stambaugh consistently stressed two core values—having a "humble heart" (compassion for others) and an "open mind" (engaging in active listening and seeking different viewpoints)—that he borrowed from the founder of Matsushita, one of Japan's leading corporations.

4. BUILD ETHICS CAPABILITIES

For companies to make effective ethical decisions, it is vital that they have both the right leadership and core values and the capabilities to implement these values in a wide range of challenging ethical situations. The case-study firms adopted a variety of approaches to building these capabilities, from the use of external consultants or advisory boards who could provide both guidance and tutoring for internal staff, to hiring full-time staff who had prior ethics training or experience. Ethics Advisory Boards (EABs) in some firms, such as Affymetrix and PharmaSNPs, deal with the full range of ethical issues facing the company; other companies, like Genzyme, have set up disease-specific advisory boards of physicians. Explains Genzyme's Therapeutics Executive Vice President: "We are trying to develop guidelines and standards of who to treat and how to treat them. We work with that group to develop clinical trials."

A number of companies also developed some form of ethics education for both employees and company partners. These could focus on specific ethics-related or compliance issues, such as good clinical procedures, conforming to new regulatory or marketing guidelines, introducing a new code of conduct, awareness of antitrust or antibribery laws, or it could have a more general focus on raising ethical awareness and the importance of acting ethically. In some cases this training was put online, for example, for business compliance or safety, to make it universally available. Educating concerned external stakeholders (consumers, non-governmental organizations [NGOs], regulators) was also seen as an important aspect of ethics education. One of the most innovative approaches was a film series (featuring movies like *Gattaca* and *Inherit the Wind*) designed by Rahul Dhanda of Interleukin and implemented both in his own firm and at Millennium to stimulate discussion of ethical dilemmas. Highlighting the importance of training, one of Millennium's former business managers commented on how useful the company's ethics training sessions were for her: "They were fascinating; my brain would hurt with the challenges [the teacher] posed."

5. INTEGRATE ETHICS AND BUSINESS STRATEGY

Perhaps one of the more surprising results to emerge from our research was that several bioscience companies had not just created a strong internal ethics capability, but gone further to integrate ethics fully into the strategic decision making process. This had been achieved by having at least one senior leader who wore two hats: for example, head of ethics and responsibility for a key business function. This was most evident at Interleukin, where both the CEO, Phil Reilly, and Rahul Dhanda, Director of Business Development, were experts on ethical issues. The CEO explained his rationale for hiring yet another ethics expert:

> Hiring Rahul gave us a disproportionately high level of attention to bioethics in this small company. But we are in a field that has relatively weak scientific credibility… We're hampered already by fly-by-night companies. We must

distinguish ourselves from them. We can't just have an appearance of bioethics, but need to have it incorporated into product development.

At Millennium, Steve Holzman was a moral philosopher who also was the key person responsible for negotiating the firm's record-setting strategic partnerships with large pharmaceutical and bioagricultural firms. At Genzyme, Elliott Hillback had served as head of several different business units while taking charge of the ethics function. In all these cases, this blending of roles ensured that everyone in the firm recognized that ethics was a high priority, seen as core to the business, and not a separate feel-good or compliance function.

6. SHIFT FROM SECRECY TO COMMUNICATION AND DIALOGUE

Given the controversies surrounding many bioethics issues, from animal rights and stem cells to how GMOs and genetic information should be used, it is not surprising that many bioscience companies have historically chosen to try to keep a relatively low profile. However, one of the key lessons from our cases is the vital need for companies to shift from a mindset of avoiding confrontation to engaging in an active dialogue with external stakeholders as part of their ethical decision making process. A director at Monsanto explained that when they were first commercializing their GM seeds, they did not engage enough with stakeholders beyond their immediate customers. He added: "It was naïve arrogance. We had a sense that our view would be unanimously accepted. We were arrogant in making decisions about what people would or should want and what they don't want." One of Novo Nordisk's Senior Vice Presidents explained that the company used to have a very introverted stakeholder model that really only included employees and customers. She explained that this was a problem until they understood that NGOs, governments, and the general public needed to be included in their dialogues with stakeholders.

As part of its efforts to reinvent its ethical decision making process after the GMO controversy, Monsanto's then-CEO explained the importance of communicating with stakeholders: "I came to the realization that we were not sensitive to our stakeholders and we had to change our behavior. Society did not understand Monsanto…We went through an extensive process to learn how to listen, not be defensive, and not be aggressive." Novo Nordisk's former Vice President for Stakeholder Relations (now Sustainable Development Consultant), Vernon Jennings, discussed the company's now-formal communication process with NGOs, which includes inviting external stakeholders into the company's laboratories each year:

> It was successful because of the openness and because we weren't seeking consensus. What we were seeking was to understand each other and to look for areas of commonality… Some think that the dialogue is sufficient. But it's not. It requires action. There has to be a tangible outcome. There are those who feel that by engaging alone they have done the job; there has to be learning and there has to be movement.

Just as important as initiating an external dialogue is creating an ongoing internal discussion about the ethical questions a company and its employees face.

A starting point for informed internal and external discussion is creating transparency through good communication about the full range of the firm's activities. Genzyme's Chief Compliance Officer discussed the importance of communicating the firm's core values: "Our values and the company mission is just so much a guiding light for ethics. We're doing the right thing for patients... I think the mission of the company really drives the ethical nature of the company." Many of the employees we interviewed, however, were somewhat reluctant to communicate about the socially beneficial activities of their firms, for fear they would be perceived as just a public relations exercise. As Merck's Mectizan Donations Program Director observed: "I think sometimes we've done too good of a job of not tooting our own horn. There's a fine line between doing good and promoting it."

7. BUILD IN STRUCTURAL SAFEGUARDS FOR ETHICAL BEHAVIOR

Another step that bioscience firms can take to foster open dialogue about ethical issues is designing structural safeguards into the organization that make employees feel safe to raise sometimes difficult ethical questions, or to identify instances when they feel that company or individual behavior is not meeting the company's ethical standards. Across our case-study firms we identified a wide range of such approaches, tailored to the firm's particular circumstances and stage of development. Among these were:

- Setting up a 24/7 ethics helpline and/or in-house ethics office with ombudsman, where individuals who have an ethical question or concern could raise it in a confidential way if desired (Millenium, Merck)
- Creating an advisory board or a separate committee of the Board of Directors to address ethical issues, with whom employees could raise issues and seek external advice (Affymetrix, PharmaSNPs, Maxim, Genzyme, Maxim)
- Designing the reporting relationships in the company so that those charged with making the clinical and regulatory decisions that are in the best interest of the patient do not report to those responsible for the profit and loss of a business or product (Genzyme, Merck)
- As noted above, establishing a performance-management system that measures both financial results and how these results are obtained, and avoids creating financial incentives (such as large bonuses for meeting certain sales targets) that could encourage bending of ethical guidelines (Maxim, Merck)
- Establishing a separate foundation to fund longer-term charitable programs independent of short-term financial results (Merck)

8. TREAT ETHICS AS A PROCESS, NOT A PLAQUE

Although symbols—whether the prominent posting of a company's values or credo (Merck) or a mouse pad that provides an ethics quick test (Millennium)—can be important reminders of ethical behavior, all the firms we studied stressed that, for ethics to really affect decision making, it is vital for the firm to treat it as an ongoing, dynamic process, rather than a set of values or code of conduct that is carefully articulated then left to gather dust. As the Enron debacle vividly illustrated, a firm may have the best-defined set of ethical guidelines available, but if it isn't enacted by the firm's leaders and employees, it provides no assurance of ethical behavior. Genzyme's Vice President of Corporate Communication explained the limit of merely putting words on a page: "If I have to pull a laminated card out of my wallet to find our mission, we have failed." To avoid this, some companies, like Pipeline Biotech, ensured that ethics was well integrated into day-to-day decision making by delegating the responsibility for making some hard ethical decisions to the frontline employees themselves.

Keeping ethics a dynamic process is vital because the types of ethical challenges a bioscience company faces evolve and become more complex as a firm grows. Many of our case-study bioscience firms had begun with a strong, yet fairly informal, emphasis on discussing ethical issues with all employees but were concerned about their ability to sustain this as their firms grew. TGN's Director of Transgenics recognized that, as the direct contact between the firm and the founders who were the ethical visionaries of the company lessened, it would be more difficult to sustain the informal influence they had on all key decisions. Likewise, employees at several companies that had experienced acquisitions— Millennium, Interleukin, PharmaSNPs—recognized that continuing with their focus on ethics would be a challenge. When Genzyme reached 5,000 employees and recognized it could no longer rely solely on informal means to communicate the desired approach to ethical decision making, it finally took some steps to formalize its core values with a code of conduct.

9. EXTEND ETHICAL APPROACH TO PARTNERS, SUPPLIERS, AND CUSTOMERS

In today's world of highly networked organizations, where companies focus on their core competencies and rely on a myriad of partners to help them develop and produce products or services, it is not enough for a firm to ensure its immediate employees are behaving ethically. As the cases clearly demonstrate, companies will also be held accountable for the actions of their partners, be it academic collaborators, suppliers, contract research organizations or, in some cases, customers. Millennium felt it was on sound ethical ground when it partnered with a Harvard professor on a study in China that was reviewed by the University's Institutional Review Board (IRB); the company wound up on the front page of *The Washington Post* when the professor began looking at the genetic links for schizophrenia, something that was not part of the original research agreement. Novo Nordisk ran into trouble when one of its suppliers was accused of violating accepted international human rights guidelines. After performing an audit of the supplier that confirmed this, Novo Nordisk is now working with this supplier to make changes to meet these standards. At Affymetrix, the EAB discussed the extent to which the firm bore ethical responsibility for the potential uses of its products by companies to which it supplies its chips, and decided that the firm should do its best to take this into account before making a sale. As each of these examples illustrate, it is vital for a firm to have a clear process in place for screening potential partners to determine whether they share similar ethical principles, and then to monitor the partnerships carefully to ensure they are comfortable with how the partner is operating.

10. MEASURE EFFECTIVENESS

One of the hardest challenges for any bioscience company to solve is to assess how well its approach to ethics is functioning. Among the difficulties inherent in assessing ethics are:

- The intangible nature of the desired outcome, such as fostering the intended types of human behavior and enhanced decision making rather than a clear sales or financial target, which makes it hard to measure;
- That key metrics may be the absence of a relatively rare, negative event, such as an FDA warning letter, lawsuits, drug recalls, rather than a more frequent, positive occurrence, and that it is therefore difficult to demonstrate the causal link between ethics programs and hoped-for outcomes;
- The initially paradoxical effects that ethics initiatives can have: for example, Merck saw a sharp increase in employee reports of possible ethical problems after it launched a new Code of Conduct, a result that the Chief Ethics Officer took as evidence that efforts to raise internal awareness of ethical issues were succeeding.

Despite these hurdles, the experience of many of our case-study firms suggests that conducting some form of ethical assessment is well worth the effort. We identified at least three different, complementary levels at which assessment can occur. First is the individual level, incorporating ethics into assessment of employee and managerial performance: at Merck, for example, every employee is assessed annually on desired leadership behaviors, so that individuals who meet or exceed financial targets, but do it in the wrong way, are not rewarded. Said one business leader:

> There are three dimensions: 1) strategic ability, which is knowing what to do, 2) can they get it done? and 3) do they get it done the right way? That's the leadership aspect. Do they collaborate and treat people with respect? We've terminated people for leadership issues, people who have been very good performers, but lousy leaders.

Merck has also been a leader in conducting a second level of assessment: evaluating the success of specific ethical initiatives, such as its drug-donation programs or internal ethical behavior. The company has typically relied on outside experts to conduct these assessments because, as the Chief Ethics Officer explained, "I wanted to use an outside consultant to do it to get as candid as possible information. We took a cold, hard look at ourselves and asked: 'Are our deeds consistent with our words?'" Likewise, several of the earlier stage bioscience companies were relying on individual external consultants or EABs to get an unbiased, external assessment of their approaches in different areas of their businesses.

A final level of assessment involves company-wide measures that integrate ethical and corporate social responsibility objectives with bottom-line business results. These measurement tools, such as Novo Nordisk's Triple Bottom Line, Maxim's Balanced Scorecard, or the Monsanto Pledge, can elevate the profile of non-financial performance (such as the firm's impact on the environment, global health, its employees, and the surrounding community), signaling the importance that the company's leaders attach to them. They can also serve as an effective communications tools with external stakeholders about these broader activities if they become part of how the firm routinely tracks and reports its performance.

KEY FUTURE TRENDS

Having reviewed the steps that individual companies can take to improve their ethical decision making, we conclude our analysis by examining some of the major ethical trends likely to face the whole bioscience industry in the near future and how the industry may approach them.

INDUSTRY SELF-REGULATION

In the earliest days of the biotech revolution, there was a major effort by scientists to self-regulate the use of this exciting, but also potentially threatening, new technology. This resulted in the famous Asilomar Conference in 1975, where many of the pioneers responsible for developing the new technology came together to discuss ethical issues and possible risks of biotechnology, and how best to deal with them. Although many of the early fears proved unfounded, the bioscience industry that emerged after Asilomar has been much less successful in anticipating potential ethical issues and dealing with them collectively. Rather, the bioscience industry has tended to wait for a crisis to occur and then only belatedly have they adopted measures to address it. This pattern has been repeated on issues such as GMOs, use of animals in research, industry marketing practices, and most recently, the registry of all clinical trials and publication of their results.

In an industry in which the pace of globalization and technological change continues to outstrip the capacity of governmental bodies to regulate it effectively, and in which the failure to gain public trust can dramatically undermine the ability to commercialize new products, many of our case-study firms agree that effective leadership extends beyond building the right values and behavior in a company; it means bringing ethical leadership to the industry as a whole. Discussing the emergence of the new field of nutragenomics, for example, Interleukin's former Director of Business Development explained the need for industry to self-regulate: "If we are going to bring the right level of science to this field, it needs the same scientific validity as other fields. We need industry leadership...and if there isn't industry leadership, you invite regulation and you deserve it." Rosalynn Gill-Garrison, Sciona's Senior Technical Officer, reinforced this point: "We felt that it is our responsibility as pioneers to shape the regulatory environment because there are currently no clear regulations."

In addition to companies working individually to create effective industry regulation, it may be beneficial to work through industry associations to generate collective action. Thane Kreiner, Senior Vice President of Corporate Affairs at Affymetrix said, "Part of trying to take a national leadership role includes trying to keep BIO's attention focused on these issues. We want to be a leader in the field of ethics in the industry." The combination of proactively addressing ethical issues and self-regulation may provide a means for the industry to address key future issues, such as global drug access, biodiversity, and gene-based medicine that are now emerging. In addition, it can create an enforceable common floor that would allow companies to compete on the basis of quality and innovation, rather than pushing the boundaries of ethical marketing practices.

Creating a self-regulating industry should be viewed as a precompetitive issue. It gives all firms license to operate. If better self-regulation emerges, perhaps through systematic application of some of the lessons of this study, it is still possible that the recent erosion in public trust of bioscience firms could be reversed.

BIOSCIENCE COMPANIES AND GLOBAL HEALTH

The common perception that biotechnology is relevant only to the developed world because it is high-tech and expensive is untrue. A recent foresight exercise identified the top 10 biotechnologies for improving global health (see Exhibit 15.3, below) that clearly shows how developing countries may benefit[3]

Addressing global health problems such as the AIDS pandemic is undoubtedly "the right thing to do" for bioscience companies, and one of the most common reasons cited by interviewees for why they entered this industry. Many of the major diseases of developing countries remain unresolved. Carl Feldbaum, the outgoing

EXHIBIT 15.3   **Top Ten Biotechnologies to Improve Health in Developing Countries**

1. Molecular diagnostics
2. Recombinant vaccines
3. Vaccine delivery systems
4. Bioremediation
5. Sequencing pathogen genomes
6. Female-controlled protection against sexually transmitted diseases
7. Bioinformatics
8. Nutritionally enhanced genetically modified crops
9. Recombinant therapeutic proteins
10. Combinatorial chemistry

[3]Daar AS, Thorsteinsdóttir H, Martin DK, et al. Top 10 Biotechnologies for Improving Health in Developing Countries." *Nat Genet.* 2002;32(2):229.

President of the Biotechnology Industry Organization (BIO), has identified global health as a priority for BIO members. In his speech entitled "Biotechnology's Foreign Policy"[4] he noted:

> Our goal must be to ensure the widest possible dissemination of biotechnology's benefits while respecting the diversity of the world's nations and peoples... (we must)...integrate biotechnology into compelling responses to public-health crises...biotech R&D can do its part in the developing world by producing vaccines that don't require refrigeration and are nasally or orally delivered. That investment in prevention can be made alongside continuing investment in diseases that afflict wealthy societies.

As a result, BIO has launched a program with the Bill and Melinda Gates Foundation called Bioventures for Global Health (BVGH),[5] to encourage biotechnology companies, foundations, and academics to carry out research to stamp out global inequities and fight neglected diseases. Feldbaum called this a "win, win, win" situation, with benefits for companies, investors, and millions of people around the world. The BVGH Resource Center provides a range of tools that businesses need to identify viable market opportunities and formulate strategies to develop global health products.

Beyond the moral case, there are a number of other compelling business reasons why bioscience companies should address global health needs. At a time when it is increasingly difficult for biotech start-ups to raise capital to fund new innovations, the Gates Foundation and National Institutes of Health (NIH) have launched the Grand Challenges in Global Health Program to channel research funds to those with bright ideas to overcome the scientific and technological bottlenecks that are stopping us from finding solutions to diseases most prevalent in developing countries. A glance at the Grand Challenges[6] indicates just how extensive are the opportunities for bioscience companies to do world-class, cutting-edge research, perhaps in partnerships with academic institutions. And unlike the money that firms may raise from venture capitalists in return for a significant (often majority) share in the ownership of their firm, these grants do not dilute the stake of the firm's founders or employees.

There will also be opportunities to enter new markets, partner with developing-country researchers and learn how to cut costs. For bioprospecting companies, the business models may require them to seek biodiversity resources in developing countries, and it is appropriate that they consider in return how to help protect the environment and address the scientific research and health needs of those countries. China and India have been experiencing approximately 8% gross domestic product (GDP) growth rates for some time now, and their spending capacity is growing very rapidly. Scientists in these and a number of other developing countries, such as South Africa, Brazil, and Mexico, are producing world-class research that might require partnering

[4]http://www.bio.org/speeches/speeches/20020610.asp.
[5]www.bvgh.org.
[6]http://www.grandchallengesgh.org.

with western bioscience companies for full development. Additionally, as the capacity to do clinical trials in the United States and Europe is being overwhelmed, bioscience companies are increasingly looking to do their clinical trials in developing countries.

Pharmacogenomics, although still poorly understood in terms of its true potential, also holds great long-run benefits for developing countries. Allen Roses of GSK has shown how use of pharmacogenomics can make drug development faster, cheaper, and more efficient.[7] What he did not discuss is the opportunity this opens for developing-country pharmaceutical companies, and the likelihood of partnering with western bioscience companies with the capacity for whole-genome scanning, to produce orphan drugs jointly. There will also be opportunities to address chronic, non-infectious diseases, which are becoming major contributors to the global burden of disease, and to help address issues of food security.

ENGAGING THE PUBLIC: SCIENCE AND ETHICS EDUCATION

An engaged and scientifically literate public is a key aspect of the innovation system in bioscience and of any informed debate about the ethical issues raised by these innovations. Engaging the public can create a more stable business climate by identifying those aspects of science and ethics that will receive public support and those likely to create controversy. One need only think of GM foods and stem cells to recognize how the public's views can greatly influence the development of a technology.

If the bioscience industry is to engage the public effectively, it means going beyond simply issuing press releases and providing some charitable funding to schools and universities. It means listening to the perspectives of members of society and perhaps modifying what the industry intends to do. As noted above, beginning in 1990, Novo Nordisk established a systematic process of stakeholder engagement and hosted annual 2-day visits for NGOs from around the world. It learned about talking honestly and openly, and that it was acceptable to disagree while seeking to understand each other and to look for areas of commonality. But Novo Nordisk also learned that dialogue alone is insufficient. It must be accompanied by action and responsiveness, and must have tangible outcomes. The company works systematically with each stakeholder group and tailors the process to each. Another important stakeholder for Novo Nordisk is the general public; this includes local communities around company facilities as well as NGOs and activists. The company holds town hall meetings, to which the public can come and raise concerns to company representatives. Some Novo Nordisk facilities have newspapers, and send their own sustainability reports out to the community.

Town hall meetings are one of a variety of ways to engage the public; others are web surveys, citizens' juries, referenda, science museums, theater, and educational modules for students. It is not clear whether any of these is superior and

---

[7] Roses AD. Pharmacogenetics and Drug Development: The Path to Safer and More Effective Drugs. *Nat Rev Genet*. 2004;5:646–656.

there is probably scope for all of them. One innovative example of the use of theater is the work of Jeff Nisker, a Professor of Obstetrics and Gynecology and playwright. Dr. Nisker has created several plays, including *Sarah's Daughters*, which focuses on genetic testing for breast cancer. After the play, Dr. Nisker leads a discussion group with the audience, and the audience's views clearly surface.

Another approach is *Project Engage*, which focuses on the use of stem cells. It uses five 1-hour sessions of class time for senior high school students. After a didactic background on the science of stem cells, students adopt the roles of various constituency groups (scientists, industry, patients, religious groups) and present briefs to a government's legislative committee. The committee later proposes legislation, and this is compared with existing legislation on stem cells. The key to the success of the exercise is balance. The *Engage* package has been sent to every high school in Canada and is available through the web page of the stem cell network.[8]

Activities like these are pre-competitive and best done proactively and collectively. They help the public to engage with future technologies and help industry (and others concerned with the innovation system) to understand how the public sees various applications of technology and what issues might prove critical to the industry's future. The bioscience industry thus has a vital interest and role to play in supporting such educational and outreach initiatives.

PREPARING FOR AN ERA OF GENETICS-BASED HEALTHCARE

The decoding of the human genome has laid the groundwork for a fundamental healthcare revolution, creating the potential for much more personalized and preventative genetics-based medicine. Within a decade, the continued sharp decline in the costs of genetic sequencing is likely to mean that all newborns in the developed countries will leave the hospital with a full sequence of their own genomes, providing the probabilities that they may suffer from an ever-widening array of genetically linked conditions and accompanying advice on diet, lifestyle, and treatments to help avoid them. To make this revolution a reality, however, requires major changes in all aspects of the approach to healthcare, from the training of healthcare practitioners and the way drugs are developed to the legal and ethical frameworks surrounding healthcare delivery. The challenges posed by the genetic revolution are likely to be starkest for countries that lack a universal national healthcare system. In the coming era, where virtually everything will be a prior condition, Americans will face the twin threat of employment and healthcare discrimination unless much more stringent safeguards are put in place. Among the questions that will need to be addressed are: what rights to privacy do individuals have over their own genetic information? How will the rights of other family members be protected when a relative elects to have a genetic test? Will we be able to afford a range of therapies, like Genzyme's orphan drugs, that are very effective but serve much smaller patient-populations?

[8]http://www.stemcellnetwork.ca/engage/index.php.

For bioscience companies, the age of pharmacogenomics offers great potential opportunities and threats. As many of our case studies illustrate, most pharmaceutical companies have thus far been reluctant to embrace a genetics-based approach to drug development, because it threatens to carve up their blockbuster drugs into much smaller, if more effective, niche products and exposes one of the industry's less-publicized secrets: that most drugs do not work for a high percentage of patients, in some cases as high as 70 to 80%. It seems inevitable that, as genetic-based diagnostics become cheaper and more widespread, the FDA and others will require them in order to obtain regulatory approval for a new drug. Also, as noted above, the common perception is that pharmacogenomics will not benefit developing countries. At a minimum, this perception is unproven, and much more thought and research are needed to explore potential uses of this technology for 5 billion people in the developing world. Rather than resist this move, the industry could enhance its ethical reputation by embracing the technology and working collectively to come up with workable standards for how it should be used. This offers the potential benefit of helping the industry tackle two of the problems that have bedeviled it and contributed to the very high cost of developing drugs: the high failure rate of drugs in Phase III trials, because of adverse side effects in some patients, and the length of time required to develop new drugs.

CONSTRUCTIVE ENGAGEMENT: FINDING THE BEST WAYS FOR BIOETHICS EXPERTS TO WORK WITH INDUSTRY

As we noted at the outset of this volume, there are contentious questions asked about the involvement of bioethicists with industry: who qualifies as a bioethics expert and can they work with industry without having their objectivity tainted? As the cases have illustrated, to be most effective such an expert needs to possess a combination of deep knowledge of applied ethical principles and an understanding of the science and business environment in which bioscience firms operate. Yet, unlike the situation with lawyers or physicians, there is currently no accrediting body or system of oversight and accountability for bioethicists. A 1998 joint report of the American Task Force of the Society for Health and Human Values and the Society for Bioethics Consultation tried to fill this gap, promoting an ethics facilitation approach based on consensus-building in the healthcare setting. Similarly, the 2001 Canadian Bioethics Working Group on Employment Standards for Bioethics proposed a "Code of Ethics for Bioethicists."[9] The specific qualifications proposed were integrity, confidentiality, absence of conflict of interest, and honesty. It sees the profession of bioethics as being not adversarial in nature, but one of negotiation, facilitation, trust, and mediation, like the earlier American report. Until some form of standard "professionalism" is achieved, companies, as

[9]Chris MacDonald, *Model Code of Ethics for Bioethics*. Available at http://www.bioethics.ca/english/workingconditions/draftcode.pdf. Accessed August 2, 2004.

much as the bioethicists involved, may be putting their reputation on the line. Whatever the qualifications demanded of bioethicists and the companies they are working with, however, it seems fair that the same accountability, professionalism, and transparency be required of the industry naysayers.

PAY AND DISCLOSURE

A few of the companies we interviewed use ethics advisors either as consultants or as EAB members. Should bioethicists be paid for giving professional advice? This is not the same as taking an advocacy position. For example, there is a difference between an expert medical witness who agrees to evaluate a case independently and give an opinion for a fee, whether that opinion is for or against the lawyer paying her, and a doctor who is a professional witness, always testifying on one side of an issue. Members of the EABs of both Affymetrix and PharmaSNPs are (or were) paid for their time. One Affymetrix EAB member was concerned that this created a perceived conflict of interest: that the money she is paid for ethics consulting work for a company could unduly influence her ethics advisor's freedom in expressing concerns. "I am conflicted. I want to do the work... However, I have to deal with the conflict-of-interest aspect.... I feel that, so long as I disclose the gift, it should be okay and satisfy the critics. But I also recognize that others may not think so." In other cases, advisors elected to turn down their payment or to give it to charity.

A separate, but related, issue concerns whether a company's ethics advisor(s) should be disclosed. In Sciona's case, the ethics consultant decided to remain anonymous. More generally, it may make sense for a company to reveal that it has set up an EAB, but then leave it up to individual members whether they want to identify themselves publicly as members.

CONCLUSION

Ethical issues are an inherent part of doing business in the bioscience sector. As awareness of ethical issues grows along with the continuing evolution of biotechnology, the bioscience industry, and the companies that operate within it, must keep pace by enhancing their own ethical decision making abilities. This book has aimed to bring to light the implicit ethics work of some of the leading bioscience companies. We hope it will stimulate much more research and education about the ways companies can best address the ethical challenges they will surely face in the years ahead.

Behaving ethically depends on the ability to recognize that ethical issues exist by seeing events and decisions from an ethical point of view. Our findings show that this ability to see and respond ethically, whether relating more to attributes of corporate culture or to explicit ethics policies and initiatives, is vital to creating a sustainable biobusiness.

INDEX

Page numbers followed by *e* refer to exhibits.